Florian Wessendorf

Supramolecular Fullerene-Porphyrin Architectures

Florian Wessendorf

Supramolecular Fullerene-Porphyrin Architectures

Carbon-Rich Nanomolecules

Südwestdeutscher Verlag für Hochschulschriften

Impressum / Imprint
Bibliografische Information der Deutschen Nationalbibliothek: Die Deutsche Nationalbibliothek verzeichnet diese Publikation in der Deutschen Nationalbibliografie; detaillierte bibliografische Daten sind im Internet über http://dnb.d-nb.de abrufbar.
Alle in diesem Buch genannten Marken und Produktnamen unterliegen warenzeichen-, marken- oder patentrechtlichem Schutz bzw. sind Warenzeichen oder eingetragene Warenzeichen der jeweiligen Inhaber. Die Wiedergabe von Marken, Produktnamen, Gebrauchsnamen, Handelsnamen, Warenbezeichnungen u.s.w. in diesem Werk berechtigt auch ohne besondere Kennzeichnung nicht zu der Annahme, dass solche Namen im Sinne der Warenzeichen- und Markenschutzgesetzgebung als frei zu betrachten wären und daher von jedermann benutzt werden dürften.

Bibliographic information published by the Deutsche Nationalbibliothek: The Deutsche Nationalbibliothek lists this publication in the Deutsche Nationalbibliografie; detailed bibliographic data are available in the Internet at http://dnb.d-nb.de.
Any brand names and product names mentioned in this book are subject to trademark, brand or patent protection and are trademarks or registered trademarks of their respective holders. The use of brand names, product names, common names, trade names, product descriptions etc. even without a particular marking in this work is in no way to be construed to mean that such names may be regarded as unrestricted in respect of trademark and brand protection legislation and could thus be used by anyone.

Verlag / Publisher:
Südwestdeutscher Verlag für Hochschulschriften
ist ein Imprint der / is a trademark of
OmniScriptum GmbH & Co. KG
Heinrich-Böcking-Str. 6-8, 66121 Saarbrücken, Deutschland / Germany
Email: info@svh-verlag.de

Herstellung: siehe letzte Seite /
Printed at: see last page
ISBN: 978-3-8381-1570-2

Zugl. / Approved by: Erlangen, Friedrich-Alexander Universität, Diss., 2010

Copyright © 2010 OmniScriptum GmbH & Co. KG
Alle Rechte vorbehalten. / All rights reserved. Saarbrücken 2010

For my Family

Table of Contents

1 Introduction 1

1.1 *Photoinduced Charge Separation and Solar Energy Conversion* *1*

1.1.1 Natural Reaction Centers   1
1.1.2 Artificial Reaction Centers   2

1.2 *Molecular Recognition* *6*

1.2.1 Receptors, Coordination and the Lock Key Analogy   6
1.2.2 Supramolecular Lock and Key Design   7

1.3 *[60]Fullerene Chemistry* *7*

1.3.1 Structural Properties of [60]Fullerenes   7
1.3.2 Reactivity of [60]Fullerenes and Synthesis of Functional Exohedral [60]Fullerenes   9

1.4 *Porphyrins* *14*

1.4.1 Nomenclature of Porphyrins   15
1.4.2 Aromaticity, Spectroscopy and Electronic Properties   15

2 State of the Art & Aims 18

2.1 *State of the Art* *18*

2.2 *Aim of the Work* *21*

3 Results and Discussion 24

3.1 *Self-Assembly of Supramolecular Fullerene-Porphyrin Architectures* *24*

3.1.1 Synthesis   25
3.1.2 NMR-Titration Experiments   45
3.1.3 Photophysical Investigations   47
3.1.4 Conclusion   52

3.2 *Aggregation of Supramolecular Fullerene-Flavin Hybrids* *54*

3.2.1 Synthesis of Fullerene Containing Single Hydrogen Bond Receptors   55
3.2.2 Formation and Characterization of Supramolecular Complex between DMA-Fl and C_{60} Derivatives   58
3.2.3 Electrochemical Studies and Electron Transfer Driving Force of Supramolecular Complex between DMA-Fl and DRC_{60}.   62

3.2.4 Assemblies of Supramolecular Complexes between DMA-Fl and C_{60} with Hydrogen Bond Receptors as Molecular Clusters in Mixed Solvents 64
3.2.5 Electrophoretic Deposition 65
3.2.6 Properties of OTE/SnO$_2$/(DMA-Fl-SRC$_{60}$)$_n$ and OTE/SnO$_2$/ ((DMA-Fl)$_2$-DRC$_{60}$)$_n$. 65
3.2.7 Photoirradiation 68
3.2.8 Photodynamics of the Supramolecular Complex Clusters between DMA-Fl and C_{60} with Hydrogen Bond Receptors in Femtosecond Order 70
3.2.9 Conclusion 71

3.3 Further Photoactive Building Blocks for Supramolecular Aggregation 72

3.4 Supramolecular Aggregates of oligo-Phenylene-Ethynylene Wires 80

3.4.1 Synthesis 80
3.4.2 Determination of Association Constants and Cooperativity of Binding 82
3.4.3 UV/Vis- and Fluorescence Titration Experiments 88
3.4.4 Conclusion 90
3.4.5 Mass Spectrometry and Ion Beam Deposition of Hamilton Receptor bearing Rods for Surface Science in Ultrahigh Vacuum 91

3.5 Porpyhinato Phosphonium Salts as Precursor for Wittig-Horner Olefinations 94

3.6 Syntheses and Characterization of oligo-Phenylene-Vinylene Wires 97

3.7 Fullerene-Porphyrin Containing Triads for the Fine Tuning of Solar Energy Conversion 102

3.7.1 Synthesis and Characterization of a Perylene-Porphyrin-Fullerene Triad 105
3.7.2 Synthesis and Characterization of a Fluorene-Porphyrin Dyad 111
3.7.3 Synthesis and Characterization of a Pyrene-Porphyrin-Fullerene Triad 115

3.7.4	Conclusion	120
3.8	*Porphyrinato Phosphonic Acids for Applications in Printable Electronics* *121*	
4	**Summary** **126**	
5	**Zusammenfassung**.. **129**	
6	**Experimental Part**... **132**	
6.1	*Utilized Chemicals and Instruments* *132*	
6.2	*Syntheses and Spectroscopic Data* *133*	
6.2.1	Supramolecular Fullerene Building Blocks and Their Precursors	138
6.2.2	Supramolecular Porphyrin Building Blocks and Their Precursors	188
6.2.3	Building Blocks for Supramolecular Wires and Their Precursors	204
6.2.4	Porphyrin Precursors for Wittig-Horner Olefinations	209
6.2.5	OPV Rods and Their Precursors	214
6.2.6	Porphyrln-Fullerene-Triads and Their Precursors	219
6.2.7	Porphyrin-Phosphonic Acid and its Precursors	235
7	**References** **242**	

List of Abbreviations

a.u.	arbitrary units
β	attenuation factor
Boc	*tert*-Butoxycarbonyl
bs	broad singlet
conc.	concentrated
I_{Sc}	Short Circuit Photocurrent
CS	Charge Separation
CV	Cyclic Voltammetrie
d	Doublet
dba	dibenzylideneacetone
DBU	1,8-Diazabicyclo[5.4.0]undec-7-ene
dctb	*trans*-2-[3-(4-*tert*-butylphenyl)-2-methyl-propenylidene]malononitril
dd	Doublet of doublet
dhb	2,5-dihydroxybenzoic acid
DDQ	2,3-Dichloro-5,6-dicyano-1,4-benzoquinone
dq	Doublet of quartet
DCC	Dicyclohexylcarbodiimid
DCU	Dicyclohexylurea
δ	Chemical Shift [ppm]
DMA	Dimethylacetamide
DMAP	Dimethylaminopyridin
DMF	Dimethylformamide
DMSO	Dimethylsulfoxide
ES	Electron Spray
ESR	Electron Spin Resonance
eq.	Equivalent
eV	electronvolt
FAB	Fast Atom Bombardment
Fl	Flavin
fs	femtosecond
FSCV	Fast Scanning Cyclic Voltammetrie
g	Landé Factor

List of Abbreviations

GABA	γ-Aminobutyric acid
GOP	General Operation Procedure
h	hours
1-HOBT	1-Hydroxybenzotriazol
IPCE	Incident Photon to Current Conversion Efficiency
J	Scalar Coupling Constant [Hz]
K_{ass}	Association Constant
$\tilde{\nu}$	Wave Number [cm^{-1}]
λ	Wavelength [nm]
m	Multiplet
MALDI	Matrix Assisted Laser Desorption/Ionisation
min	Minute
mg	Milligram
mmol	Millimole
m/z	Mass-to-Charge Ratio
MeOH	Methanol
MS	Mass Spectrometry
η_h	Hill Coefficient
NHE	Normal Hydrogen Electrode
NMR	Nuclear Magnetic Resonance
ns	nanosecond
OD	Optical Density
ODCB	1,2 Dichlorobenzene
OF	Oligofluorene
OFET	Organic Field Effect Transistor
OLED	Organic Light Emitting Diode
OPA	Oligophenyleneacetylene
OPV	Oligophenylenevinylene
OTE	Optical Transparent Electrode
ppm	*Parts per Million*
ps	picosecond
PS	Photosystem
pTSA	Para Toluenesulfonic acid
r	Occupency

List of Abbreviations

rt	Room Temperature
s	Singlet
SCE	Standard Calomel Electrode
sin	Sinapinic acid
STM	Scanning Tunnelling Microscopy
t	Triplet
τ	Fluorescence Life Time
TBAF	Tetrabutylammonium fluoride
TEA	Triethylamine
TEM	Transmission Electron Microscopy
TFA	Trifluoroacetic acid
THF	Tetrahydrofurane
TOF	Time Of Flight
TPP	Tetraphenylporphyrin
UHV	Ultra High Vacuum
UV/Vis	Ultraviolet/Visible
V_{OC}	Open Circuit Voltage

1 Introduction

Sunlight is beside wind, geothermic energy and biomass our ultimate energy source. The economic development of the world business and society leads to an extreme increasing demand for energy.[1,2] At the same time the availability of fossil energy sources like oil, gas, coal and uranium recedes continuously.[3-5] On the other hand, we are not yet able to use the whole potential of the extraordinary amount of energy the sun supplies every day. This paradox was first pointed out by Giacomo Ciamician in his lecture entitled "The Photochemistry of the Future" in New York at the VIII International Congress of Pure and Applied Chemistry (1912): *"So far human civilization has made use almost exclusively of fossil solar energy. Would it not be advantageous to make a better use of radiant energy?"*[6] In order to increase the efficiency of solar energy conversion humanity needs first to understand in detail how sunlight is used by nature to power life. Indeed light excitation can induce a far range of chemical processes. For energy conversion purposes photoinduced electron transfer is by far the preferred reaction in nature. In this process a charge-separated state is created which is then used to prepare various high-energy molecules which fuel the organism.[7-9] When the mechanism of natural photosynthesis has been appreciated the mission of chemists is to arrange an array of artificial molecules to mimic this natural processes. The aim in this context is to convert and store solar energy for further applications. In the last few years many research groups in various current research projects were inspired by the power and effectiveness of natural processes. Scientist all over the world started to mimic natural systems and to examine artificial systems with high efficiency.[7-23] Unfortunately, up to now none of them was able to get even close to nature's efficiency.

1.1 Photoinduced Charge Separation and Solar Energy Conversion

1.1.1 Natural Reaction Centers

Nature widely uses antenna to solve the problem of light harvesting. Natural antenna collect large amounts of solar energy and redirect it as electronic excitation energy to reaction centers where subsequent conversion into redox chemical energy follows.

Introduction

The simplest and best understood reaction center is that found in purple bacteria which are often used as model for the photosynthetic centers.[7-9,24,25]

However, the most important solar energy conversion process is that explored in green plants.[26-28] There the PS II (Figure 1.1) has an electron-acceptor side similar to that of bacterial reaction center and a very peculiar donor side which can use water as an electron source and produces oxygen as "waste" product. PS II performs all the processes needed for photosynthesis–light absorption in antenna: energy transfer to a reaction center, charge separation and charge stabilization. To do that, PS II must:

• reach potentials high enough to oxidize water (>+0.9 V relative to NHE),
• handle such a high oxidation potential in fragile biological structures,
• couple the one-photon/one-electron charge-separation process to the four-electron water oxidation process.[29,30]

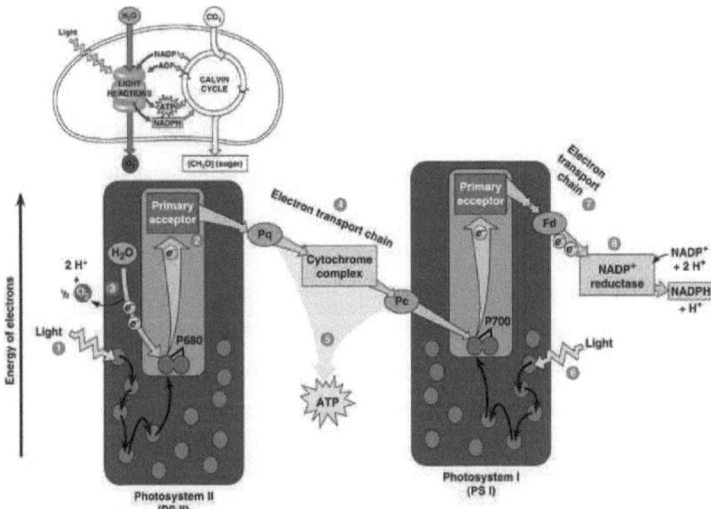

Figure 1.1: Schematic representation of the two reaction centers in naturally occurring photosynthesis. Photosystem II and photosystem I (PS II and PS I) are pigment/protein complexes that are located in specialized membranes called thylakoids.[31-33]

1.1.2 Artificial Reaction Centers

The complexity of natural photosynthetic systems is widely beyond the reach of the synthetic chemist. However this complexity is largely related to their living nature and it is entirely plausible that a single photosynthetic function, e.g. photo induced charge separation, can be duplicated by relatively simple artificial systems. The transport of energy and electrons within nanoscale-ordered materials is critical to the realization of systems for artificial photosynthesis as well as engineering molecular electronics and optoelectronics.[34] Studying multifunctional nanostructures – their design, synthesis, characterization and performance – mandates a careful control over composition, inter-chromophore-separation/angular relationship, overall dynamical and stimulus-induced reorganization, and electronic coupling. The overall goal is to achieve control over both the organization of the assemblies and their physical and chemical properties (i.e., electronic structures, etc.) and, thereby, enhancing desired functionalities, through simple external parameters or variables, with the intent of creating new tailored materials. Among the tools exploitable for the creation of new assemblies are organization principles, such as biomimetic methodologies, that help regulate size, shape and function down to the molecular scale.[35-37] Nature has long used these interactions to create a formidable array of structures, where a great level of control over the organization and an increased flexibility in replacing individual building blocks has been realized.

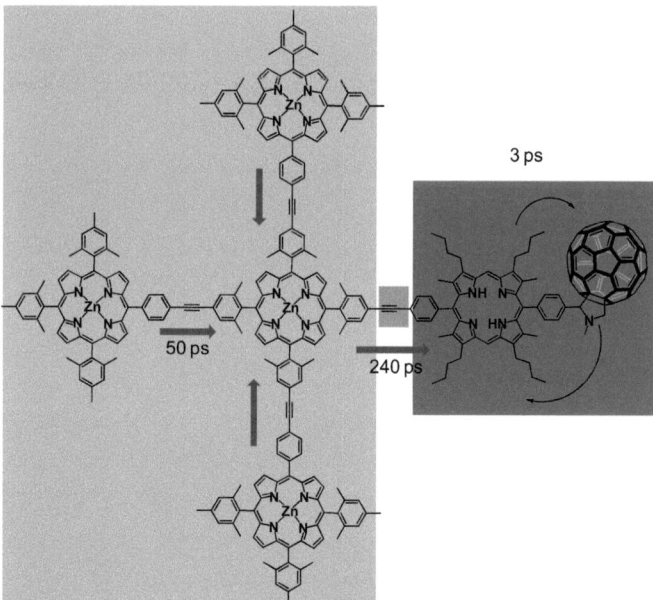

Figure 1.2: Integration of a light harvesting array and a reaction center.[38]

Nanohybrid systems that combine the favorable features of, for example, fullerenes and porphyrins as electron acceptors and electron donors, respectively, have received considerably interest (Figure 1.2).[38-51] As a consequence, new materials with a wide range of unique and spectacular physico-chemical properties have resulted in noteworthy advances in the areas of light induced electron transfer chemistry and solar energy conversion. It is mainly the small reorganization energy, which fullerenes exhibit in electron transfer reactions, that is accountable for a noteworthy breakthrough.[16,52,53] In particular, ultrafast charge separation together with very slow charge recombination features lead to unprecedented long-lived radical ion pair states formed in high quantum yields. This aspect is crucial towards the use of solar energy conversion as a clean, abundant and economical energy source. One of the commonly used strategies to achieve long-lived charge-separated states during photo induced electron transfer in model compounds (Figure 1.2) involves promoting multi-step electron transfer reactions along well-defined redox gradients (e.g. triads, tetrads, pentads, etc.).[39,41,54-67]

To control the rates and yields of electron transfer reactions and to eliminate the energy wasting charge recombination reactions, a better control over separation, angular relationship, electronic coupling and composition in donor-acceptor assemblies on a molecular level is desired. Self-assembled donor-acceptor nanohybrids (Figure 1.3) are considered to be a viable alternatively for the covalently linked molecular polyads in order to achieve an increased rate and yield of the charge-separation process and prolongation of the lifetime of the charge separated state.

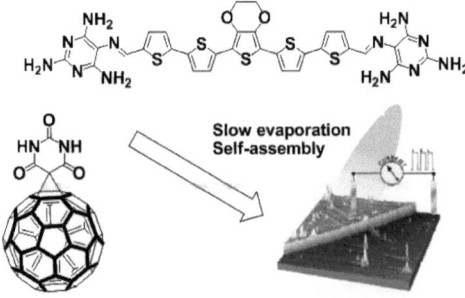

Figure 1.3: Hydrogen bonded all-organic device.[68]

Introduction

Biomimetic methodologies such as hydrogen bonding, metal-ligand complexation, electrostatic interactions and π–π stacking can be used to control the composition and architecture of photo and redox active complexes (Figure 1.4).
In literature several examples and methodologies to self-assemble fullerenes[42,69-75] and fullerene bearing donor-acceptor systems in solution and on electrode surfaces[76-87] are reported. In the case of modified electrodes containing porphyrin-fullerene systems high yields of photoelectrochemical currents (up to nearly 2%) are reported.[88-90]

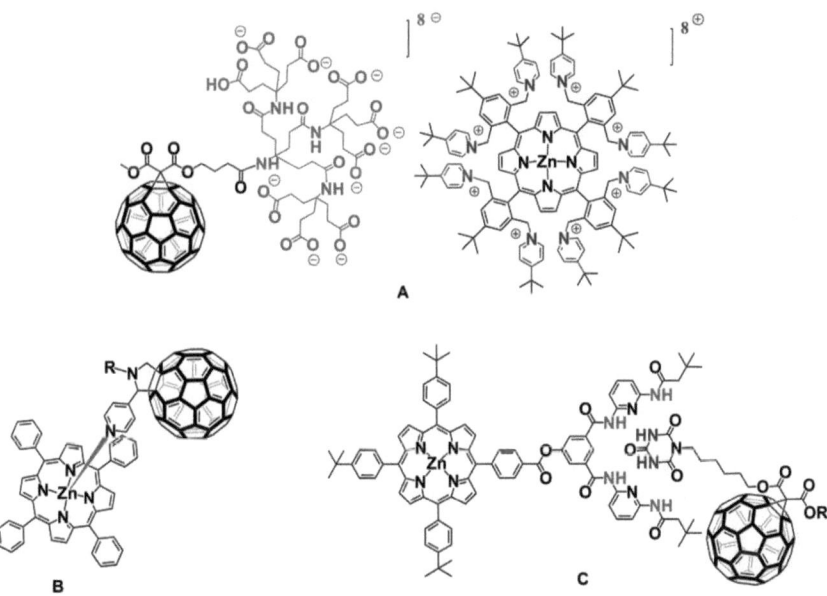

Figure 1.4: Photo and redox active compounds with electrostatic interaction (A),[90] Metal-ligand bonding (B)[48] and hydrogen bonding (C)[91] motif.

There is little doubt that the most potent supramolecular binding motif is hydrogen bonding. Despite its tremendous potential for the realization of highly directional self-assembly only a few examples of fullerene-porphyrin donor-acceptor nanohybrids have been constructed in this way so far.[91-95]

Introduction

1.2 Molecular Recognition

1.2.1 Receptors, Coordination and the Lock Key Analogy

Chemistry based on a host-guest (or receptor-substrate or lock-key) interaction is rest upon three historical concepts:

- The recognition by Paul Ehrlich in 1906 that molecules do not act if they do not bind, "Corpora non agunt nisis fixata". In this way Ehrlich introduced the concept of a biological receptor.
- The discovery by Emil Fischer in 1894 that binding must be selective. He found this by studying receptor-substrate binding in the case of enzymes. Fischer describes this as a lock-key image of steric fit. The guest has the shape or size complementary to the receptor or host (Figure 1.5). This concept represents the basis for molecular recognition, the discrimination by a host between a number of different guests.
- The fact that selective binding has to involve attraction or mutual affinity between receptor and substrate is in effect a generalization of Alfred Werner's theory of coordination chemistry from 1893. In this theory a metal ion is coordinated by a sphere of ligands.

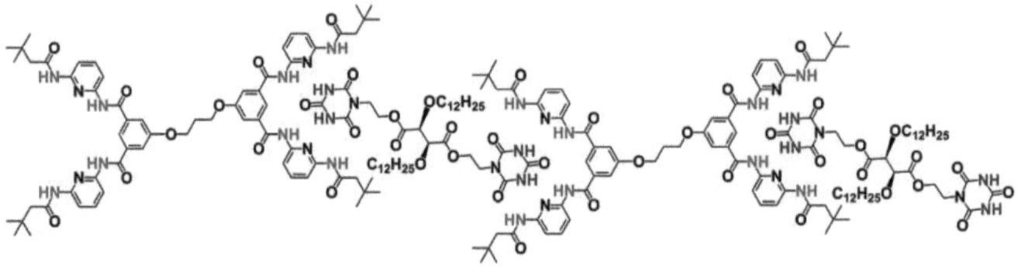

Figure 1.5: Lock and key principle of geometric fit between host (Hamilton receptor) and guest (cyanurate).[96]

These three concepts were developed independently and it took many years to associate them and to give birth to the new interdisciplinary field of supramolecular chemistry.[37]

1.2.2 Supramolecular Lock and Key Design

An important area of supramolecular chemistry is the assembly of multiple components in a defined way to perform diverse specific functions, such as photo induced electron or energy transfer.[97-102] In general self-assembly and molecular recognition involves the design and use of mono- or oligofunctionalized host or guest. In artificial systems multibinding within the same molecule is very rare. Especially when hydrogen bonding is involved the design is often complicated.

Dendrimers often have been proven to be suitable supramolecular hosts for guest molecules or guests for host molecules, respectively.[103-112] Due to their monodispersity and highly branched three-dimensional structure a microenvironment is created that encapsulates the host or guest. This entrapment is possible via hydrophilic/hydrophobic interactions.[104-111] Such non-bonding interactions are unspecific and even encapsulation of solvent molecules can be followed as a form of topological entrapment.

The integrated binding sites might recognize their targets based on acid-base, electrostatic or hydrogen-bonding interaction. The use of backbones which contain chromophoric units is extremely appealing since there is a great need for systems that can perform sensor functions and immobilization of biological substrates.[113]

Several publications have reported the ability of Hamilton receptors[114] to bind barbiturates from serum,[115,116] its use as a model for enzyme catalysis,[117-119] and as receptors in photoactive hydrogen-bond-based assemblies.[75,91,94,120-122]

1.3 *[60]Fullerene Chemistry*

1.3.1 Structural Properties of [60]Fullerenes

Fullerenes build closed carbon clusters which can be described by the empirical formula C_{20+2n}. In these molecules n is the number of hexagons.[123] This is the conclusion of the Euler Theorem which predicts that for each closed network of n hexagons exactly twelve pentagons are necessary.[124] Arrays of hexagons are planar like the structure of graphite. In the case of introducing pentagons into the array a bending attends which affords a closed carbon frame. Additionally all stable fullerenes satisfy the "isolated pentagon rule" (IPR).[125,126] This rule demands that all pentagons are surrounded by hexagons. This avoids pentalene analogue structures.

These structures are energetically disfavored due to their 8 π antiaromatic electron system and to a significant stronger bending.[123] The smallest fullerene that fulfills the IPR is [60]fullerene, also named buckminsterfullerene.[125,126]

Figure 1.6: Model of C_{60}

C_{60} exhibits I_h symmetry like a regular soccer ball which is shown in Figure 1.6. Each carbon atom of this icosahedron is identical. This can be concluded from the ^{13}C-NMR spectrum which shows only one peak at 143.2 ppm.[127] Due to the bending of the surface a pyramidalization of the carbon orbitals appears. The additional bending energy of 10.2 kcal/mol[128] is allocated to all carbon atoms. This results in an advanced p-character of the sp^x-hybrid orbitals from sp^2 in an exactly planar graphite layer to sp$^{2.278}$ in the fullerene frame. This leads to greater s-proportion in the π-symmetrical orbitals. Together with the minor overlap due to the bending a clear change in reactivity in comparison to planar conjugated systems appears. The electron affinity of [60]fullerene is decisive for its reactivity. That is why C_{60} reacts predominantly with nucleophiles.

Furthermore the localization of the double bonds at the cooperated edges of the hexagons can be investigated. These bonds are named [6,6] bonds. Their length of 1.36 Å is significantly shorter than the 1.47 Å of the [5,6] bonds between pentagons and hexagons. Due to this alternans and to the high electron affinity C_{60} is not treated as an aromatic but as an electron poor polyolefine. The structure is described as an alliance of 1,3,5-hexatriene and [5]radialene unities (Figure 1.7). Every [6,6] bond and every [5,6] bond is identical to each other due to symmetry.[123,129,130]

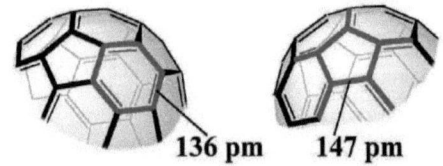

Figure 1.7: 1,3,5-hexatriene and [5]radialene substructure of C_{60}.

Introduction

The molecular frontier orbitals in Figure 1.8 proof these conclusions. The t_{1u} symmetrical LUMO is energetically relatively low and shows a great opportunity to gain electrons. The first reduction potential of C_{60} is between -0.3 to -0.4 V (against SCE).[39] Altogether as far as six electrons can be reversibly accepted.[131] This reflects the high electron affinity. The differences in comparison to aromatic compounds like benzene can also be proofed with the molecular frontier orbitals in Figure 1.8. Especially the dramatically smaller HOMO-LUMO gap in the case of C_{60} in comparison with benzene is outstanding. Due to their large HOMO LUMO gap aromatics are relatively inert towards nucleophilic reactions and prefer electrophilic substitutions. C_{60} in contrast prefers nucleophiles to react with.

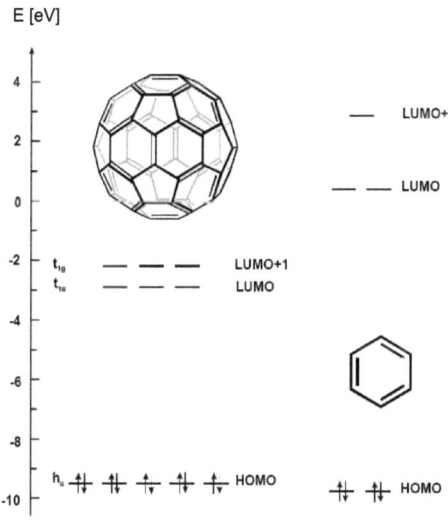

Figure 1.8: Molecular frontier orbitals of C_{60} and benzene.

1.3.2 Reactivity of [60]Fullerenes and Synthesis of Functional Exohedral [60]Fullerenes

The electronic and structural parameters and properties of C_{60} described in the chapter above demonstrate clearly that the buckminsterfullerene is a quite electronegative system. These informations imply that C_{60} behaves like an electron-poor conjugated polyolefine which consists of fused [5]radialene and cyclohexatriene units. Consequently C_{60} is predestinated to undergo nucleophilic addition reactions with carbon, nitrogen (Scheme 1.1), phosphorus and oxygen nucleophiles. After the

Introduction

initial attack of the nucleophiles to C_{60} the intermediate $Nu_nC_{60}^{n-}$ is formed (Scheme 1.2). This intermediate can be stabilized by:

• the addition of electrophiles E^+, e. g. H^+, or carbocations to give $C_{60}Nu_nE_n$
• the addition of neutral electrophiles EX such as alkylhalides to give $C_{60}Nu_nE_n$
• a S_Ni or internal addition reaction to give methanofullerenes and cyclohexenofullerenes
• an oxidation to give $C_{60}Nu_2$

Although statistically many isomers are possible the 1,2-addition pattern is preferred. In the case of sterically hindered nucleophiles/electrophiles 1,4-additions and even 1,6-additions are reported.[132]

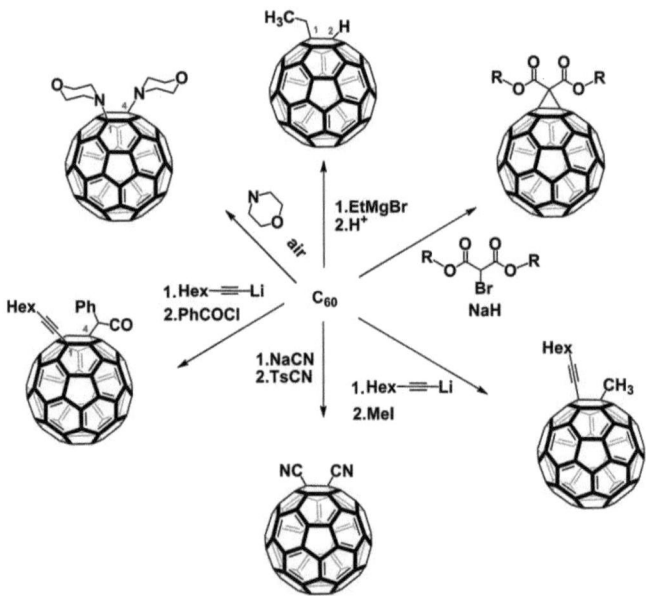

Scheme 1.1: Nucleophilic addition reactions on C_{60}.[123]

Typical reactions of C_{60} with carbon nucleophiles are reactions with organolithium (Scheme 1.1) and Grignard compounds containing alkyl or aryl groups (Scheme 1.2)[133-136] and cyclopropanation reactions (Scheme 1.3).[137-141] In the first case the reactions are very fast and the corresponding salts $C_{60}R_nM_n$ (M= Li or Mg) precipitate instantaneously. The protonation yields the hydrofullerene derivative $C_{60}R_nH_n$ (Scheme 1.2 B).

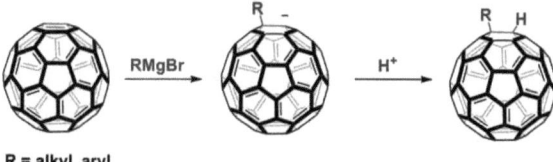

R = alkyl, aryl

Scheme 1.2: Nucleophilic addition on C_{60} leads to the intermediate. After adding an electrophile (H^+ in this case) the stabilized monoadduct can be isolated.

In the case of the cyclopropanation reaction the intermediate $C_{60}R^-$ can be stabilized by an intramolecular nucleophilic substitution (S_Ni) if R contains a good leaving group. As shown by BINGEL,[137] the generation of a carbon nucleophile by deprotonation of α-halo esters or ketones leads to clean cyclopropanation of C_{60}. Cyclopropanation of C_{60} with dimethyl bromomalonate in toluene with DBU as auxiliary base (Scheme 1.3) proceeds smoothly at room temperature. By-products are unreacted C_{60} and higher adducts. Modified reaction protocols are based on the *in situ* preparation of the halomalonate starting from iodine or CBr_4 as halogen source, malonate and DBU as base. The monoadduct can be isolated easily using column chromatography.

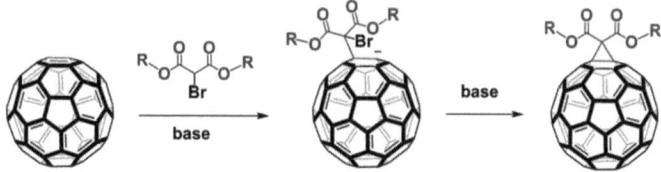

Scheme 1.3: Nucleophilic cyclopropanation of C_{60}. R is a wide variety of organic rests (e. g. Me), base is in most cases NaH or DBU.[142]

Quite similar protocols for introducing cyanides,[123] amines,[143] hydroxides,[144] alkoxides,[145] phosphorus nucleophiles,[146] silicon,[147] germanium nucleophiles[148] and macromolecular nucleophiles[149] exist.

Another interesting and versatile approach to functionalize fullerenes is to use its [6,6] double bonds for cycloadditions. C_{60} reacts as dienophile which leads to a large variety of compounds, mainly monoadducts (Scheme 1.4). Almost any functional group can be covalently linked to C_{60} by cycloaddition of suitable addends. Some of these addends show a remarkably stability and can for example be heated up to 400 °C without decomposition. This seems to be a prominent requirement for further side chain chemistry as well as for possible application of new fullerene derivatives as new materials or in pharmacy.

The most reported cycloaddition reactions of C_{60} are [4+2]-cycloadditions (Diels-Alder reactions).[150,151] The dienophilic character of C_{60} is comparable with that of maleic anhydride[152] or N-phenylmaleimide.[153] The reaction conditions strongly depend on the reactivity of the diene. Most of the [4+2]-cycloadditions of C_{60} are accomplished under thermal conditions, but photochemical reactions have also been reported, and in various additions microwave irradiation can be used efficiently.[154]

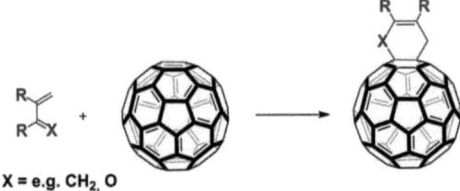

X = e.g. CH_2, O

Scheme 1.4: Typical [4+2]-cycloaddition at C_{60}.

Not only [4+2]-cycloadditions are reported in the literature, [3+2]- and [2+2]-cycloadditions are versatile tools to functionalize fullerenes at their periphery. In case of [3+2]-cycloadditions often azomethine ylids[155] (Scheme 1.5), diazoacetates,[156] diazoamides,[157] azides,[158] nitrile oxides[159] or carbonyl ylids[160] are involved as reactants.

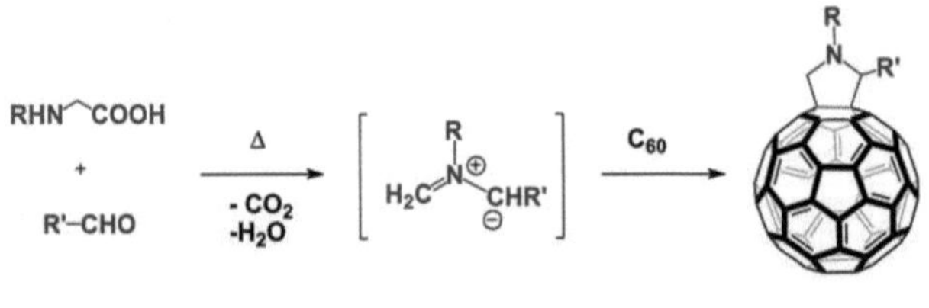

Scheme 1.5: [3+2] cycloaddition by azomethine ylids.[155]

Introduction

In most of these cases stable heterocycles are built directly at the fullerene [6,6]-bond (Scheme 1.5). Beside closed fullerene derivatives even opened structures are accessible via this route (Scheme 1.6).

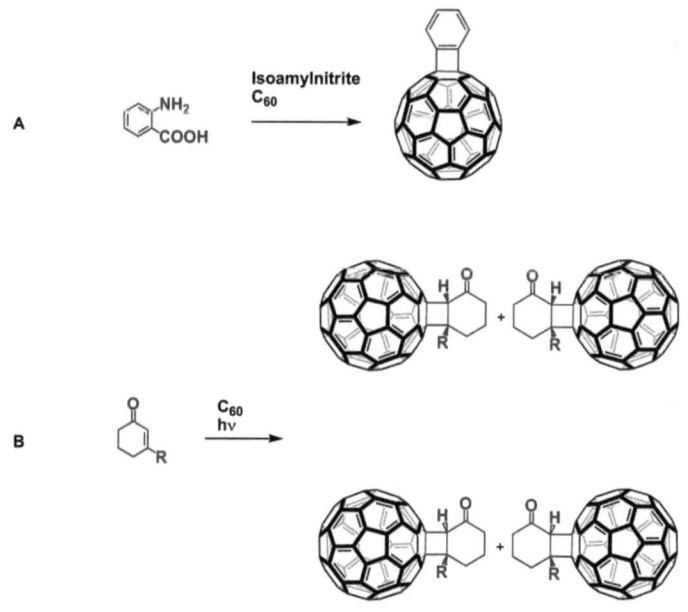

Scheme 1.6: [3+2]-cycloaddition of an organo azide at C_{60}. At first a [6,6]-triazoline intermediate is formed which can be detected or even be isolated in some cases.

Scheme 1.7: A) [2+2] cycloaddition of a benzyne derivative with C_{60}. The benzyne was prepared *in situ* from anthranilic acid with iosamyl nitrite. B) photoinduced [2+2]-cycloaddition of an enone with C_{60}. The mechanism is described as a stepwise addition of the enone triplet excited state to the C_{60} via an intermediate triplet 1,4-biradical.

[2+2] cycloadditions of C_{60} are reported with benzyne,[161] enones,[162] electron-rich alkynes[163] and alkenes,[164] ketenes[165] and ketene acetals.[165] Two examples are shown in Scheme 1.7.

The comparison of the reaction conditions show that monoadducts which are synthesized by a nucleophilic approach are often more stable than cycloadducts. In most cases the reaction conditions in nucleophilic reactions are easier to control and lead to higher yields. Side products and isomers can be prevented or minimized mostly in nucleophilic routes in contrast to cycloadditions.

1.1 Porphyrins

Porphyrins and other closely related tetrapyrrole pigments are found in a great number in natural systems. They often play a tremendous role in a wide range of biological processes. The analysis of haemoglobins (Figure 1.9) and myoglobins showed that heme, which is the iron(II)-protoporphyrin-complex, is responsible for the oxygen transport and storage in living tissues.[166] Heme was also isolated in the enzyme peroxidase, which catalyzes the oxidation of different substrates with hydrogen peroxide. The related enzyme catalase, containing heme as well, catalyzes the disproportionation of hydrogen peroxide to water and oxygen.[167,168] The family of cytochromes is another group of heme containing enzymes that is of great importance for the physiology and pharmacology of human beings. Whereas many cytochromes are crucial for electron transferring processes cytochrome P450 is important for the metabolism of naturally and exogenous substances occurring in the body.[169]

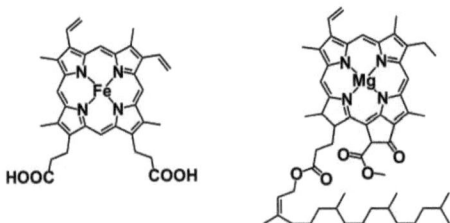

Figure 1.9: The two most common natural metal-porphyrins, haemoglobin (left) and chlorophyll-a (right).

Introduction

Closely related to porphyrins are chlorins which are assigned to one reduced pyrrole unit compared to the porphyrin core. The probably most famous representors of this class are the chlorophylls (e.g. chlorophyll-a, figure 1.9) which are found abundantly in green plants and phytobacteria. Chlorophyll systems play an important role in the processes of photosynthesis and are therefore essential for organic life on earth as we know it. Nearly all life on earth relies directly (autotrophic organisms) or indirectly (heterotrophic organisms) on the central role of the chlorophylls in photosynthesis by means of which sunlight energy is converted and stored as chemical energy.

1.1.1 Nomenclature of Porphyrins

The term porphyrin is derived from the Greek πορψυρα which means purple. Porphyrins are a huge class of organic molecules characterized by their tetrapyrrolic core structure. Therein four pyrroles are bridged by methine units under formation of an aromatic macrocycle. KÜSTER was the first who described the basic porphyrin structure in 1912,[170] before FISCHER & ZEILE[171] verified the postulated structure experimentally. The basic porphyrin core is shown in figure 1.10 including the denotation of the different positions of this system.

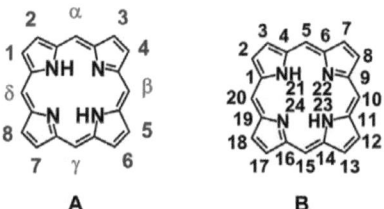

Figure 1.10: Structure of the basic porphyrin structure with numbering system according to FISCHER[172,173] A and IUPAC[174-176] B.

1.1.2 Aromaticity, Spectroscopy and Electronic Properties

Most of the porphyrins are planar, π-conjugated systems which consist of C-C double bonds and nitrogen lone pairs. Although the aromatic system consists of 22 π-electrons just the 18 π-electrons of the C-C double bonds are directly involved in the aromaticity.[177-179] This circumstance in mind the porphyrin core can be discussed as [18]annulene, respectively its bis-etheno bridged aza-analogue. Computational molecular dynamics assume an aromatic inner cross (Figure 1.11). Independently both considerations fulfil the HÜCKEL's rule (4n+2 π-electrons) and are considered as aromatic.[180,181] This structural anomaly strongly influence the spectroscopic (NMR, absorption, emission) and chemical behavior of porphyrins.

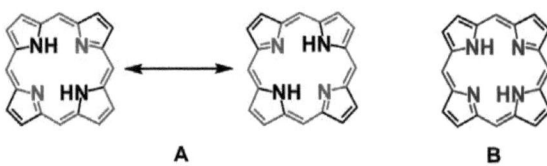

Figure 1.11: Mesomeric forms of the porphyrin core. Analogy to diaza-[18]annulene (A) and inner cross approach (B).

The aromatic character of porphyrins can be monitored by NMR spectroscopy very convincingly. Due to the anisotropic effect of the ring current, the NMR signals for the deshielded meso protons (protons on the bridging methine carbons) are shifted to low field (8 to 10 ppm), whereas the signals for the shielded protons on the inner nitrogen atoms occur at very high field (-2 to -4 ppm).[182]

The absorption of light in the UV-visible region is another important characteristic of porphyrins (Figure 1.12). The highly conjugated porphyrin macrocycle shows intense absorption (extinction coefficient > 200,000) at around 420 nm (the "SORET" band), followed by several weaker absorptions (Q-bands) at higher wavelengths (500 to 700 nm). Variations of the peripheral substituents on the porphyrin ring often lead only to minor changes as far as intensity and wavelength of these absorptions are concerned.

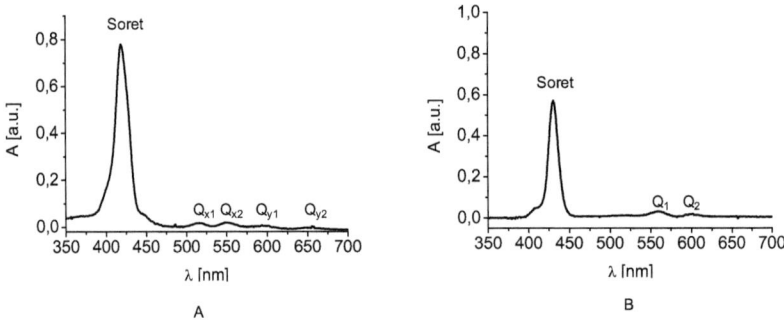

Figure 1.12: Characteristic absorption spectrum of a free base porphyrin A and zinc porphyrin B.

To explain the absorption behavior of porphyrins GOUTERMAN introduced the Four-Orbital Model[183,184] in the 1960s. According to this theory the absorption can generally be divided in two main parts. The first part is below 500 nm in which the main absorption band (SORET band) of porphyrins is localized. This band arises from an allowed π-π^* transition ($S_0 \rightarrow S_m$, m>2). In the region above 500 nm the Q-bands represent also π-π^* transitions ($S_0 \rightarrow S_1$ and $S_0 \rightarrow S_2$). In free base porphyrin usually four Q-bands were detected. Protonation of two of the inner nitrogen atoms or insertion of different metals into the porphyrin cavity typically changes the visible absorption. In the case of metal insertion the system achieves greater symmetry which is clearly proven by the reduction of the number of the Q bands.[185,186]

2 State of the Art & Aims

This thesis is majorly focused on the development of novel carbon-rich compounds, their photophysical properties and their application for solar energy conversion. In this context supramolecular fullerene-porphyrin architectures are highlighted. Accessorily to further supramolecular nanohybrids (i.e. fullerene-flavin hybrids and supramolecular rods), triads containing fullerene, porphyrin and perylene/fluorene subunits and molecular wires are synthesized and analyzed towards their properties.

2.1 State of the Art

Concerning the supramolecular setup of fullerene-porphyrin nanohybrids D'SOUZA et al.[95] reports a Hamilton type hydrogen bond fullerene-porphyrin complex (Figure 2.1) where amino pyridine functionalized fulleropyrrolidine **1** binds to carboxylic acid porphyrin **2**. Kinetic studies, namely time-resolved emission and transient absorption, show efficient photoinduced charge separation (k_{cs} = 1.4 x 10^{-9} s^{-1}) and slow charge recombination.

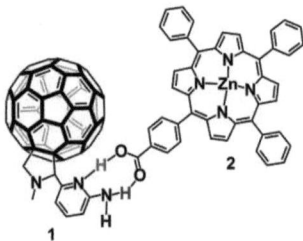

Figure 2.1: Schematic representation of Hamilton receptor type fullerene-porphyrin hybrid.

In this architectures even two folded hydrogen bonding leads to a rather stable 1:1 complex (K_{ass} = 4.4 x 10^3 M^{-1}). Surprisingly ionic interactions in this supramolecular array are not discussed.

In contrast to this features SANCHEZ et al[93] described supramolecular fullerene-porphyrin architectures assembled via a combination of hydrogen bonding and electrostatic attraction. In this particular light carboxylate modified fullerene derivative **3** complexes amidinium bearing porphyrins **4** (Figure 2.2).

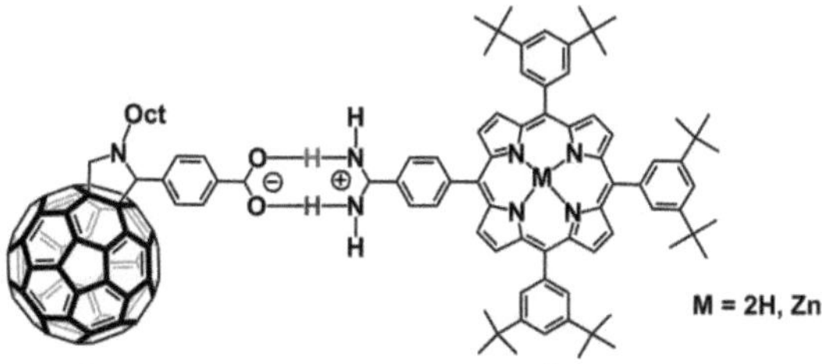

State of the Art & Aims

Figure 2.2: Amidinium-carboxylate fullerene-porphyrin assemblies.

The complexation constants K_{ass} reaches values of 10^7 M^{-1} which is extremely high in comparison to systems in which an electrostatic interaction additional to the hydrogen bonding is missing. The very fast formation, the high efficiency and the long-lived (10 µs in THF) charge separation are attributed to extremely strong electronic couplings of those systems.

In the light of hydrogen bonded supramolecular fullerene-porphyrin architectures intensive preliminary work during my masterthesis was done.[91] In this context we have investigated for the first time the self-assembly and the photophysical properties of supramolecular fullerene-porphyrin donor-acceptor nanohybrids connected *via* one Hamilton-receptor based hydrogen binding motif. In this light, we have synthesized an entire modular set of new C_{60} monoadducts **5** (Figure 2.3) carrying a cyanuric acid terminus within the malonate addend connected via hexylene spacers.

These new derivatives complement the structure of previously reported adducts bearing propylene spacers. To overcome the overall poor solubility, especially in solvents where no interference with the hydrogen bonding is expected, we introduced as the second malonate termini a variety of dendritic groups. The Hamilton-receptor counterpart was coupled to a library of porphyrin derivatives involving either tin or zinc as central metals leading to new porphyrin building blocks **6** (Figure 2.3). The corresponding association constants were determined by NMR- and fluorescence titration experiments. They were found to be in the range between 3.7×10^3 and 7.9×10^5 M^{-1}. Interestingly the association constants decrease when hexylene instead of propylene spacers are used. In response to visible light irradiation, the corresponding **6d·5** complexes give rise to a fast charge separation evolving from the photoexcited ZnP chromophores. Time-resolved fluorescence and transient absorption measurements were employed to corroborate this pathway. Most importantly, while the oxidized zinc porphyrin π-radical cation (ZnP$^{•+}$) was identified through its fingerprint absorption in the 550 – 800 nm range, the signature of the reduced fullerene π-radical anion (C$_{60}$$^{•-}$) appeared at 1040 nm (Figure 2.4). In stark contrast, in the analogous SnP complexes (**6a·5**) energy transfer – instead of electron transfer – occurs from the initially excited SnP excited state to C$_{60}$. This is due to the shifted oxidation potential when comparing ZnP and SnP.

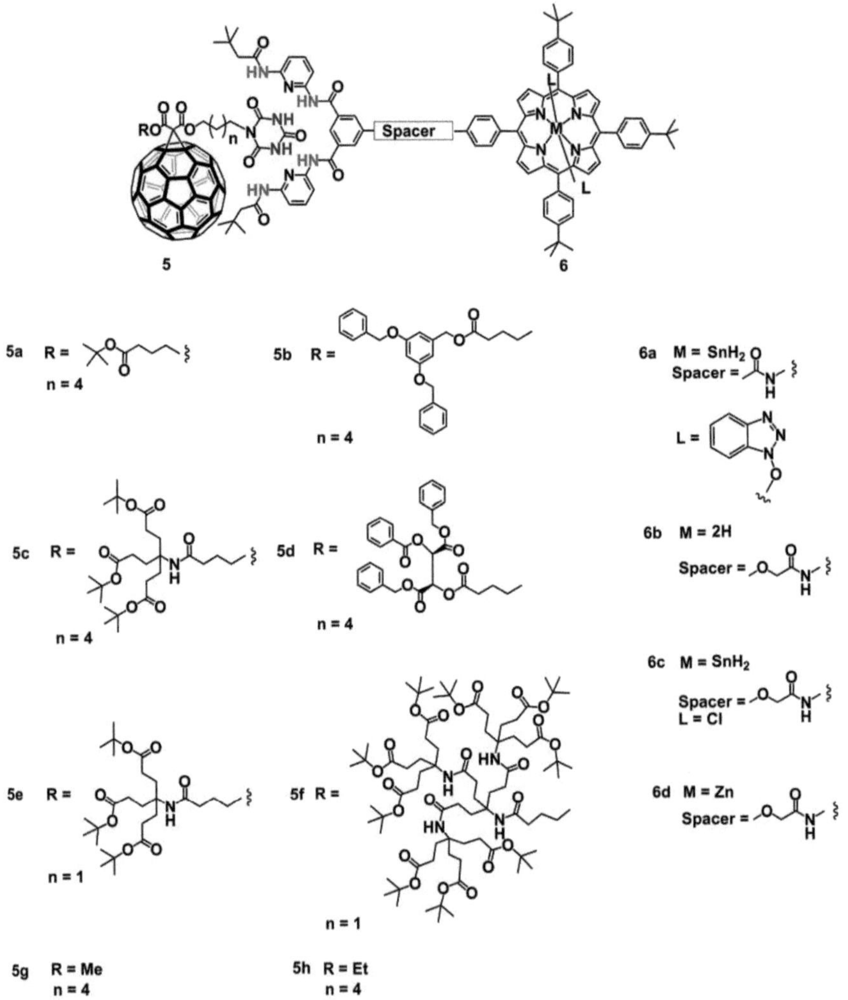

Figure 2.3: Schematic representation of the 5x6 complexes between fullerene cyanurates and porphyrin-Hamilton receptors associated by six-fold hydrogen bonding.

The most important and unprecedented finding is that electronic communication between the porphyrin donor and fullerene acceptor is even possible through a considerable number of σ- and hydrogen bonds. This is expected to open new avenues of approaches for the inexpensive and efficient construction of new prototypes of photovoltaic devices.

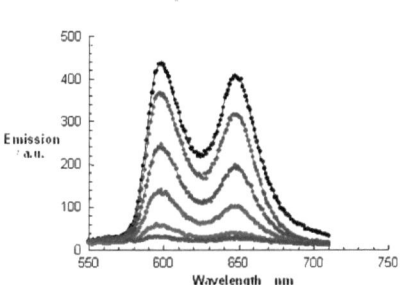 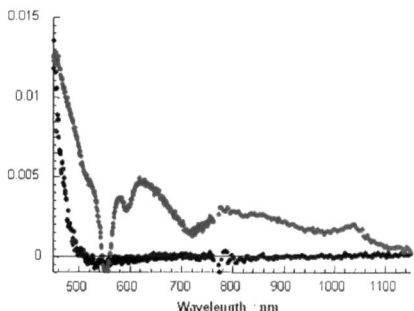

Figure 2.4: Left part: Fluorescence intensity of Zn porphyrin 2d (1.2×10^{-6} M) in toluene upon addition of increasing amounts of C_{60} compound 1e ($0-7.1 \times 10^{-6}$ M). Right part: Differential absorption spectra (visible and near-infrared) obtained upon femtosecond flash photolysis (550 nm) of Zn porphyrin 2d / C_{60} compound 1e in nitrogen saturated toluene with several time delays (i.e., 0 ps – black spectrum; 500 ps – grey spectrum).

2.2 Aim of the Work

The intention of this work is to develop and to analyze novel electron donor-acceptor systems for solar energy conversion. In respect to their well known role in photoactive devices the focus of these studies is on fullerene and porphyrin systems. Starting from our well established supramolecular approaches, directed hydrogen bonding is chosen to self-assemble the fullerene-porphyrin architectures. Due to their strongly interaction the Hamilton receptor/cyanuric acid key-lock system is applied in this context.

In order to study electron transfer systems and their properties linear, rigid systems seem to be favorably due to the exact localization of electron donor and acceptor in solution. Beyond that spacers that include conjugated unsaturated carbons (e.g. oligophenyl, oligoacetylene, oligophenyleneacetylene (OPA), oligophenylenevinylene (OPV) and oligothiophene) allow electron transfer processes through bonds. Electron

State of the Art & Aims

transfer processes in saturated carbon chains instead facilitate only through space. The circumstances in mind unsaturated systems are introduced as spacers between fullerene and Hamilton receptor (Figure 2.5). As key cyanuric acid bearing porphyrin **7** is an interesting possibility. Beside synthetic aspects the photophysics of the hybrids are discussed.

Additionally the electron donating probability of flavin compound **8** is been taken into account. Therefore Hamilton receptor analogues for three-folded hydrogen bonding are developed and the self-assembled fullerene-flavin hybrids (Figure 2.5) should be analyzed towards their photophysical and electrochemical properties.

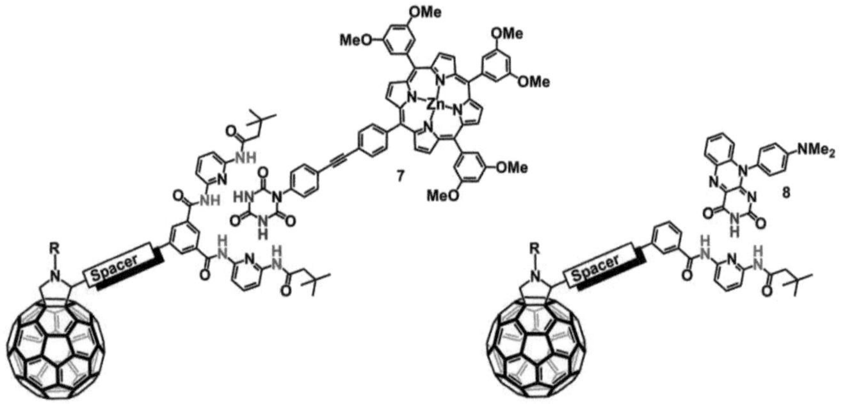

Figure 2.5: Self-assembled nanohybrids for solar energy conversion.

Meanwhile the synthesis of supramolecular linear rigid oligomers containing phenyleneacetylene building blocks is interesting for the examination of oligophenyleneacetylenes' (OPA) role in electron transfer processes. In this context bis Hamilton receptors connected *via* OPAs shall be developed and the photophysical interaction after complexation with chromophores (i.e. porphyrin **7**) be investigated.

Complementary to supramolecular donor-acceptor systems molecular triads where an additional light absorbing chromophore (i.e. perylene **9**, pyrene or fluorene) is included in the fullerene-porphyrin architecture (Figure 2.6 left) are of great interest. The conceptual aim of these hybrids is to broaden the absorption spectrum of the light harvesting antennas and therewith a fine tuning of electron transfer processes.

State of the Art & Aims

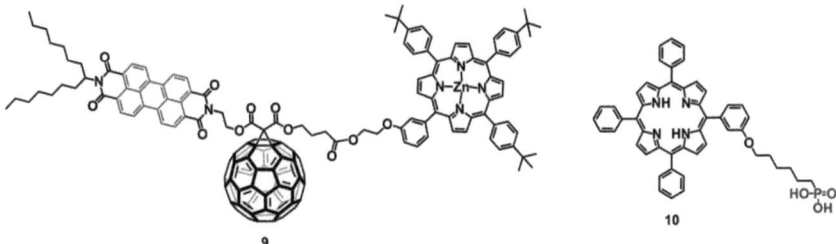

Figure 2.6: Triad **9** containing an additional chromophore for broadening the absorption spectrum (left). Porphyrinato porphyrin **10** for the application in printable electronics (right).

This work should be completed by the synthesis and characterization of porphyrinato phosphonic acids **10** (Figure 2.6 right) for applications in nanoelectronic devices. The combination of a suitable anchoring group and an electroactive subunit is a promising assembly for printable electronics (e.g. nanostructured transistors, etc.).

3 Results and Discussion

3.1 Self-Assembly of Supramolecular Fullerene-Porphyrin Architectures

In the case of the described supramolecular fullerodendrimer-porphyrin assemblies **5 6** (see figure 2.3 on page 20) flexible alkyl chains as spacer between the key-lock systems were introduced what might result in an inter/intramolecular folding. Conjugated spacers (i.e., *p*-phenylene-ethynylene, *p*-phenylenevinylene, *p*-ethynylene and fluorine) provide wire-like behavior in terms of electron transfer processes and restrict the flexibility of the Hamilton receptor and cyanuric acid functionalities at the same time. The efficient electronic coupling between electro active units -donor and acceptor termini- guarantees in this context a wire-like behavior. Therefore, different factors play a key role for an ideal molecular wire: 1) matching between the donor (acceptor, respectively) and bridge energy levels, 2) good electronic coupling between the electron donor and acceptor units by means of the bridge orbitals[187-190] and 3) a small attenuation factor (β).[191,192] Intramolecular electron transfer along π-conjugated oligomers, such as o-phenylene-vinylene (oPV), has been tested in several donor-acceptor conjugates involving porphyrins,[193,194] anilines,[195,196] or ferrocenes[197] as electron donors and C_{60} as electron acceptor. In fact, recent studies show that electron transfer along oPVs is much more efficient as in o-fluorenes (oF) and o-phenylene-ethynylenes (oPE) which is caused by a complete delocalization of the π-electrons in oPVs, whereas the electron density in oPE is localized at the phenyl rings. The electron delocalization in oFs can be found between both extremes.[198] Here, an electron transfer along π-conjugated spacers in a novel series of supramolecular assembled fullerene-porphyrin hybrids is reported. In this particular light *p*-phenylene-ethynylene, *p*-phenylene-vinylene, *p*-ethynylene, phenylene, thiophene and fluorine are introduced as spacer between [60]fullerene and Hamilton receptor and a *p*-phenylene-ethynylene bridged porphyrinato cyanuric acid.

3.1.1 Synthesis

Hamilton Receptor Modified Fullerenes

The target Hamilton receptor derivatives are synthesized starting from commercially available 5-bromo-xylene **11** and 5-iodo-xylene **12**, which can be easily oxidized by $KMnO_4$ to 5-bromo isophthalic acid **13** or 5-iodo isophthalic acid **14**, respectively. Treatment of the diacids **13** and **14** with thionyl chloride ($SOCl_2$) at reflux and under dry conditions leads to the dichlorides **15** and **16**, which then can be coupled with aminopyridine derivative **17**[199] affording the bromo Hamilton receptor **18** in 92 % and the iodo Hamilton receptor **19** in 94 % isolated yield (Scheme 3.1).

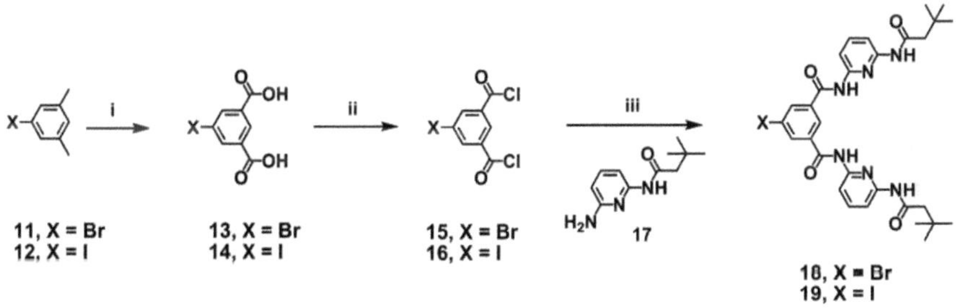

Scheme 3.1: Synthesis of bromo Hamilton receptor 18 and iodo Hamilton receptor 19. i) $KMnO_4$, tert-butanol/H_2O 1:1, Δ; ii) $SOCl_2$, DMF, Δ; iii) NEt_3, THF, 0°C→rt.

Iodo Hamilton receptor **19** is a versatile building block for subsequent C-C cross coupling reactions. In particular SONOGASHIRA cross coupling reactions[200,201] between terminal alkynes and aryl halides are expected to be suitable. Due to further functionalization reactions (i.e. PRATO reaction[202] with C_{60}) formyl groups, e.g. in para position to the cross coupled aryl groups are inevitable. To combine both synthetic accounts 4-ethynyl benzaldehyde **20** is used as a promising reagent for the derivatization of the Hamilton receptor **19**. The coupling of both compounds is nearly quantitative at room temperature (Scheme 3.2). As Pd^0 source, $Pd(PPh_3)_2Cl_2$ (bis (triphenylphosphine)palladium(II)dichloride) and as Cu^I catalyst, CuI are chosen. Elongation of the p-phenylene-ethynylene chain is possible by using 4-bromo benzaldehyde **23** and (2-(4-ethynylphenyl) ethynyl)triisopropylsilane **22,** which are coupled using Pd^0/Cu^I catalysis at 75 °C (Scheme 3.2). Deprotection of the resulting elongated p-phenylene-ethynylene wire **24** using TBAF (tetrabutylammonium fluoride) as fluoride source leads to the terminal ethynyl derivative **25**, which is subsequently coupled with the Hamilton receptor building block **19** as described above for 4-ethynyl benzaldehyde **20** to give the molecular wire **26** (Scheme 3.2).

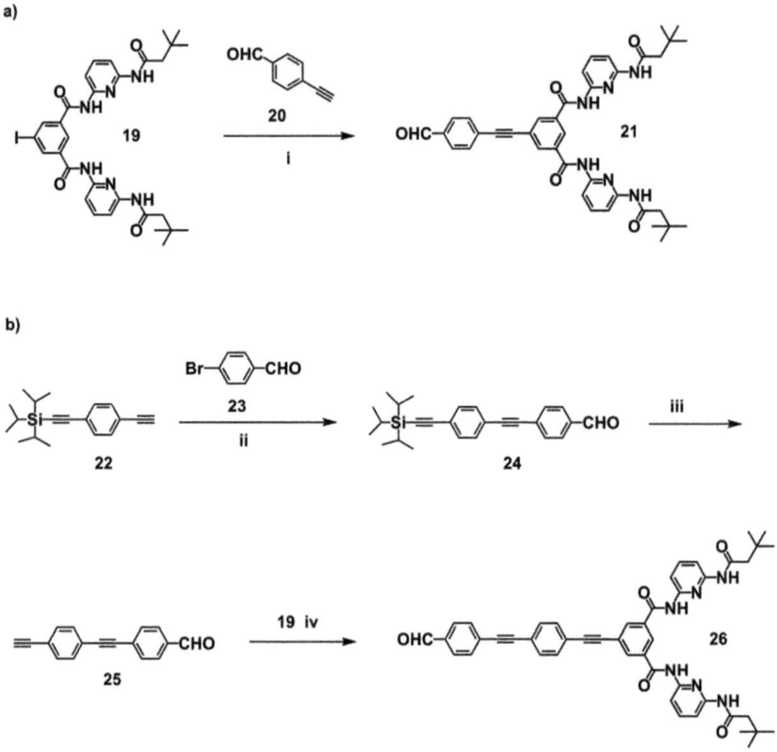

Scheme 3.2: Syntheses of Hamilton receptor bearing compounds 21 and 26. i) Pd(PPh$_3$)$_2$Cl$_2$, CuI, NEt$_3$, THF, rt; ii) Pd(PPh$_3$)$_2$Cl$_2$, CuI, HNEt$_2$, toluene, 75 °C; iii) TBAF, THF, 0 °C; iv) Pd(PPh$_3$)$_2$Cl$_2$, CuI, NEt$_3$, THF, rt.

To investigate the influence of the bridge's nature on the photophysical properties Hamilton receptor bearing fullerenes that are not only connected by *p*-phenylene-ethynylene chains, but also by *p*-ethynylene, *p*-phenylene-vinylene, phenylene, thiophene and fluorene spacers are designed. In case of the short ethynyl bridge *p*-ethynyl precursor compound **28** can be easily synthesized using 3,3-diethoxyprop-1-yne **27** and the iodo Hamilton receptor **19** under Pd^0/Cu^I catalysis in tetrahydrofurane and diethylamine. Deprotection of the acetal by adding TFA and subsequent

Results and Discussion

neutralization with sodium bicarbonate afforded the ethynyl Hamilton receptor **29** (Scheme 3.3).

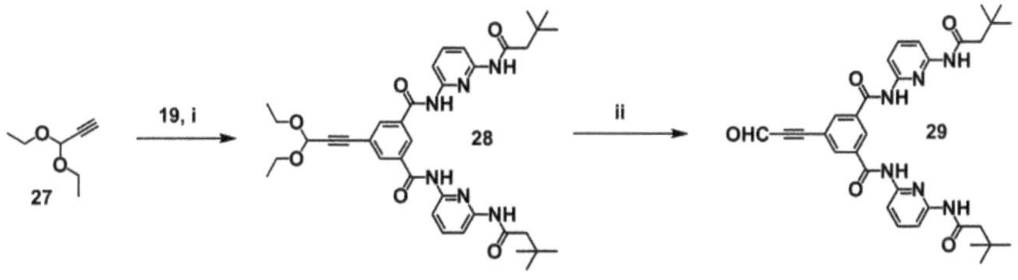

Scheme 3.3: Synthesis of Hamilton receptor bearing compound 29. i) Pd(PPh$_3$)$_2$Cl$_2$, CuI, HNEt$_2$, THF, rt; ii) 1. TFA, CH$_2$Cl$_2$, rt, 2. H$_2$O/NaHCO$_3$, rt.

The introduction of a *p*-phenylene-vinylene spacer is managed by the synthesis of 4-ethenyl benzaldehyde **31**. Therefore n-BuLi and DMF are added to 4-bromo styrene **30** at -78 °C. HECK conditions[203] are then applied for the coupling of 4-ethenyl benzaldehyde **31** and the iodo Hamilton receptor **19** to give the *p*-phenylene-vinylene Hamilton receptor **32** using Pd$_2$dba$_3$ (tris(dibenzylideneacetone)dipalladium(0)) as Pd^0 source and triphenyl arsine as additional ligand (Scheme 3.4).

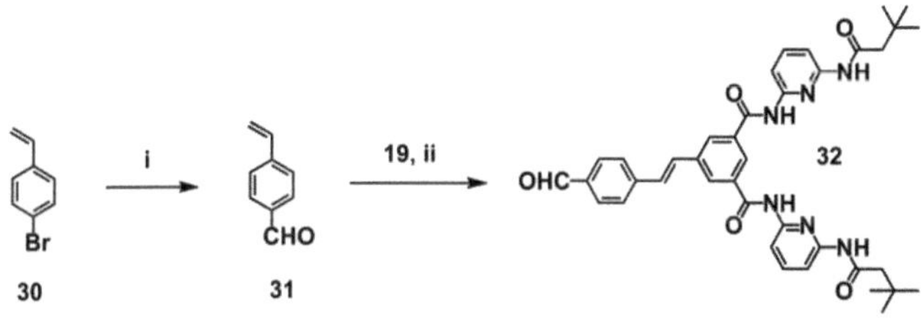

Scheme 3.4: Synthesis of vinylene Hamilton receptor 32: i) 1. n-BuLi, THF, -78 °C, 2. DMF, -78 °C, 3. H$_2$O/NH$_4$Cl, rt; ii) Pd$_2$dba$_3$, AsPh$_3$, NEt$_3$, THF, 70 °C;

Besides HECK and SONOGASHIRA coupling techniques SUZUKI coupling[204-206] can be used for further functionalization of the Hamilton receptor motif. In this light 4-formylphenyl boronic acid **33** and 3-formylthiophene boronic acid **35** are coupled with iodo Hamilton receptor **19** using tetrakis(triphenylphosphine)palladium(0) as catalyst and potassium carbonate as base (Scheme 3.5). Both π-conjugated systems can be isolated in good or moderate yield (phenylene Hamilton receptor **34** in 71 % and thiophene Hamilton receptor **36** in 51 %).

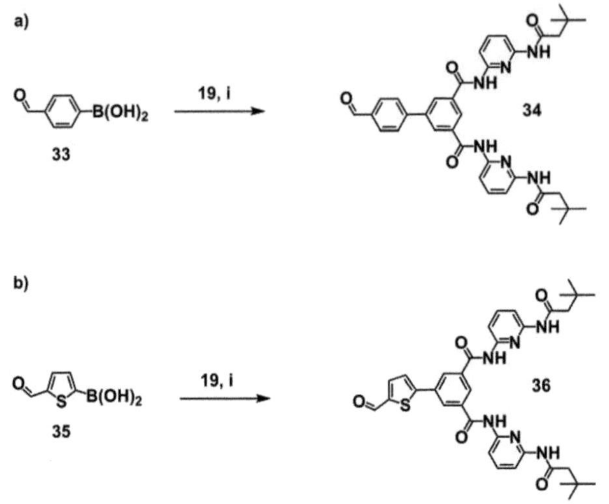

Scheme 3.5: Syntheses of π-conjugated Hamilton receptors **34** and **36**. i) Pd(PPh₃)₄, K₂CO₃, DMF, rt.

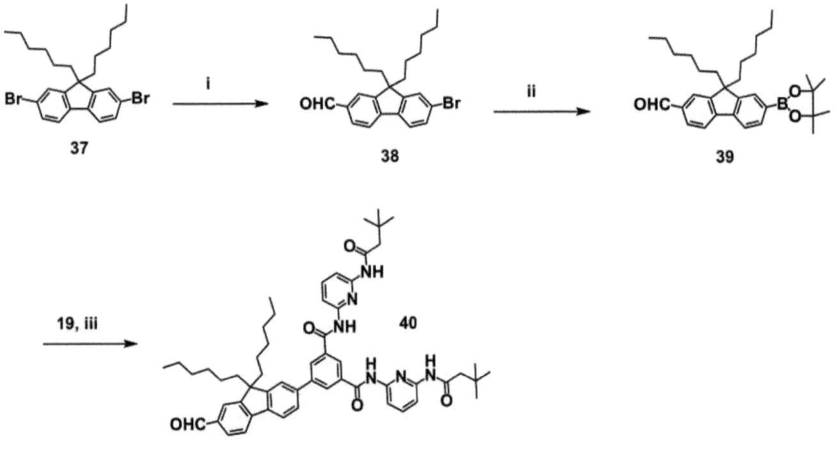

Scheme 3.6: Synthesis of fluorene Hamilton receptor **40**. i) 1. n-BuLi, THF, -78 °C, 2. DMF, -78 °C, 3. H₂O/HCl, rt; ii) pinacolyldiborane, Pd(OAc)₂, KOAc, DMF, 90 °C; iii) Pd(PPh₃)₄, K₂CO₃, DMF, rt.

Starting from 9,9-dihexyl-2,7-dibromofluorene **37** monoformyl fluorene boronic ester **39** can be achieved easily in a two step synthesis according to the literature known procedure.[207] The fluorene containing precursor **40** was obtained by SUZUKI coupling[204-206] of the boronic acid ester **39** and Hamilton receptor **19** under the action of Pd(PPh$_3$)$_4$ (tetrakis(triphenylphosphine)palladium(0)) and potassium carbonate (Scheme 3.6).

The wire like Hamilton receptor fullerene hybrids **41-47** are then synthesized by applying standard PRATO conditions[202] and allow C$_{60}$ to react with the aldehydes **21, 26, 29, 32, 34, 36** and **40** in THF/toluene at reflux (Scheme 3.7). To increase the yield of the desired monoadduct a slight excess of C$_{60}$ and the adequate dilution (1 mg C$_{60}$ per 1 mL solvent) is crucial.

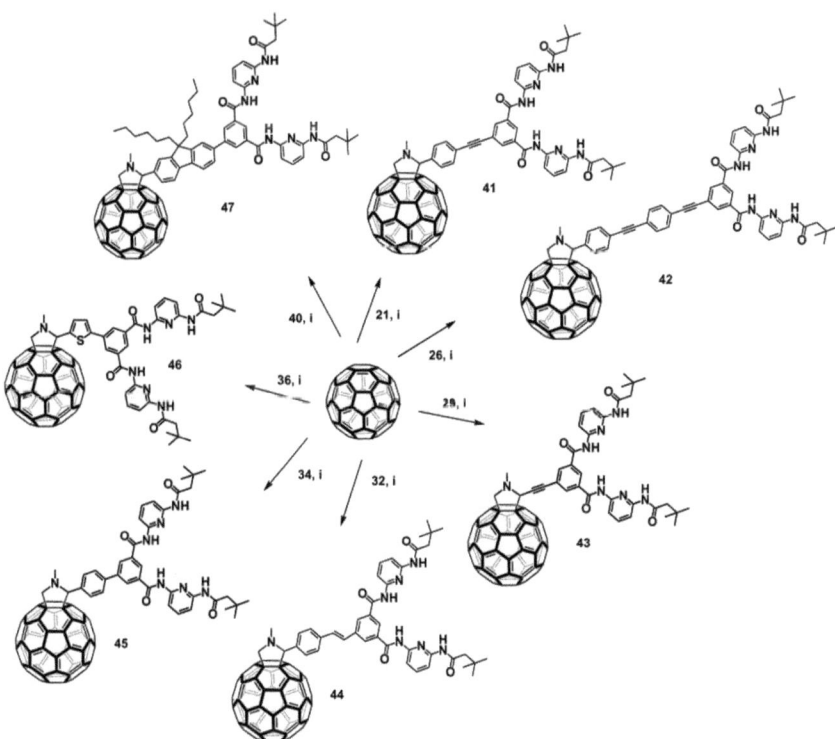

Scheme 3.7: Syntheses of the Hamilton receptor bearing fullerene derivatives **41-47**. i) C$_{60}$, sarcosine, THF/toluene, Δ.

All Hamilton receptor modified fulleropyrrolidines are characterized by ^{1}H/^{13}C NMR, UV/Vis and IR spectroscopy and mass spectrometry. As typical example for NMR spectra of Hamilton receptor bearing fullerene monoadducts the ^{1}H NMR spectrum (Figure 3.1) and the ^{13}C NMR spectrum (Figure 3.2) of compound **41** are depicted.

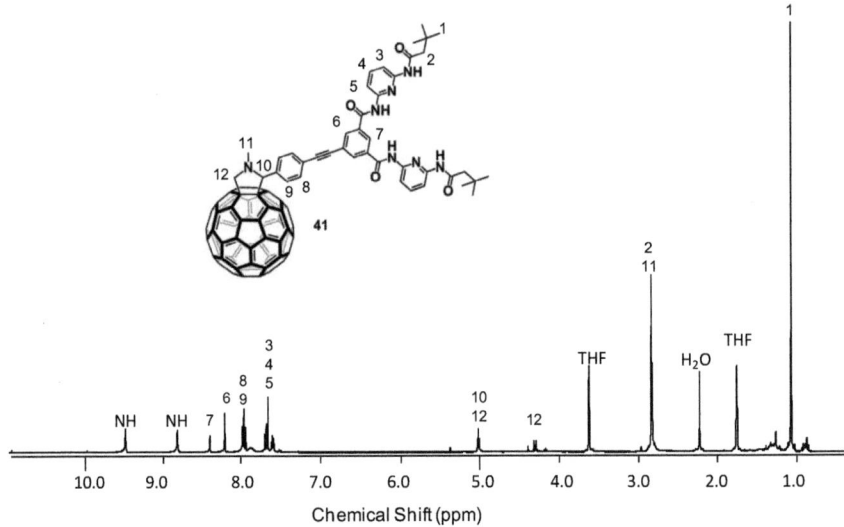

Figure 3.1: ^{1}H NMR (400 MHz) spectrum of fulleropyrrolidine 41 in THF d$_8$/CS$_2$ at rt.

Typical for the ^{1}H NMR spectrum (Figure 3.1) of product **41** in comparison to the precursor **21** are the signals which can be assigned to the pyrrolidine and the loss of the signal of the formyl proton. The singlet at 2.81 ppm belongs to the *N*-methyl group of the pyrrolidine. The singlet for the CH-group of the pyrrolidine appears at 4.97 ppm. The signals for protons of the CH$_2$-group split into two doublets which occur at 5.00 ppm (d, ^2J = 9.23 Hz, 1H) and 4.28 ppm (d, ^2J = 9.40 Hz, 1H). The other signals can clearly be related to the Hamilton receptor moiety. The most characteristic signals are certainly the one of the NH groups which occur as singlets at 9.45 (s, 2H) and 8.79 ppm (s, 2H) and the singlet of the *tert*-butyl group which can be found at 1.04 ppm (s, 18H).

Characteristic for the ^{13}C-NMR spectrum of monoadduct **41** (Figure 3.2) are the signals for the sp^2 carbons of C$_{60}$. There are 27 signals which appear in the region of 147.88 ppm to 138.88 ppm. Most monoadducts have a C$_s$ symmetry (one vertical mirror plane) which let us expect 29 signals for the sp^2 carbon atoms of C$_{60}$ between 148 and 138 ppm. Actually only 27 signals appear in the ^{13}C-NMR spectra of the synthesized monoadduct. By closer examination not all signals show the same

Results and Discussion

structure. Some are very broad in comparison to other, some nearly twice intense than other. This can only be explained by an overlap of some signals to a more intense or a broader signal. The signals for the quaternary C_{60} carbons appear at 70.47 and 69.79 ppm. Typical signals which differ from the ^{13}C-NMR spectrum of the precursor compound are the signals of the pyrrolidine ring. The *N*-methyl carbon causes a signal at 40.14 ppm, the CH_2-group at 77.52 ppm and the CH-group at 124.80 ppm. All other signals are very similar to compound **21**. The carbonyl signals occur at 170.77 and 164.52 ppm, whereas the α-pyridine carbons produce signals at 151.08 and 150.62 ppm. The signals of the β-pyridine carbons arise at 110.47 and 110.24 ppm. The ethynyl carbon signals can be found at 91.69 and 89.02 ppm. Characteristic for the neopentyl group are the signals at 50.98 (CH_2), 31.59 ($C(CH_3)_3$) and 30.11 (CH_3) ppm. Notable is the loss of the formyl carbon signal.

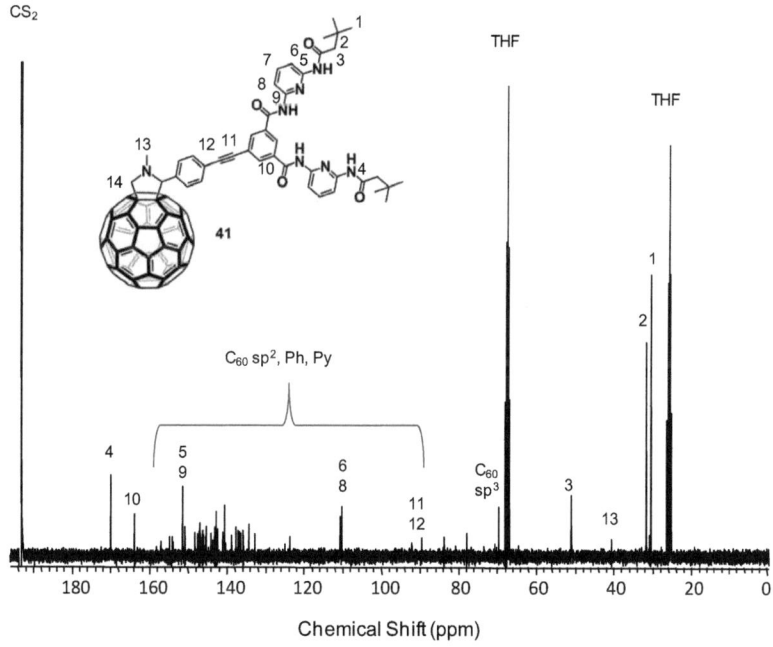

Figure 3.2: ^{13}C NMR (100.5 MHz) spectrum of fulleropyrrolidine 41 in THF d_8/CS_2 at rt.

Results and Discussion

Barbituric/Cyanuric Acid Bearing Porphyrins

As key-system to the lock-containing [60]fullerene derivatives **41-47** porphyrin building blocks that contain barbituric or cyanuric acid moieties are designed. In the first approach the synthesis of a tetraphenylporphyrin core with three *tert*-butyl groups for increasing the solubility and one bromo or iodo group for further derivatization is planned. The idea is to create a mono-substituted porphyrin which can be easily formylated and transferred into further derivatives. 4-bromo-benzaldehyde **23** and 4-iodo-benzaldehyde **49** are chosen as educts which are commercially available and easy to handle. Another reason is that it provides a certain rigidity in the desired final product. The synthesis is performed with a statistical batch under modified LINDSAY-conditions.[208-211] Four equivalents pyrrole **50**, one equivalent 4-bromo-benzaldeyhde **23** or 4-iodo-benzaldehyde **49** and three equivalents 4-*tert*-butylbenzaldehyde **48** are dissolved in dichloromethane under addition of ethanol and tetraphenyl phosphonium chloride (Scheme 3.8).

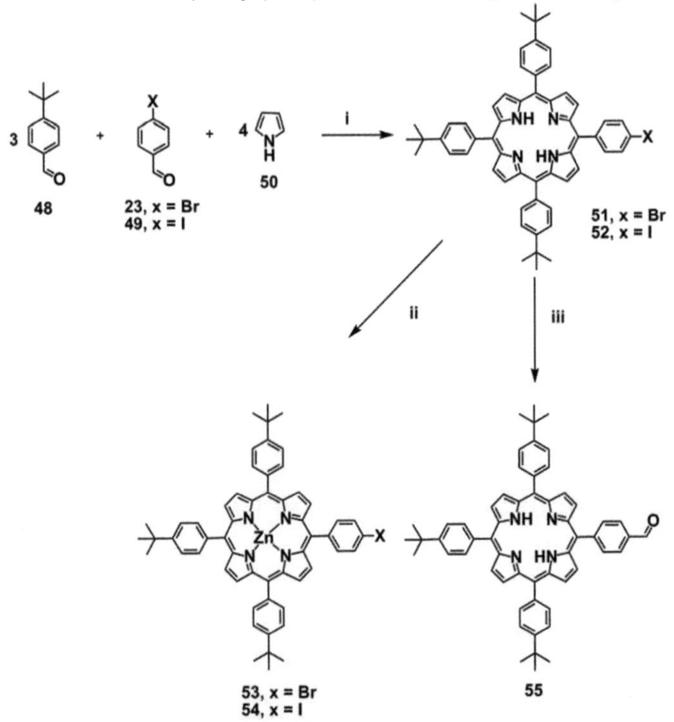

Scheme 3.8: Synthesis of porphyrins 53-55. i) 1. BF$_3$*OEt$_2$, PPh$_4$Cl, CH$_2$Cl$_2$, 2. DDQ; ii) Zn(OAc)$_2$, THF, Δ; iii) 1. n-BuLi, THF, -78 °C, 2. DMF, -78 °C, 3.HCl/H$_2$O, rt.

Boron trifluoride etherate seemed to be the more prosperous catalyst in comparison to trifluoro acetic acid to build up the porphyrin indicated by higher yields. The final oxidation is performed with DDQ. The purification on silica gel with dichloromethane/hexanes 2:1 as eluent made it possible to separate the different products.

Zinc porphyrins are well analyzed towards their photophysical properties.[91,212-214] Therefore metalloporphyrin systems with zinc as metal are very attractive for electron transfer investigations. The metallation is carried out in refluxing THF using zinc(II)acetate as zinc(II)source (Scheme 3.8). Formylation of porphyrin **51** and **52** can be carried out in dry THF at -78 °C. At first n-BuLi is added dropwise. After two hours of stirring at -78 °C DMF then is added at once. Hydrolysis leads to the formylated TPP **55** (Scheme 3.8).

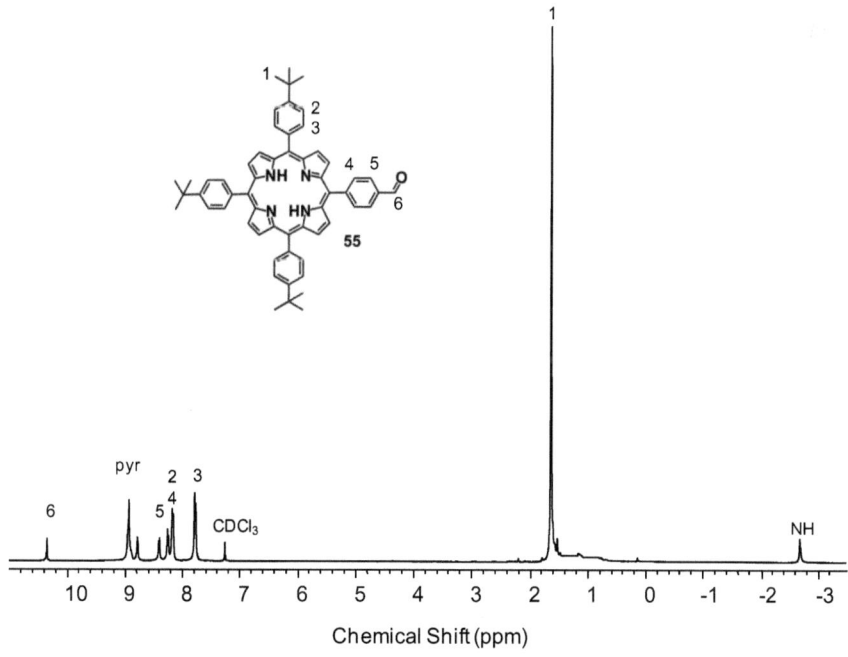

Figure 3.3: ^{1}H NMR (400 MHz) spectrum of formylporphyrin 55 in CDCl$_3$ at rt.

Characterization of formyl porphyrin **55** is done by ^{1}H/^{13}C NMR, UV/Vis and IR spectroscopy, mass spectrometry and elementary analysis. The ^{1}H-NMR spectrum in

CDCl$_3$ shows the expected chemical shifts (Figure 3.3). Characteristic for free base porphyrins in the proton NMR spectrum is the broad singlet at -2.71 ppm. The singlet for the *tert*-butyl groups at the porphyrin periphery occurs at 1.61 ppm. The signals for aromatic protons can be easy differentiated by integration: the doublets of the phenylene protons of *tert*-butyl substituted phenyl rings occur at 7.76 (d, ^3J = 7.95 Hz, 6H) and 8.15 ppm (d, ^3J = 7.95 Hz, 6H). The doublets for the phenylene protons of the formylated phenyl ring appear at 8.23 (d, ^3J = 6.96, 2H) and 8.37 (d, ^3J = 7.32, 2H) ppm. By the same approach the pyrrolic protons can be identified. The singlet for the pyrrolic protons between different substituted phenyl rings occurs at 8.76 ppm whereas the singlet of the other pyrrolic protons appears at 8.90 ppm. The formyl proton causes a singlet at 10.32 ppm.

KNOEVENAGEL condensation[215] of porphyrin **55** and barbituric acid **56** in pyridine with piperidine as base (Scheme 3.9) unfortunately does not lead to an isolable amount of barbituric porphyrin **57**. Although porphyrin **57** can be detected by mass spectrometry a separation from excessive porphyrin **55**, barbituric acid and side products is not successful. Neither crystallisation nor chromatography leads to the clean product.

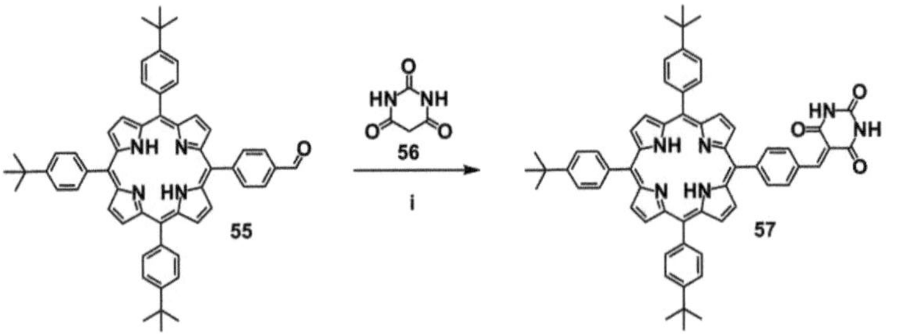

Scheme 3.9: Synthetic pathway to porphyrin **57**. i) piperidine, pyridine, 100 °C.

To solve this problem the concept of the barbituric acid key-system is altered to cyanuric acid systems (see for instance porphyrin **7** in figure 2.5 on page 22) which show similar behaviour in complexing Hamilton receptor derivatives.[91,96,114] In this context the main concept is to synthesize a tetraphenylporphyrin (TPP) which has just one cyanuric acid functionality and three additional substituents which should provide a good separation of the occurring isomers by chromatography besides a sufficient solubility of the porphyrin building blocks. Therefore 4-trimethylsilylethynylbenzaldehyde **58** and 3,5-dimethoxy benzaldehyde **59** are

Results and Discussion

chosen as starting materials. The synthesis is performed under statistical LINDSEY-conditions (Scheme 3.10).[208-211] Four equivalents of pyrrole **50**, one equivalent of 4-trimethylsilylethynylbenzaldehyde **58** and three equivalents of 3,5-dimethoxy benzaldehyde **59** are dissolved in dichloromethane. Then TFA is added and the solution is stirred for 1 h before NEt$_3$ is added to neutralize the reaction mixture. The final oxidation is then performed with DDQ. Separation of all statistically formed porphyrin isomers is possible by careful column chromatography (hexanes/EtOAc 4:1). Mono substituted 4-trimethylsilyl-ethynyl TPP **60** is isolated in 5 % yield after precipitating with n-pentane. The yield is typical for this type of reaction.

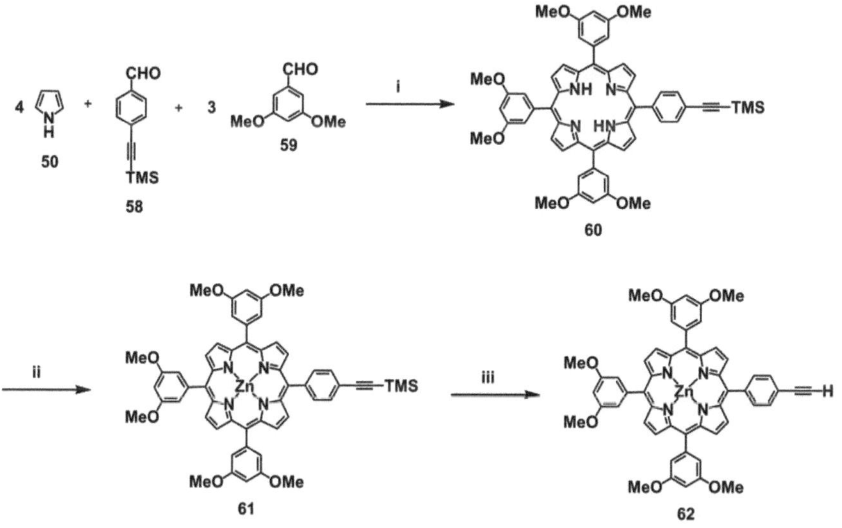

Scheme 3.10: Synthesis of zinc porphyrin **62**: i) 1. TFA, CH$_2$Cl$_2$, 60 min, 2. NEt$_3$, 8 min, 3. DDQ, rt; ii) Zn(OAc)$_2$, THF, Δ; iii) TBAF, THF, rt.

Metallation with zinc is straight forward in the case of tetraphenylporphyrins. Zinc porphyrins have been intensively investigated with respect to their photophysical properties and are very attractive building blocks for architectures enabling photoinduced electron transfer.[91,212-214] The metallation is carried out in refluxing THF using zinc(II)acetate as zinc(II) source (Scheme 3.10).

Mild deprotection can be promoted by addition of TBAF to a solution of **61** in THF. Porphyrin **62** is accessible in nearly quantitative yield by this route (Scheme 3.10). SONOGASHIRA coupling[200,201] of the free acetylenic porphyrin **62** with 4-iodophenyl isocyanuric acid **63**[216] using CuI, Pd$_2$dba$_3$ and triphenyl arsine as catalyst-ligand systems in a mixture of NEt$_3$ and THF (Scheme 3.11) leads to the desired cyanuric acid modified porphyrin **7** in good yield (81 %). As side product porphyrin dimer **64** (Scheme 3.11) can be isolated by column chromatography. To purify target porphyrin **7** dissolving in THF and precipitation with n-pentane is necessary to get rid of solvent inclusion. Porphyrin butadyine **64** can then be transformed into the corresponding 2,5-thiophene compound **65** (Scheme 3.11) by applying a reaction protocol developed by BÄUERLE et al.[217,218]

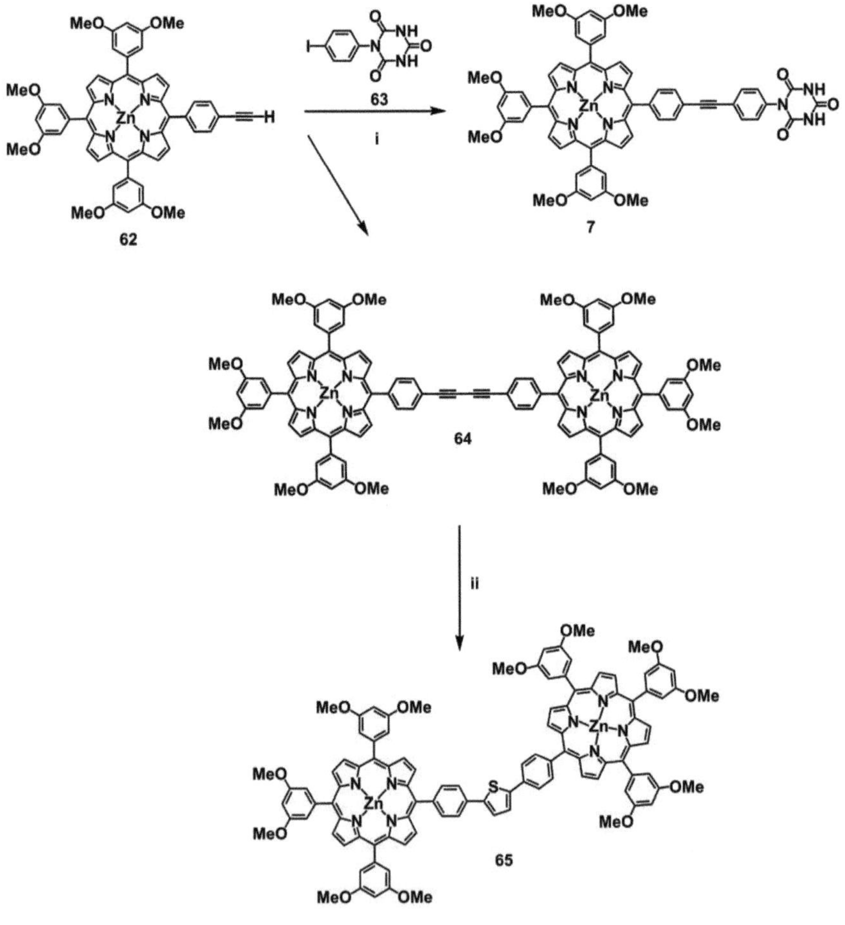

Scheme 3.11: Synthesis of target porphyrin 7 and porphyrin dimer 65: i) TBAF, THF, rt, 2 h; ii) Pd$_2$dba$_3$, AsPh$_3$, THF, NEt$_3$, rt; ii) Na$_2$S*9 H$_2$O, DMF, 100 °C, 2 h, then rt.

Results and Discussion

Porphyrinato thiophene **65** can be isolated in a good yield (79 %) and is characterized by ¹H NMR, UV/Vis and IR spectroscopy, mass spectrometry and elementary analysis. The proton NMR of compound **65** is depicted in Figure 3.4 and posseses characteristic signals in the aromatic region as well as two singlets for the methoxy ethers at 3.96 ppm (24H) and 3.93 ppm (12H). A splitting of the methoxy signal can be explained convincingly by the angulated structure of the 2,5 substituted thiophene ring. The two thiophene protons cause a signal at 7.69 ppm which occurs as singlet. Typical for the porphyrin substitution pattern of derivative **65** are two doublets at 8.30 (d, ^3J = 7.91 Hz, 4H) and 8.10 (d, ^3J = 7.91 Hz, 4H) and the singlets at 7.43 (s, 8H), 7.40 (s, 4H), 6.89 (s, 4H) and 6.86 ppm (s, 2H). The signals of the pyrrolic protons arise at 9.10 (d, ^3J = 3.77 Hz, 4H) and 9.06 ppm (s, 12H).

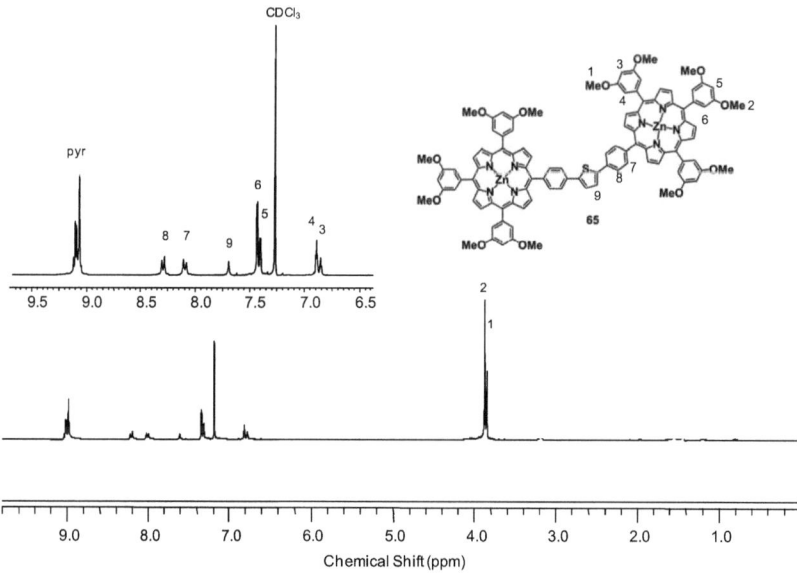

Figure 3.4: ¹H NMR (400 MHz) spectrum of porphyrinato thiophene 65 in CDCl₃ at rt.

Dendrimers

The main problem of Hamilton receptor modified fullerene monoadducts is their relatively poor solubility in common organic solvents. To overcome this problem dendritic subunits are a powerful tool. We recently described dendrofullerenes where the dendrimers are introduced at asymmetric malonates (compounds **5** in figure 2.3 on page 20).[75,91,94,121] These fullereneated compounds show excellent solubility in

CH$_2$Cl$_2$, CDCl$_3$ and toluene. A transfer of these properties to fulleropyrrolidines is possible by introducing dendritic building blocks into glycine derivatives. These compounds can then be condensed with formylated Hamilton receptor analogues to reactive azomethines for the cycloaddition at C$_{60}$.

In a first concept Boc-GABA **66** is coupled under Steglich conditions[219-221] with the Frechet dendron of the 1st generation **67**.[222,223] After acidic deprotection of the obtained Boc-GABA dendron **68**, neutralization of isolated triflate **69** with triethylamine and conversion with iodo acetic acid is performed (Scheme 3.12). Unfortunately it is not possible to isolate the *N*-glycine dendron **70** from the amine excess due to a very similar dissolution behavior and nearly exact retention times in chromatographic approaches on silica gel or AlOx.

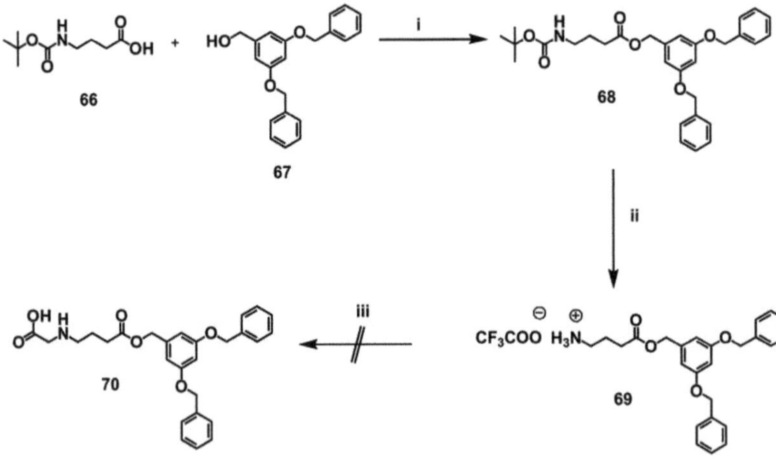

Scheme 3.12: Synthetic path for Frechet dendron containing *N*-glycine **70**. i) DCC, 1-HOBT, DMAP, CH$_2$Cl$_2$, 0 °C→rt; ii) TFA, CH$_2$Cl$_2$, rt; iii) 1. NEt$_3$, CH$_2$Cl$_2$, 2. iodo acetic acid, EtOH/H$_2$O, 0 °C→rt.

A similar approach for the synthesis of dendritic *N*-glycines is tried with (*R,R*)-depsipeptide of the 1st generation **71**[224] under the same conditions mentioned above (Scheme 3.13). Also this approach turned out to be not successful due to failing isolation possibility. After evaluation of the reaction parameters it should be mentioned that not only the failing separation potential, but also the large excess of dendrimer containing amine (amine/iodo acetic acid 4:1) in this synthetic approach is an enormous limitation. A further development of the dendrimer concept has to consider the great effort to synthesize a noteworthy amount of dendrimers and therefore prefer more effective approaches.

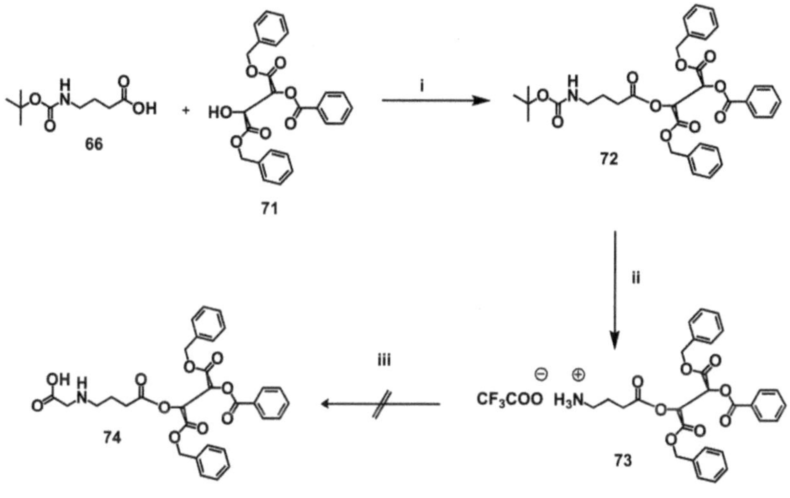

Scheme 3.13: Synthetic path for dendron containing N-glycine **74**. i) DCC, 1-HOBT, DMAP, CH_2Cl_2, 0 °C→rt; ii) TFA, CH_2Cl_2, rt; iii) 1. NEt_3, CH_2Cl_2, 2. iodo acetic acid, $EtOH/H_2O$, 0 °C→rt.

To avoid these problems a different synthetic route is developed for the introduction of dendritic structures to glycine. Starting from three different dendrimers (Frechet 1st generation **67**[222,223], depsipeptide 1st generation **71**[224] and Newkome 1st generation **76**[225]) the formyl group containing dendroesters/-amide **77-79** (Scheme 3.14) are synthesized using the well established DCC-coupling techniques (Scheme 3.14).[219-221]

The formyl dendrons can then be used for the *in situ* synthesis of the methylglycine ester imines by neutralizing methylglycine ester hydrochloride with triethylamine and adding the corresponding dendroaldehydes **77-79**. Mild reduction with sodium cyanoboronhydride which selectively reduces the imine in presence of the ester and amide groups leads to the *N*-dendroglycine methylesters **80-82** (Scheme 3.15). Deprotection of the *N*-dendroglycine methylesters **80-82** proves to be not as simple as planned. Standard methods using sodium hydroxide or lithium hydroxide in mixtures of THF and MeOH fail as well as using lithium iodide in refluxing pyridine (Scheme 3.16). Besides the cleavage of the glycine methyl ester, which should probably be the most reactive ester in the substrates, the other esters/amide are cleaved at least partially.

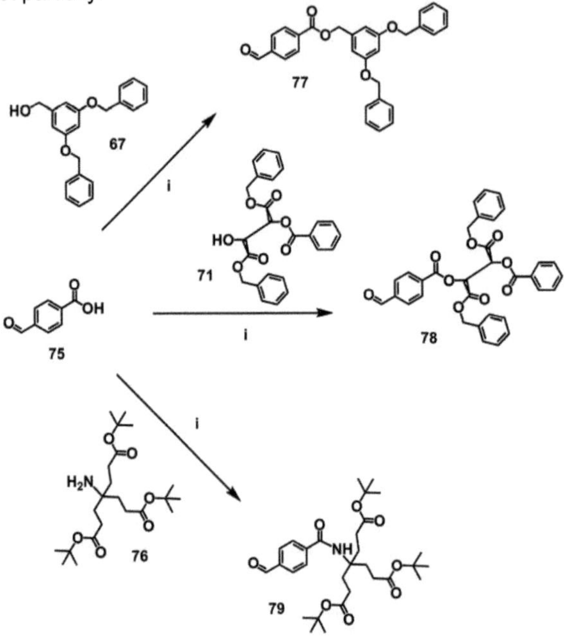

Scheme 3.14: Syntheses of dendroesters/-amids **77-79**. i) DCC, 1-HOBT, DMAP, CH$_2$Cl$_2$, 0 °C→rt.

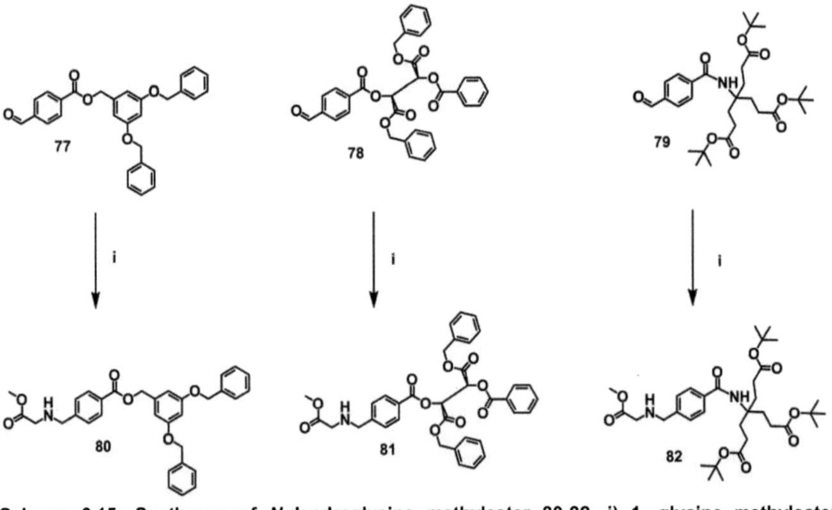

Scheme 3.15: Syntheses of *N*-dendroglycine methylester **80-82**. i) 1. glycine methylester hydrochloride, NEt$_3$, THF/MeOH, rt, 2. NaBH$_3$CN.

Results and Discussion

So the usage of methyl glycine esters as precursors for dendron glycine compounds **83-85** is discarded for this reason.

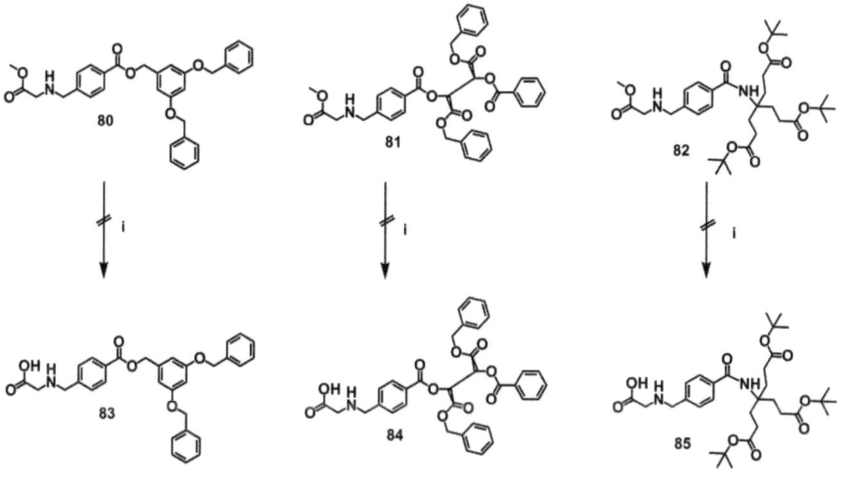

Scheme 3.16: Deprotection of the methyl glycine esters 80-82. i) 1 N NaOH or 1 N LiOH, THF/MeOH, rt or LiI, pyridine, Δ.

To overcome the problem of basic mediated methyl glycine ester cleavage the concept is expanded towards the introduction of *tert*-butyl glycine esters which then should be cleaved under mild acidic conditions. Therefore the formyl dendrons **77** and **78** are converted into the corresponding *tert*-butyl glycine esters applying the same conditions described above for the methyl glycine esters but using *tert*-butyl glycine hydrochloride (Scheme 3.17). Naturally Newkome dendrimer **79** cannot be used in this approach due to the multiple *tert*-butyl esters at its periphery which are deprotected under acidic conditions surely as well.

Unfortunately this approach is not successful as well. Although adding trifluoroacetic acid to a rapid stirring solution of the *tert*-butyl glycine esters **86** and **87** in dichloromethane at room temperature and neutralization by adding sodium bicarbonate leads to the cleavage of the ester, the planned dendroglycine derivatives **83** and **84** cannot be isolated. ^{1}H NMR analysis suggests that even in the case of mild acidic deprotection the dendritic esters or confocal points in the dendrimers are attacked. Concluding the glycine ester approach it must be mentioned that the complications of the ester cleavage do not lead to promising precursor compounds for dendrofulleropyrrolidines. Therfore alternative routes must be established.

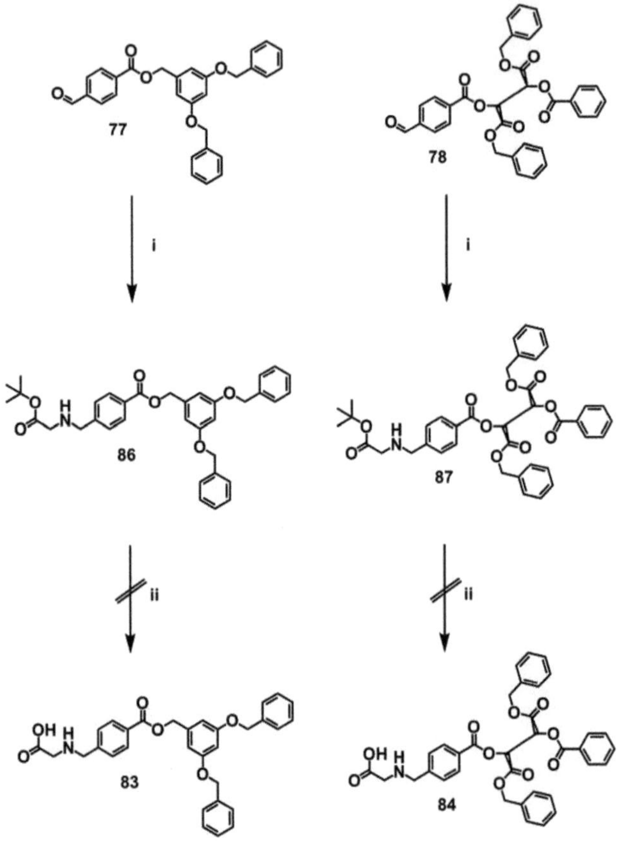

Scheme 3.17: Syntheses of *N*-dendroglycine methylesters **86** and **87** and their cleavage. i) 1. glycine *tert*-butylester hydrochloride, NEt$_3$, THF/MeOH, rt, 2. NaBH$_3$CN; ii) 1. TFA, CH$_2$Cl$_2$, rt, 2. NaHCO$_3$, rt.

PEREZ et al. described the synthesis of dendrofullerenes in which the dendritic side chain is connected to the pyrrolidine ring by carboxylic chloride coupling with the according fulleropyrrolidine.[226] Inspired by this approach Hamilton receptor fulleropyrrolidine **88** is synthesized using the above mentioned PRATO reaction protocol[202] with Hamilton receptor derivative **21**, glycine and [60]fullerene in refluxing o-dichlorobenzene (Scheme 3.18). As dendron 2nd generation Frechet dendrimer **89**

Results and Discussion

is chosen and converted into the carboxylic chloride **90** by stirring in oxalyl chloride at room temperature. Fulleropyrrolidine **88** can then be further functionalized by adding chloride **90** in tetrahydrofurane in the presence of pyridine (Scheme 3.18). Dendrofulleropyrrolidine **91** can be isolated in a moderate yield (70 %) by this approach.

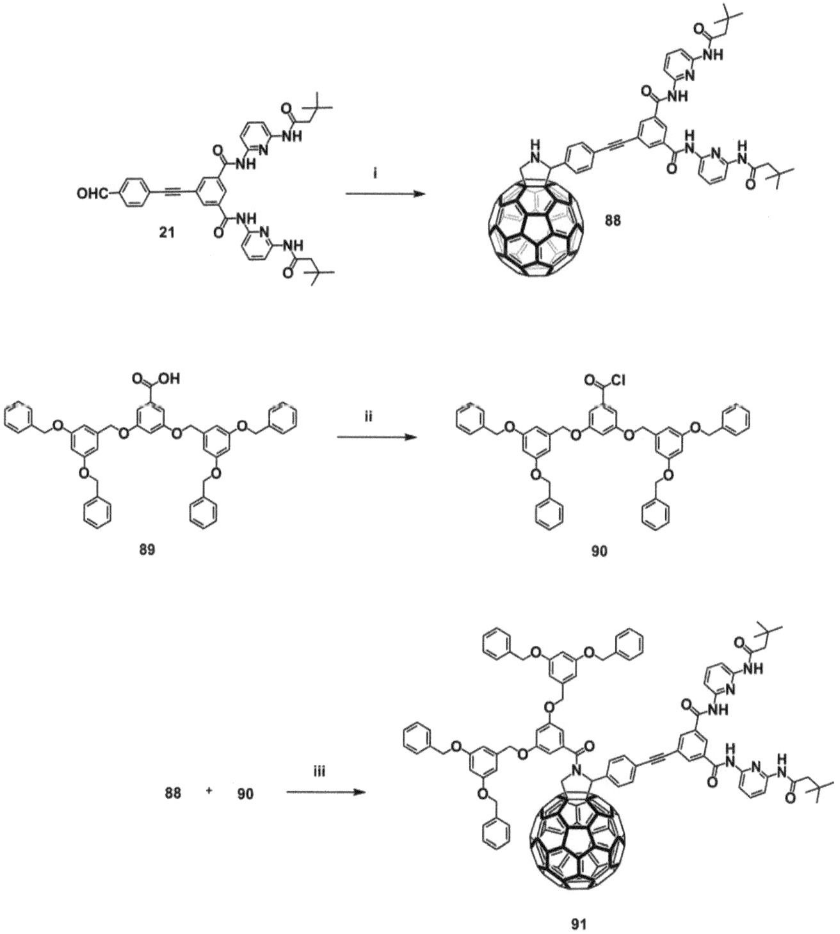

Scheme 3.18: Synthesis of dendrofulleropyrrolidine 91. i) glycine, C$_{60}$, ODCB, Δ; ii) oxalyl chloride, rt; iii) pyridine, THF, rt.

In comparison to the above described approaches the introduction of the dendrimer in the last step seems to be by far the best. Even if a suitable selective glycine ester cleavage method would be found the PEREZ way should be favored towards its short synthetic pathway and higher effectiveness. Due to the acidic conditions of the chlorination reaction the usage of Newkome type dendrimers is not suitable for this approach. For Depsipetide bearing dendrimers this pathway is in principle possible, but was not persued as a matter of time and lacking starting material.

The Alternative to Dendrimers: Alkylchains

As mentioned above in detail dendritic systems in the light of Hamilton receptor bearing fulleropyrrolidines are used to achieve a higher solubility in organic solvents. But all dendritic systems exhibit the disadvantage of a relatively complex multiple step synthesis. Therefore it is tested whether the application of an additional alkyl chain instead of a dendrimer enhance the solubility sufficiently. The two hexyl chains at the fluorene spacer of fullerene derivative **47** increase the solubility in CH_2Cl_2 and $CHCl_3$ clearly in comparison to the monoadducts **41-46**.

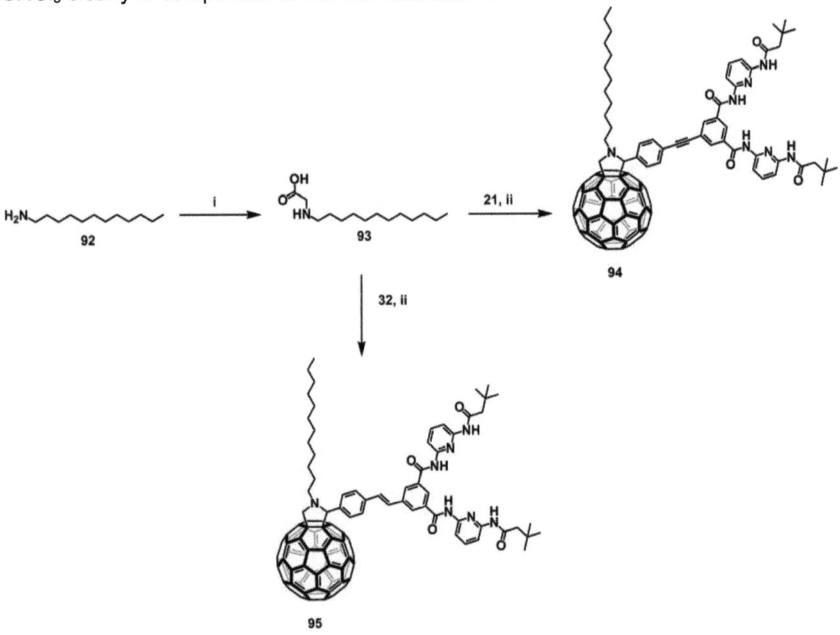

Scheme 3.19: Synthesis of *N*-dodecyl fulleropyrrolidines **94** and **95**. i) iodo acetic acid, EtOH/H_2O, 0 °C→rt, ii) C_{60}, toluene, Δ.

So the next concept should be to displace the commercially available and cheap N-Methylglycine (sarcosine) by a long chain N-alkyl glycine derivative. The best opportunity of course is to use relatively cheap starting materials and a reaction protocol that guarantees a high yield. Therefore a dodecyl chain is introduced at the glycine which is quite easily done adding iodo acetic acid to large excess of dodecylamine in ice cooled EtOH/H$_2$O (Scheme 3.19).

N-dodecyl glycine can then be applied in the PRATO reaction protocol as it is depicted in Scheme 3.19. Hamilton receptor containing fulleropyrrolidine **94** and **95** show much higher solubility in nonpolar solvents (i.e. CH$_2$Cl$_2$, CHCl$_3$ and toluene) than their N-methyl counterparts **41** and **44**.

3.1.2 NMR-Titration Experiments

The formation of fullerene-porphyrin complexes (i.e. Hamiltonreceptor modified fullerrenes **47**, **94** and **95** and cyanuric acid bearing prophyrin **7**) is first investigated by ^{1}H NMR spectroscopy. In particular, the determination of the association constant K_{ass} required a series of titration experiments in CDCl$_3$. The characteristic downfield shifts of the NH^1- and NH^2-protons of the Hamilton-receptor moieties[91,103,114,120,121,227] in the fullerene compounds **47**, **94** and **95** are used to analyze the complex formation.

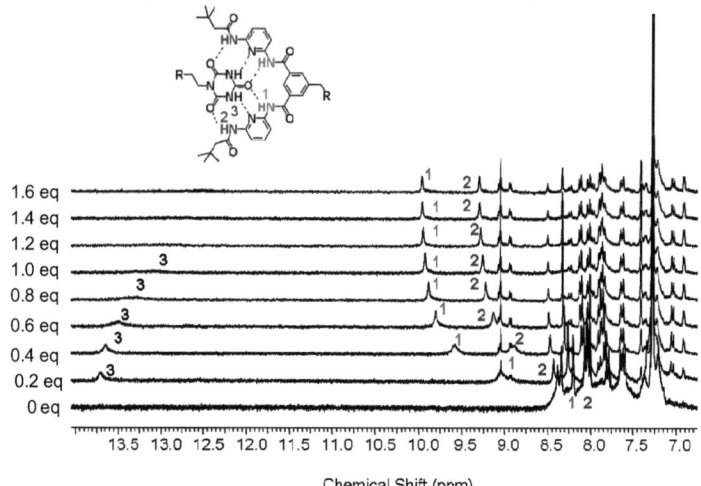

Figure 3.5: Binding motif between **47**, **94**, **95** and **7** with indication of the NH protons NH^1, NH^2 and NH^3 (top) and 300 MHz ^{1}H NMR spectra of **94** at a concentration of 3.3 mM in CDCl$_3$ in the presence of various equivalents of **7** (bottom).

Figure 3.5 depicts the corresponding shifts of the protons of **94** as a function of added porphyrin derivative **7**. Additionally, the shifts of the NH^3-protons within the cyanuric acid moiety of **7** were investigated. In a typical experiment, 0.5 mL of a 2 mM solution of fullerene derivative **47**, **94** and **95** were titrated with 100 µL of a 2 mM solution of porphyrin derivative **7**. The ^{1}H-NMR spectra were recorded approximately 30 min after mixing the solutions. In each case, the spectrum remained unchanged after 60 min indicating a fast equilibrium formation.

Whereas the NH^1- and NH^2-protons undergo a shift to lower fields, the NH^3-protons of the cyanuric acid moieties at 13 ppm are subject to an opposite effect. The corresponding signals undergo a high field shift during addition of the porphyrin cyanuric acid derivative **7**. Furthermore, they broaden and, finally, they disappear. The reason for the disappearance in the presence of an excess of **7** is a fast equilibrium (coalescence regime) between bound and free cyanurate. Based on these titration experiments the association constants K_{ass} (Table 1) can be determined with the help of the program Chem-Equili.[228,229]

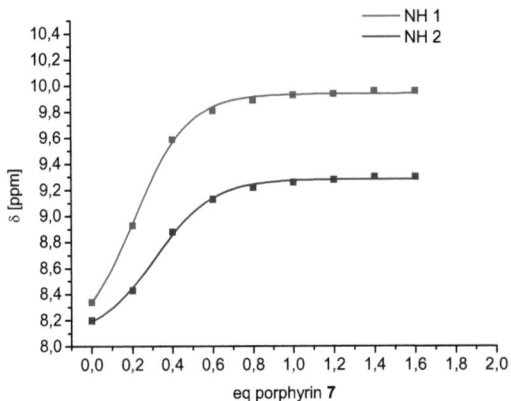

Figure 3.6: Plot of the chemical shift (δ [ppm]) of the NH^1 and NH^2 as function of mole fraction X of titrated cyanuric acid bearing porphyrin **7**.

A plot of the chemical shift (δ [ppm]) of the NH protons as function of mole fraction X of titrated cyanuric acid bearing porphyrin **7** shows the characteristic sigmoidal shape with plateau at X = 1 (Figure 3.6) which is a clear hint for the formation of a 1:1 complex.

Results and Discussion

	94:7	95:7	47:7
log K_{ass}	5.32	5.18	5.16

Table 1: Association constants K_{ass} of the fullerene porphyrin hybrids (determined by NMR-titration experiments).

3.1.3 Photophysical Investigations

Further information on the supramolecular complexation was obtained by optical absorption spectroscopy and steady state fluorescence spectroscopy. To this end, variable amounts of the different [60]fullerene Hamilton receptors **42-44, 47, 94** and **95** were added to porphyrinato cyanuric acid **7** containing o-dichlorobenzene (oDCB) or dichloromethane solution. Upon increasing concentration of [60]fullerene **43, 44, 47, 94** and **95**, respectively, the SORET band of **7** is slightly decreased with no spectral shifts (Figure 3.7). Isosbestic points (i.e. 415 nm and 429 nm) in each titration assay confirm the formation of a supramolecular complex for the before mentioned entities. We note however that the strength of the spectral changes relates to the nature of the spacer. For the supramolecular complexes **44, 47, 94** and **95 : 7** spectral changes are rather marginal in contrast to complex **43 : 7** noticeable.

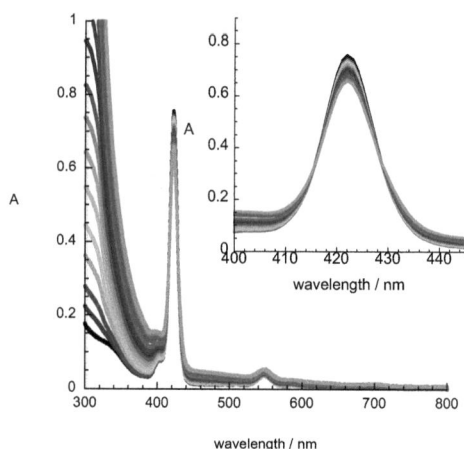

Figure 3.7: Optical absorption spectra of **7** (2.83 x 10^{-6}M) on increasing addition of **44** (0 – 2.10 x 10^{-5}M) in CH_2Cl_2.

Opposing to that, titration of **7** with [60]fullerene **42** results in the digital addition of the individual spectra of both entities. Comparing the obtained results from absorption

Results and Discussion

spectroscopy to previous studies,[91] the noticed spectral changes are rather negligible and imply that introducing rigid spacer moieties constrains ground state interaction between porphyrin and fullerene.

To gain quantitative details on the binding strength, complementary fluorescence titration experiments were carried out. In line with earlier work, the prominent porphyrin emission of **7** (Φ_F = 0.04) was monitored while distinct amounts of [60]fullerene **42-44** and **94** were added. Figure 6 shows an exemplary fluorescence titration of **7** with **44** in CH_2Cl_2. The gradual quenching at the long wavelength maximum of the emission was used to estimate the binding constant according to equation 1:

$$\frac{I_f}{I_0} = 1 - \frac{1}{2c_D}\left[\frac{1}{K_s} + c_0 + c_D - \sqrt{\left(\frac{1}{K_s} + c_0 + c_D\right)^2 - 4c_0 c_D}\right] \quad (1)$$

whereas I_0 refers to the initial fluorescence intensity, c_0 is the total porphyrin concentration, c_D is the total concentration of the added fullerene, and K_S is the binding constant.[230]

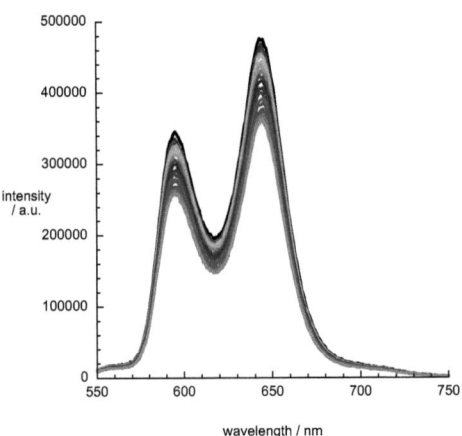

Figure 3.8: Steady state fluorescence titration of **7** (2.3 x 10^{-6}M) with **44** (0 – 2.10 x 10^{-5}M) in CH_2Cl_2 (λ_{exc} = 416 nm).

In contrast to the binding constants obtained by ^{1}H NMR, binding constants deduced by fluorescence quenching show some interesting trends. On one hand, a weakening

Results and Discussion

in binding was obtained when increasing the donor-acceptor distance of the p-phenylene-ethynylenes wire systems. On the other hand, [60]fullerenes with incorporated fluorine and p-phenylene-vinylene spacers (i.e. **44**, **47** and **95**) show a higher affinity in binding than [60]fullerenes with p-phenylene-ethynylene spacers (i.e. **94** and **42**). Moreover, [60]fullerene **43** with p-ethynylene spacers bears the highest binding constant to the porphyrinato cyanuric acid moiety, owing to the shortest donor acceptor distance (Table 2).

	log K_{ass}			
	oDCB	CH_2Cl_2	τ_F, ns	k_{CS}, s^{-1}
42:7	2.182	2.698	1.470	not detectable
43:7	---	4.667	1.480, 0.260	1.064×10^{10}
44:7	4.089	4.332	1.550, 0.640	4.348×10^{9}
47:7	4.104	4.201	1.490, 0.660	4.065×10^{9}
94:7	3.528	3.719	1.450, 0.470	7.576×10^{9}

Table 2: Obtained data from the photophysical measurements.

Next we turned to time-resolved fluorescence. To this end, the zinc porpyhrin bearing cyanuric acid **7** and different ratios of **7** and **42-44, 47, 94** and **95** were excited at 403 nm and the resulting emission analyzed by mono-, bi- or polyexponential decay fit functions. The fluorescence of the pristine zinc porphyrin **7** was well fitted with a monoexponential fit function yielding lifetime of 1.55 ns in oDCB and CH_2Cl_2. By adding [60]fullerene **43, 44, 47, 94** and **95** to **7** lifetimes of ca. 1.50 ns, which can be assigned to the unbound Zinc porphyrin **7**, and, in addition, in the range of hundreds of picoseconds, which can be attributed to the complexed species, were obtained.

Interestingly, the presence of [60]fullerene **42** did not result in any appearance of a second fluorescence lifetime at all. In a separate experiment we increased incrementally the concentration of [60]Fullerene **43, 44, 47, 94** and **95**. Analysis of the preexponential factors of both lifetimes clearly demonstrated that an increase of [60]fullerene concentration causes an increase of the preexponential factor of the short and - simultaneously – causes a decrease of the preexponential factor of the longer lifetime. Importantly, the two lifetimes remained unchanged throughout the experiments.

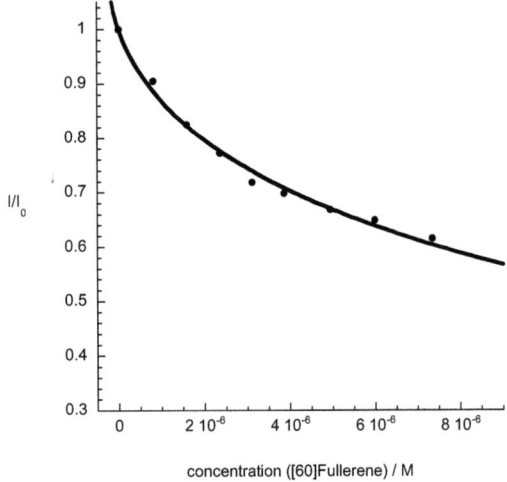

Figure 3.9: Fluorescence intensity at 645 nm of 7 and 44:7 with non-linear fit (according to equation 1) to obtain the binding constant.

Results from steady state und time-resolved experiments point to a deactivation of the photoexcited zinc porphyrin **7** *via* charge transfer to the [60]fullerene. Fs transient absorption spectroscopy measurements were performed to indentify the expected charge transfer process and – in addition – to analyze the kinetics of the involved processes. The fs transient spectra of zinc porphyrin **7**, taken after a 150 fs laser pulse at 420 nm in oDCB, show in absence of any [60]fullerene significant transient features of the singlet excited state of the zinc porphyrin **7** in the region between 600 nm and 1100 nm. Besides these characteristic transient bands, bleaching features of the Q ground state absorption bands emerge at 550 nm and 600 nm shortly after the laser pulse. The zinc porphyrin singlet excited state ($E_{Singlet}$ = 2.00 eV) deactivates slowly *via* intersystem crossing to energetically low lying triplet excited state ($E_{Triplet}$ = 1.53 eV) with a rate constant of 4.446 x 10^8 s^{-1}.

When adding [60]fullerene **43, 44, 47, 94** and **95** to the cyanuric bearing zinc porphyrin **7**, the tentatively drawn conclusion of an electron transfer in the before mentioned systems could be confirmed. For instance, the presence of [60]fullerene **94** changes the porphyrin excited state deactivation dynamics towards a charge separation, indicated by the radical ion fingerprint at 680 nm and 1010 nm, respectively (Figure 3.10). Interestingly, introduction of a second repetition unit (i.e., **42:7**) hampers the electron transfer process and no evidence for any charge separated species could be found (Figure 3.11).

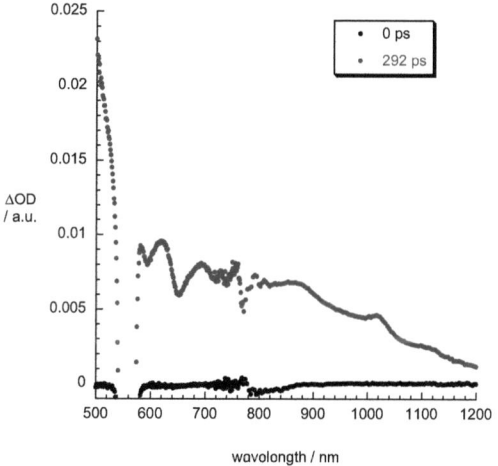

Figure 3.10: Fs transient absorption spectra of 94:7 obtained after 150 fs laser pulse at 550 nm in Ar saturated oDCB.

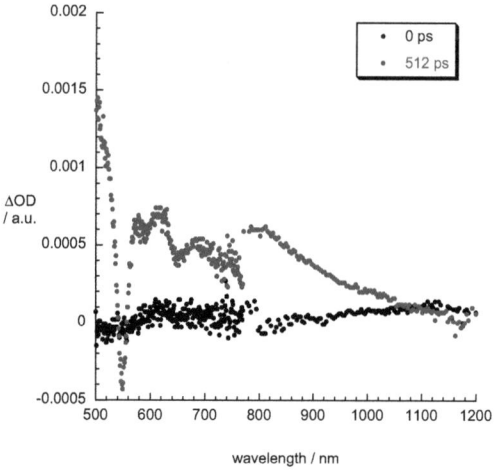

Figure 3.11: Fs transient absorption spectra of 42:7 obtained upon laser flash photolysis (420 nm) in Ar saturated oDCB.

This finding is well in line with our observations from steady state and time-resolved fluorescence measurements and supports the assumption that the photoexcited porphyrin in system **42:7** deactivates *via* intrinsic pathways, as for example, fluorescence and internal conversion. The rates of charge separation k_{CS} were calculated for the systems **43, 44, 47, 94** and **95:7** and are shown in table 2.

Nevertheless, we were not able to shed light on the process of charge recombination, because transient spectra beyond the ns time regime were all characterized by strong triplet excited state features of the unbound [60]fullerene. Complementary ns transient absorption spectra of **43:7** after 552 nm laser flash show the characteristic triplet-triplet absorption of [60]fullerene at 710 nm (Figure 3.12).

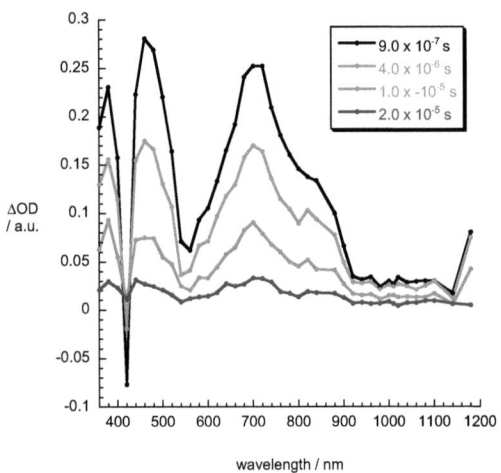

Figure 3.12: Differential absorption spectra of 43:7 upon ns flash photolysis (532 nm) in Ar saturated oDCB.

3.1.4 Conclusion

In this work the self-assembly and the photophysical properties of novel supramolecular wire-like donor-acceptor nanohybrids are investigated. The examined fullerene-porphyrin complexes are constructed *via* the six-folded hydrogen bonding Hamilton receptor motif. In this particular light the fullerene derivatives are connected *via* rigid spacers, i.e. *p*-phenylene-ethynylene, *p*-phenylene-vinylene, *p*-ethynylene and fluorene, with the Hamilton receptor. To overcome the poor solubility, especially in solvents where no interference with the hydrogen bonding is expected, one dodecyl or two hexyl chains are introduced. The zinc porphyrin **7** is bridged *via p*-phenylene-ethynylene with the cyanuric acid. The corresponding binding constants

K_{ass} were determined *via* comparative ^{1}H NMR and fluorescence titration experiments. Interestingly both methods show different binding strengths. The ^{1}H NMR experiments reveal K_{ass} in the range of 10^5 mol^{-1}dm^3, means while the binding quantified by fluorescence titration experiments show a clear trend from 10^2 (elongated *p*-phenylene-ethynylene spacer) to 10^4 mol^{-1}dm^3 (*p*-phenylene-vinylene and *p*-ethynylene). These circumstances correlate to the different methods in characterization. In the case of ^{1}H NMR shifts the key lock interaction of the Hamilton receptor and the cyanuric acid is fundamental, whereas the fluorescence quenching relates to non-emissive deactivation pathways, i.e. electron transfer or internal conversion. Herein the differences in binding point to increments in electron transfer efficiency of the included spacers (*p*-phenylene-vinylene > fluorene > *p*-phenylene-ethynylene). However, the highest binding affinity was obtained in the *p*-ethynylene spacer based [60]fullerene:porphyrin system. Here, the deactivation rate of the photoexcited porphyrin is almost completely characterized by the short donor acceptor distance. By means of time-resolved techniques (i.e. time-resolved fluorescence and transient absorption), the before mentioned conclusions could be further confirmed and gave clear evidence for spacer mediated electron transfer processes. For instance, on one hand complex **94**:**7** shows a clear electron transfer characteristic, on the other hand elongation with a second *p*-phenylene-ethynylene repetition unit impedes the photoinduced electron transfer.

Results and Discussion

3.2 Aggregation of Supramolecular Fullerene-Flavin Hybrids

Besides porphyrins, flavins (formed by the tricyclic heteronuclear organic ring isoalloxazine) play an important role as electron mediators.[231,232] Their redox reactivity is finely controlled by hydrogen bonding in biological redox systems.[233,234] In addition the flavin chromophore, like porphyrins, has a strong absorption band in the visible region which let it act as an efficient photoreceptor.[235] Therefore photoinduced electron transfer reactions of flavins have been the subject of intense research in photocatalyst for photobiological redox processes.[236-238] Although supramolecular electron donor-acceptor complexes linked with hydrogen bonding have merited current interest as functional ensembles to build molecular machines and optoelectronic devices,[91,239-246] there has been no report on formation of hydrogen bonded complexes of flavins with electron donors or/and acceptors.

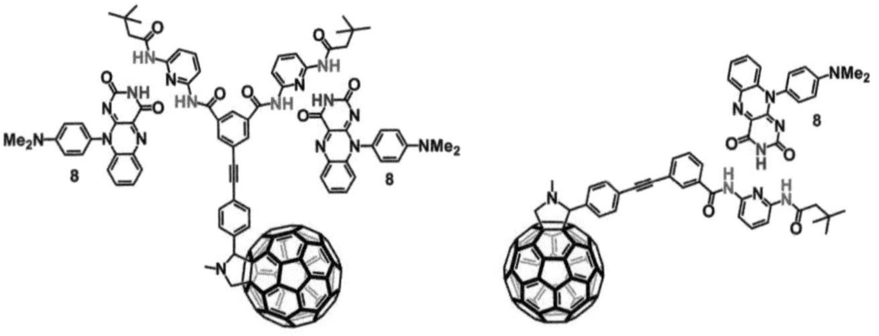

Figure 3.13: Supramolecular Hybrids of DMA-Flavin 8 with fullerene 43 and 109.

Herein hydrogen bonded supramolecular assemblies composed of *N,N*-dimethylaniline-substituted flavin (DMA–Fl)[244] as electron donor and fullerene (electron acceptor) are examined, which contain single and double hydrogen bonded receptors (SRC$_{60}$ **109** and DRC$_{60}$ **43**) as shown in Figure 3.13. Fullerene is suitable as an electron-acceptor component since electron-transfer reduction of fullerene is highly efficient because of the minimal changes of structure and solvation associated with the electron-transfer reduction.[245,246]

In these supramolecular systems, directional electron transfer from the terminal electron donor (*N,N*-dimethylaniline) moiety to the terminal electron acceptor (fullerene) is made possible by photoexcitation of the flavin moiety. Importantly for

Results and Discussion

the first time the evidence for a multistep electron transfer as opposite to a hole transfer along a fine-tuned and fine-balanced redox gradient is provided in this work. The detailed photodynamics were examined by laser flash photolysis measurements. The DMA-Fl-SRC$_{60}$ and (DMA-Fl)$_2$-DRC$_{60}$ complexes were assembled as clusters on a optically transparent electrode (OTE) of nanostructured SnO$_2$ (OTE/SnO$_2$). The photoelectrochemical behavior of the nanostructured SnO$_2$ films of DMA-Fl-SRC$_{60}$ and (DMA-Fl)$_2$-DRC$_{60}$ clusters are reported in comparison with reference systems without flavin.

3.2.1 Synthesis of Fullerene Containing Single Hydrogen Bond Receptors

For the synthesis of the single arm receptors (SRC$_{60}$) the starting points are the commercially available 3-iodo benzoic acid **96** and 2-iodo benzoyl chloride **99**, respectively 4-iodo benzoyl chloride **100**. Iodo compound **96** is transformed by DCC / 1-HOBT coupling with aminopyridine **17** to the iodo single arm component **97**. Coupling of the carboxylic chlorides **99** and **100** with single arm amine **17** is used for the synthesis of the corresponding iodo single arm derivatives **101** and **102**.

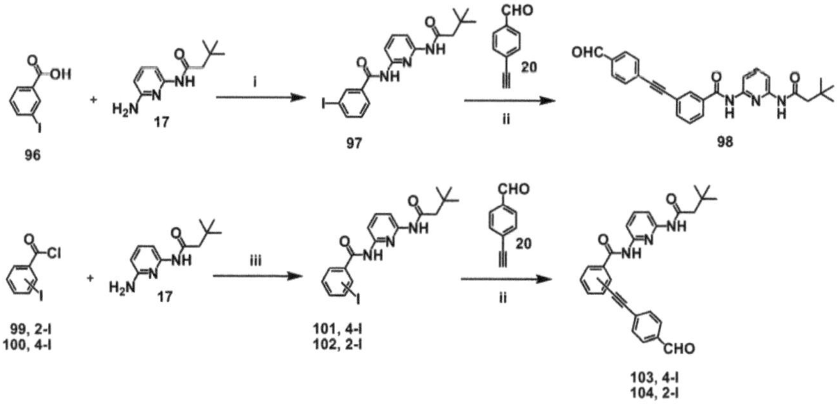

Scheme 3.20: Syntheses of the formyl single arm receptors 98, 103 and 104. i) DCC, 1-HOBT, DMAP, CH$_2$Cl$_2$, 0 °C→rt; ii) Pd(PPh$_3$)$_2$Cl$_2$, CuI, NEt$_3$, THF, rt; iii) NEt$_3$, CH$_2$Cl$_2$, 0 °C→rt.

The aldehyde functionalized precursors **98**, **103** and **104** are then obtained by SONOGASHIRA coupling[200,201] of the iodo receptors **97**, **101** and **102** with 4-ethynyl benzaldehyde **20** in nearly quantitative yield (Scheme 3.20).

Furthermore single arm receptor **97** can be used as a basic precursor for several C-C cross coupled carbon rich receptors (Scheme 3.21 and 3.22).

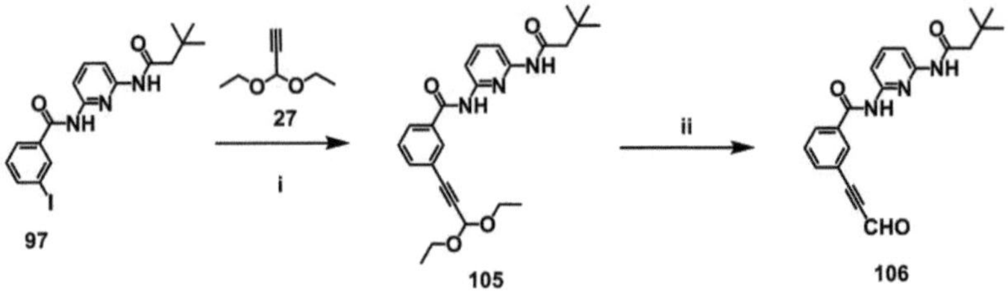

Scheme 3.21: Synthesis of compound **106**. i) Pd(PPh$_3$)$_2$Cl$_2$, CuI, HNEt$_2$, THF, rt; ii) 1. CH$_2$Cl$_2$, TFA, rt; 2. NaHCO$_3$, rt.

Different spacers between the fullerene moiety and the single arm receptor should show interesting behavior towards their photophysical properties after irradiation with light. Similar to the considerations mentioned in the section about fullerene-porphyrin architectures (3.1) ethynylene **106**, phenylene **107** and thiophene **108** elongations are promising towards the electron transfer fine tuning possibilities.

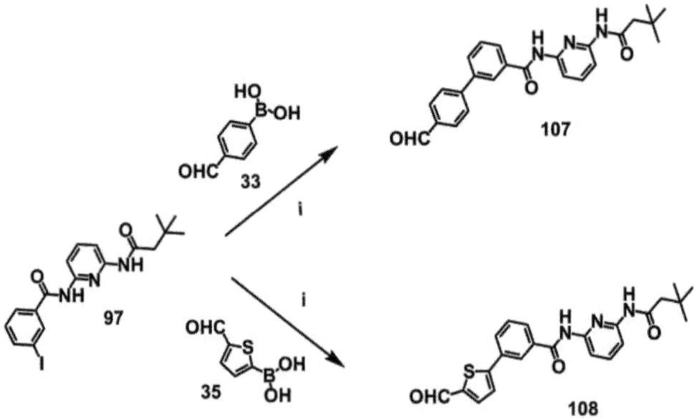

Scheme 3.22: S$\scriptsize{UZUKI}$ coupling for further functionalization of **97**. i) Pd(PPh$_3$)$_4$, K$_2$CO$_3$, DMF, rt.

In this context 3,3-diethoxyprop-1-yne **27** is connected via S$\scriptsize{ONOGASHIRA}$ coupling to single arm compound **97**. Acidic cleavage of the acetal **105** leads to the formyl compound **106** (Scheme 3.21). S$\scriptsize{UZUKI}$ coupling[204-206] techniques are used for functionalization of iodo derivative **97** with 4-formylphenyl boronic acid **33** and 5-

formylthiophene boronic acid **35** to the corresponding carbon rich rods (Scheme 3.22).

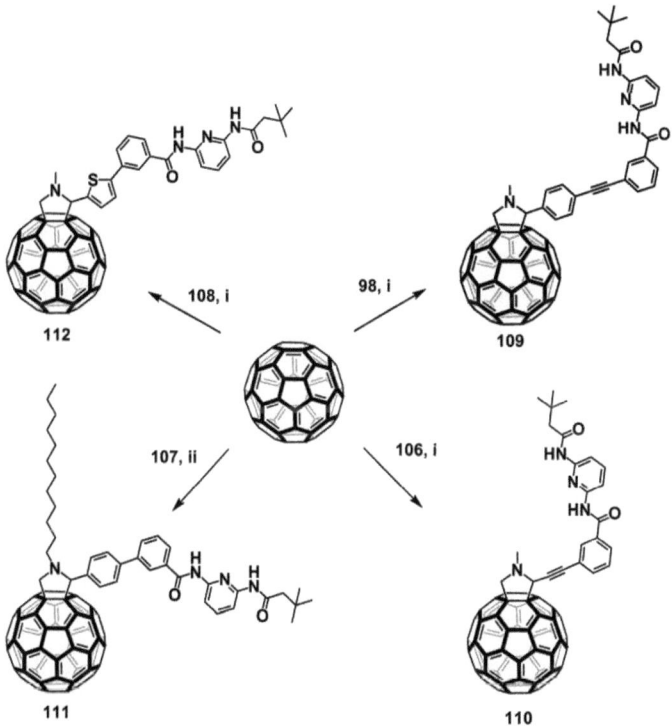

Scheme 3.23: Syntheses of single arm receptor containing fullerenes. i) C_{60}, sarcosine, toluene/THF, Δ; ii) C_{60}, *N*-dodecylglycine, toluene/THF, Δ.

Conversion into the C_{60} monoadducts **109-112** can be afforded by a PRATO reaction protocol[155,202] using the aldehydes **98, 106-108**, an excess of C_{60} and sarcosine (Scheme 3.23). Conversion of the aldehydes **103** and **104** into the corresponding monoadducts fails serval times and is not followed further.

3.2.2 Formation and Characterization of Supramolecular Complex between DMA-FI and C_{60} Derivatives

The formation of the supramolecular complexes can be analyzed using ^{1}H NMR titration experiments.[91,94,103,120,121,227,247] Starting from a 3.3 mM solution of either DMA-FI **8** or DRC$_{60}$ **43** in CDCl$_3$ the corresponding compound is added in consecutive steps (0.3 mM - 3.3 mM). The characteristic downfield shifts of the NH^1- and NH^2-protons of the Hamilton-receptor moieties[91,94,103,120,121,227] are used to analyze the complex formation.

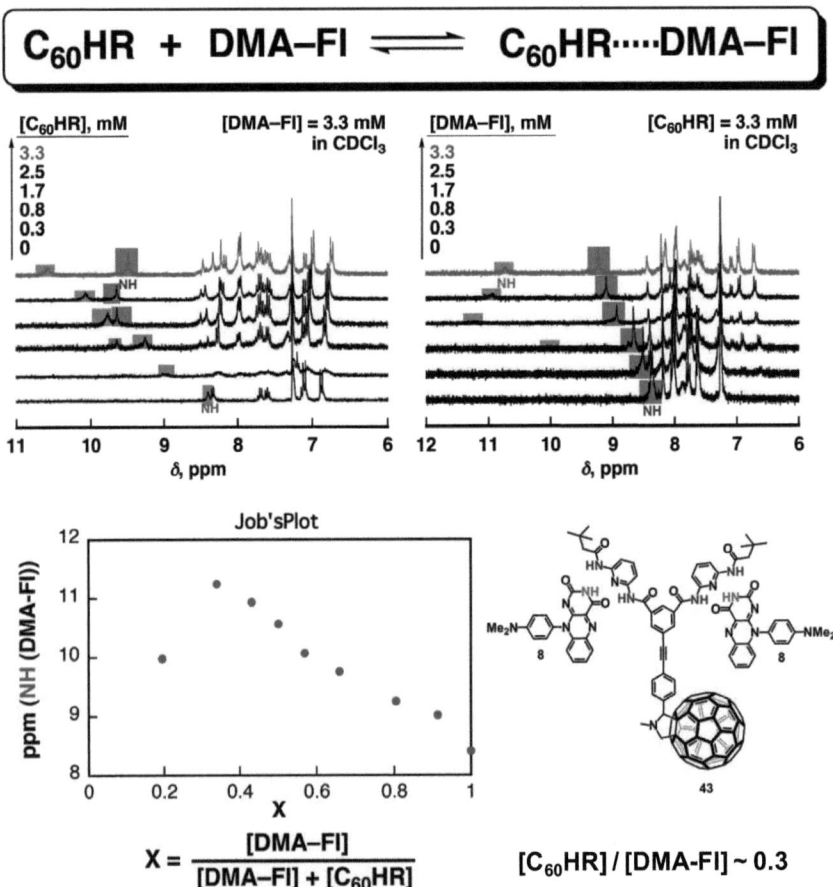

Figure 3.14: ^{1}H NMR titration experiments for determination of the stoichiometry of the formed nanohybrid of DRC$_{60}$ **43** and DMA-FI **8** in CDCl$_3$ (above). Job's plot analysis suggests the formation of a 1:2 complex (botton).

Results and Discussion

Additionally, the shifts of the NH-protons within the DMA-Fl **8** are investigated starting either with DMA-Fl **8** in CDCl$_3$ and addition of Hamilton receptor modified fulleropyrrolidine **43** (Figure 3.14 above) or the otherway around (Figure 3.14 bottom) In each case, the spectrum remained unchanged after 60 min indicating a fast equilibrium formation. Whereas the NH^1- and NH^2-protons undergo a shift to lower fields, the NH-protons of the flavin moieties at 13 ppm are subject to an opposite effect (Figure 3.14, bottom). The corresponding signals undergo a high field shift during addition of the DMA-Fl **8**. Furthermore, they broaden and, finally, they disappear. The reason for the disappearance in the presence of an excess of **8** is a fast equilibrium (coalescence regime) between bound and free flavin. Based on this titration experiments Job's plot analysis[248] to clarify the stoichiometry is possible. The plot of the chemical shifts as function of the mole fraction of DMA-Fl **8** shows a global maximum at 0.3 (Figure 3.14, right) which is a clear hind for the formation of a 1:2 complex.

The UV-Vis absorption spectrum of fullerene with single hydrogen bond receptor (SRC$_{60}$) spreads broadly through the visible region. Thus the reference compound **97** (I-SR) of SRC$_{60}$ additionally is titrated to N,N-dimethylaniline-substituted flavin **8** (DMA–Fl) to investigate the formation of supramolecular complex between DMA–Fl and hydrogen bond receptor moiety. The absorption spectrum is changed upon addition of I-SR **97** as shown in figure 3.15, where the broad absorption band is observed from 480 nm to 600 nm. The absorption change exhibits a saturation behavior with increasing I-SR concentration. This indicates that I-SR forms a supramolecular complex with DMA–Fl. Formation of the supramolecular complex was confirmed by UV-Vis absorption spectral measurements in tetrahydrofurane. According to equation 2,

$$\mathbf{8} + \mathbf{97} \overset{K}{\rightleftharpoons} \mathbf{8} \cdot \mathbf{97} \qquad (2)$$

$$[DMA\text{–}Fl]_0/(A_0 - A) = (\varepsilon_c - \varepsilon_p)^{-1} + (K_1[I\text{-}SR](\varepsilon_c - \varepsilon_p))^{-1} \qquad (3)$$

the absorption change is given by eq 3[249], which predicts a linear correlation between [DMA–Fl]$_0$/(A_0 – A) and [I-SR] **97**, where A_0 and A are the absorbance of DMA–Fl at 510 nm in the absence and presence of I-SR **97**, K_1 is formation supramolecular complex formation constant of I-SR-DMA–Fl, and ε_p and ε_c are the molar absorption coefficients of DMA–Fl at 510 nm in the absence and presence of I-SR, respectively.

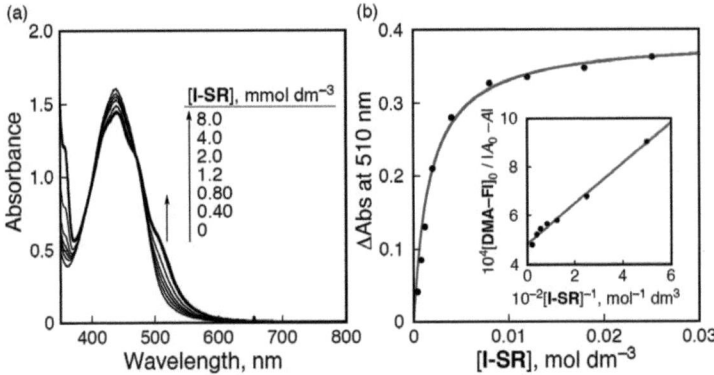

Figure 3.15: (a) UV-vis spectral change of DMA–Fl **8** (1.9 × 10^{-4} M) in the presence of various concentrations of I-SR **97** (0-8 mM) in THF. (b) Plots of Absorbance change vs [I-SR] **97** at 510 nm: Plot of [DMA–Fl]$_0$/|A_0 – A| vs [I-SR]$^{-1}$ at 510 nm for the supramolecular complex formation between DMA–Fl **8** and I-SR **97**.

From a linear plot of [DMA–Fl]$_0$/(A_0 – A) vs [I-SR]$^{-1}$ the formation constant is determined to be 6.8 × 10^2 M^{-1} in THF. Similarly, the reference compound **18** (Br-DR) of DRC$_{60}$ **43** is titrated to DMA–Fl **8** additionally to investigate the formation of supramolecular complex between DMA–Fl and Br-DR (Br-DR-(DMA–Fl)$_2$). The absorption spectrum is also changed and observed a saturation behavior upon addition of Br-DR **18** in THF as shown in Figure 3.17. According to equation 4,

$$\text{[DMA-Fl] 8} + \text{Br-DR 18} \underset{}{\overset{K}{\rightleftharpoons}} \text{Br-DR-(DMA-Fl)}_2 \quad (4)$$

the absorbance change is given by equation 5[249],

$$[DMA–Fl]_0/|A_0 – A| = (\varepsilon_c – \varepsilon_p)^{-1} + (K_2[Br\text{-}DR]^2(\varepsilon_c – \varepsilon_p))^{-} \quad (5)$$

which predicts a linear correlation between $[DMA-Fl]_0/|A_0 - A|$ and $[Br-DR]^{-2}$. From a linear plot of $[DMA-Fl]_0/(A_0 - A)$ vs $[Br-DR]^{-2}$ the formation constant can be determined to be 6.1×10^6 M^{-2} in THF. The difference in binding strength is tentatively ascribed to the number of the hydrogen bonds involved. This led us to optimize the structure of the supramolecular complexes by semiempirical method PM3[250,251] calculation with Gaussian 03. In the case of SRC$_{60}$ **109** with DMA–Fl **8** interacts on triple hydrogen bonds moiety, in contrast, DRC$_{60}$ **43** with DMA–Fl **8** interacts not only triple hydrogen bonds but also double DMA–Fl as π interaction.

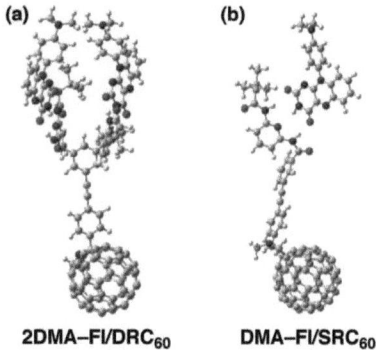

Figure 3.16: Optimized structures of supramolecular complex between DMA–Fl **8** and (a) DRC$_{60}$ **43** and (b) SRC$_{60}$ **109** calculated by a PM3MM method with Gaussian 03.

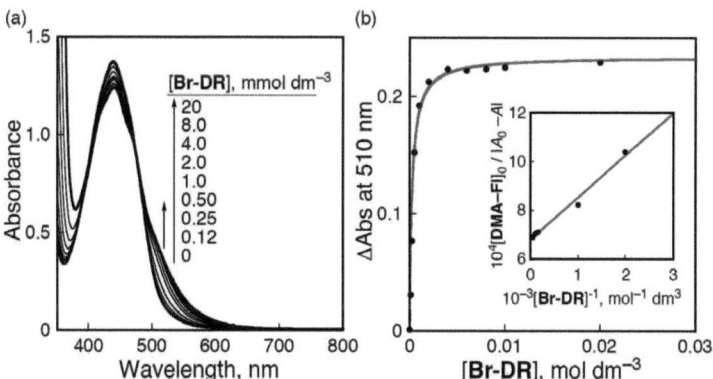

Figure 3.17: (a) UV-Vis spectral change of DMA–Fl **8** (1.6×10^{-4} M) in the presence of various concentrations of BrDR **18** (0-8 mM) in THF. (b) Plots of Absorbance change vs [Br-DR] at 510 nm: Plot of $[DMA-Fl]_0/|A_0-A|$ vs $[Br-DR]^{-1}$ at 510 nm for the supramolecular complex formation between DMA–Fl **8** and Br-DR **18**.

Results and Discussion

The stoichiometry of the complex formation between DMA–Fl **8** and Hamilton receptor **18** was examined by Job's plot analysis[248] using the absorption changes of DMA–Fl **8** at 470 nm. The absorbance change is given by equation 6, in which A_T and $(A_T)_R$ are the absorbances of total and the reference solution, $(A_T)_{max} - (A_T)_{R,\,ex}$ is the maximum absorbance term, respectively.

$$y = (A_T - (A_T)_R) / ((A_T)_{max} - (A_T)_{R,\,ex}) \qquad (6)$$

The Job's plot showes maxima at a mole fraction of 0.67, indicating a 2:1 complex formation. Structures of the complexes can be depicted as shown in equation 4. The same strategy can be extended to the stoichiometry of the 1:1 complex formation between DMA-Fl **8** and I-SR **97**.

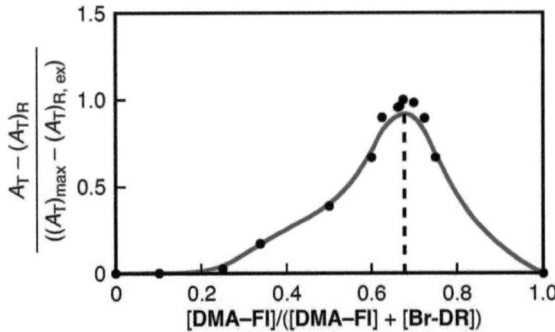

Figure 3.18: Job's plot obtained by absorption band 470 nm for the supramolecular complex formation between DMA–Fl 8 and Br-DR 18 in THF ([DMA–Fl] + [Br-DR] = 9.0 × 10⁻⁵ M).

3.2.3 Electrochemical Studies and Electron Transfer Driving Force of Supramolecular Complex between DMA-Fl and DRC$_{60}$.

Determination of the redox potentials of the newly formed donor-acceptor dyad is important to evaluate the energetics of electron transfer reactions and to probe the existence of electron transfer interactions between the donor and acceptor in the ground state. With this in mind, we have performed a study to evaluate the redox potential of the supramolecular complex between DMA–Fl **8** and DRC$_{60}$ **43** by using the cyclic voltammetric (CV) and fast scanning CV (FSCV) technique.

In deaerated THF containing 0.5 M LiClO$_4$, (DMA–Fl)$_2$-DRC$_{60}$ revealed one-electron oxidation and one-electron reduction were determined by FSCV (50 Vs^{-1}) and CV

(0.1 Vs^{-1}) as 1.02 and −0.46 V vs SCE, respectively (Figure 3.19a). The one-electron oxidation and reduction potentials of (DMA–Fl)$_2$-DRC$_{60}$ are assigned to one-electron oxidation potential of DMA–Fl **8** and one-electron reduction potential of DRC$_{60}$ **43**, as indicated by the reference FSCV and CV (Figure 3.19b and 3.19c), respectively. By comparison with reference redox potential, one-electron oxidation potential shifts to lower potential and one-electron potential shifts to higher potential by forming the supramolecular complex. This result suggests photoinduced ET could occur easier in the supramolecular complex than intermolecular photoinduced ET. Similarly, the comparison of redox potentials of the DMA–Fl-SRC$_{60}$, DMA–Fl, and SRC$_{60}$ are determined (Figure 3.20).

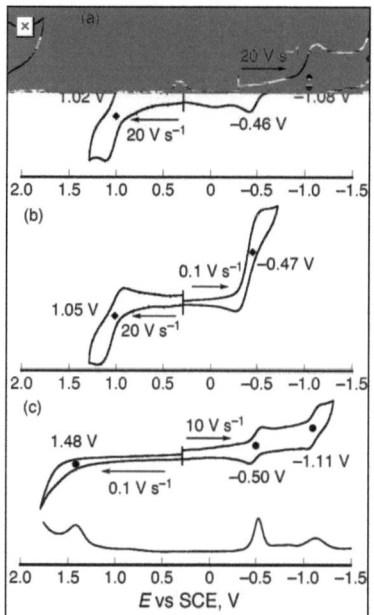 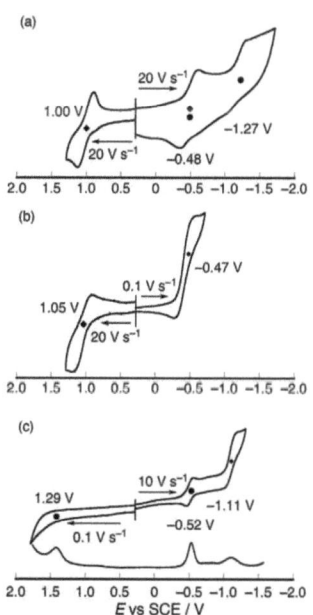

Figure 3.19: a) Fast scanning cyclic voltammogram (FSCV) (sweep rate 20 V s^{-1}) with a platinum microelectrode (i.d. 100 μm) of DMA–Fl (2.0 × 10^{-3} M) and DRC$_{60}$ (2.0 × 10^{-3} M), b) FSCV (sweep rate 20 V s^{-1}) and cyclic voltammogram (CV) (sweep rate 0.1 V s^{-1}) of DMA–Fl (2.0 × 10^{-3} M), and c) FSCV (sweep rate 10 V s^{-1}) and CV (sweep rate 0.1 V s^{-1}) of DRC$_{60}$ (2.0 × 10^{-3} M) in deaerated THF containing LiClO$_4$ (0.5 M) at 298 K.

Figure 3.20: a) Fast scanning cyclic voltammogram (FSCV) (sweep rate 20 V s^{-1}) with a platinum microelectrode (i.d. 100 μm) of DMA–Fl (2.0 × 10^{-3} M) and SRC$_{60}$ (2.0 × 10^{-3} M), b) FSCV (sweep rate 20 V s^{-1}) and cyclic voltammogram (CV) (sweep rate 0.1 V s^{-1}) of DMA–Fl (2.0 × 10^{-3} M), and c) FSCV (sweep rate 10 V s^{-1}) and CV (sweep rate 0.1 V s^{-1}) of SRC$_{60}$ (2.0 × 10^{-3} M) in deaerated THF containing LiClO$_4$ (0.5 M) at 298 K.

3.2.4 Assemblies of Supramolecular Complexes between DMA-Fl and C_{60} with Hydrogen Bond Receptors as Molecular Clusters in Mixed Solvents

DMA-Fl **8** and C_{60} with hydrogen bond receptors (SRC_{60} **109** and DRC_{60} **43**) are soluble in polar solvents such as tetrahydrofurane and benzonitrile, but sparingly soluble in nonpolar solvents such as hexane. When a concentrated solution of DMA-Fl **8** or SRC_{60} **109** or DRC_{60} **43** in THF is mixed with hexane by a fast injection method, the molecules aggregate to form stable clusters.[243,252-255] The final solvent ratio of mixed solvent employed in the present experiments was 3:1 (v/v) hexane/THF. The same strategy can be extended to prepare the supramolecular clusters of DMA-Fl-SRC_{60} and (DMA-Fl)$_2$-DRC_{60}. The supramolecular clusters aggregates in the present investigation were prepared by mixing the solution of DMA-Fl **8** (1 mM) and SRC_{60} **109** (1 mM) or DMA-Fl **8** (2 mL) and DRC_{60} **43** (1mM) in THF (0.5 mL) and then inject them into a pool of hexane (1.5 mL). The obtained optically transparent composite clusters are stable at room temperature, and they can be reverted back to their monomeric forms by diluting the solution with toluene.[243,252-255]

Figure 3.21 shows transmission electron microscopy images of the composite clusters, (DMA-Fl-SRC_{60})$_n$ and ((DMA-Fl)$_2$-DRC_{60})$_n$. DMA-Fl **8** moieties are self-assembled with DRC_{60} **43** moieties to yield controlled D-A assemblies with an interpenetrating network [((DMA-Fl)$_2$-DRC_{60})$_n$]. The diameter of ((DMA-Fl)$_2$-DRC_{60})$_n$ is ~40 nm in diameter, whereas (DMA-Fl-SRC_{60})$_n$ forms uniformed nanoparticles (diameter: ~ 200-500 nm). The size and shape is largely dependent on the different number of receptors.

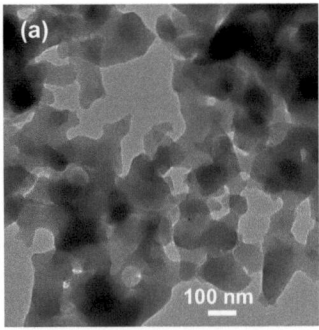 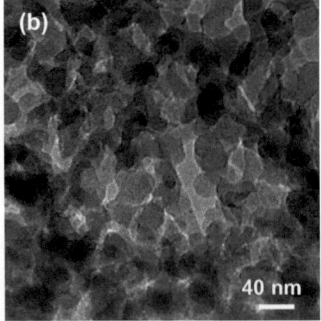

Figure 3.21: TEM images of clusters prepared with (a) [DMA-Fl] = [SRC_{60}] = 0.25 mM, and (b) [DMA-Fl] = 0.50 mM and [DRC_{60}] = 0.25 mM in hexane/THF (3/1, v/v).

3.2.5 Electrophoretic Deposition

Electrophoretic deposition was applied to fabricate films of clusters ((DMA-Fl-SRC$_{60}$)$_n$) onto OTE and nanostructured SnO$_2$ films casted onto OTE (OTE/SnO$_2$).[256-258] A suspension of the clusters (~2 mL) in THF was transferred to a 1 cm cuvette. Two OTEs cutted from conducting glass were inserted and a dc electric field (~100 V/cm) was applied. The clusters from the suspension were driven to the positive electrode surface and a robust thin film (abbreviated as OTE/SnO$_2$/(DMA-Fl-SRC$_{60}$)$_n$) was deposited within 2 min. An analogous process was adopted to deposit the clusters of (DMA-Fl)$_2$-DRC$_{60}$)$_n$ onto OTE/SnO$_2$ (abbreviated as OTE/SnO$_2$/(DMA-Fl)$_2$-DRC$_{60}$)$_n$). The steady-state electron absorption spectra of the OTE/SnO$_2$/((DMA-Fl)$_2$-DRC$_{60}$)$_n$ and OTE/SnO$_2$/(DMA-Fl-SRC$_{60}$)$_n$ electrodes in comparison to the absorption spectrum of ((DMA-Fl)$_2$-DRC$_{60}$)$_n$ in THF/hexane (3/1, v/v) are shown in Figure 3.22.

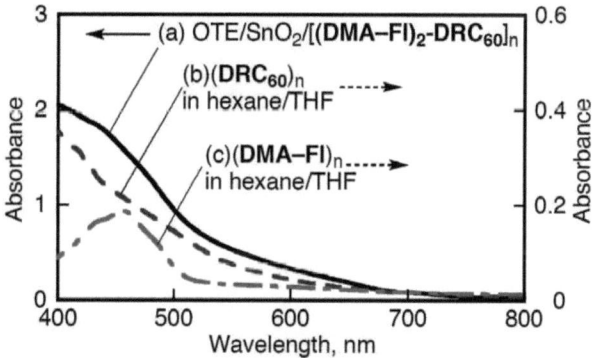

Figure 3.22: Absorption spectra of (a) OTE/SnO$_2$/((DMA-Fl)$_2$-DRC$_{60}$)$_n$ film prepared from cluster solutions of ([DMA-Fl] = 0.50 mM, [DRC$_{60}$] = 0.25 mM), (b) OTE/SnO$_2$/(DMA-Fl-SRC$_{60}$)$_n$ film prepared from cluster solutions of ([DMA-Fl] = [SRC$_{60}$] = 0.25 mM) , and (c) ((DMA-Fl)$_2$-DRC$_{60}$)$_n$ solution in hexane/THF (3/1, v/v).

3.2.6 Properties of OTE/SnO$_2$/(DMA-Fl-SRC$_{60}$)$_n$ and OTE/SnO$_2$/ ((DMA-Fl)$_2$-DRC$_{60}$)$_n$.

Photoelectrochemical measurements were performed in acetonitrile containing 0.5 M LiI and 0.01 M I$_2$ as the redox electrolyte with OTE/SnO$_2$/((DMA-Fl)$_2$-DRC$_{60}$)$_n$ and OTE/SnO$_2$/(DMA-Fl-SRC$_{60}$)$_n$ as the working electrode and a Pt wire as the counter electrode. The photoelectrochemical performance of the OTE/SnO$_2$/((DMA-Fl)$_2$-DRC$_{60}$)$_n$ electrode was examined by employing a standard two-compartment cell with

Results and Discussion

a Pt wire gauge counter electrode.[259] The photocurrent and photovoltage responses upon the excitation of the OTE/SnO$_2$/(DMA-Fl-SRC$_{60}$)$_n$ electrode in the visible region ($\lambda > 400$ nm) are shown in Figure 3.23 and 3.24, respectively. The photocurrent response is prompt, steady, and reproducible during repeated on/off cycles of the visible light illumination. The short circuit photocurrent (I_{sc}) of the OTE/SnO$_2$/((DMA-Fl)$_2$-DRC$_{60}$)$_n$ electrode is 0.28 mA cm^{-2} (Figure 3.23 a) under white light illumination (AM 1.5 condition; input power) 100 mW cm^{-2}), and this value is nearly two times greater than that (0.16 mA cm^{-2}) obtained with the OTE/SnO$_2$/(DRC$_{60}$)$_n$ electrode (Figure 3.23b). The open circuit voltage (V_{oc}) of the OTE/SnO$_2$/((DMA-Fl)$_2$-DRC$_{60}$)$_n$ electrode is 220 mV (Figure 3.24a), and this value is nearly the same (200 mV) obtained with the OTE/SnO$_2$/(DRC$_{60}$)$_n$ electrode (Figure 3.24b).

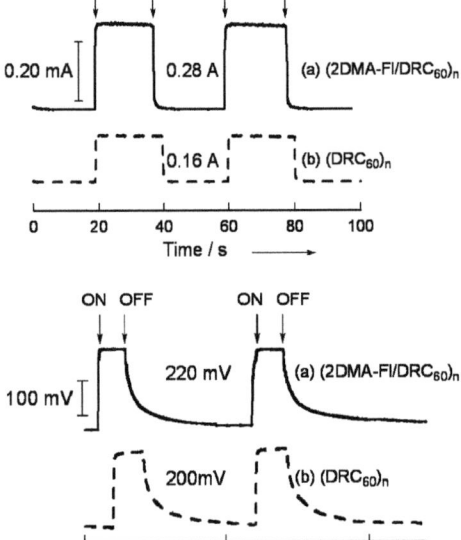

Figure 3.23: Photocurrent response of (a) OTE/SnO$_2$/((DMA-Fl)$_2$-DRC$_{60}$)$_n$ and (b) OTE/SnO$_2$/(/DRC$_{60}$)$_n$ electrode prepared from cluster solution of ([DMA–Fl] = 0.50 mM; [DRC$_{60}$] = 0.25 mM to visible light illumination ($\lambda > 400$ nm); electrolyte: 0.5 M LiI and I$_2$ 5 mM in acetonitrile; input power 100 mW/cm^2.

Figure 3.24: Photovoltage response of (a) OTE/SnO$_2$/((DMA-Fl)$_2$-DRC$_{60}$)$_n$ and (b) OTE/SnO$_2$/(/DRC$_{60}$)$_n$ electrode prepared from cluster solution of ([DMA–Fl] = 0.50 mM; [DRC$_{60}$] = 0.25 mM to visible light illumination ($\lambda > 400$ nm); electrolyte: 0.5 M LiI and I$_2$ 5 mM in acetonitrile; input power 100 mW/cm^2.

Furthermore, an evaluation of the photocurrent action spectrum of the OTE/SnO$_2$/((DMA-Fl)$_2$-DRC$_{60}$)$_n$ electrode and OTE/SnO$_2$/(DMA-Fl-SRC$_{60}$)$_n$ was performed by examining the wavelength dependence of the incident photon to current conversion efficiency (IPCE). The IPCE values are calculated by normalizing the photocurrent densities for incident light energy and intensity and by use of the following expression:[260]

Results and Discussion

$$\text{IPCE (\%)} = 100 \times 1240 \times i/(W_{in} \times \lambda) \qquad (6)$$

where i is the photocurrent density (A cm^{-2}), W_{in} is the incident light intensity (W cm^{-2}), and λ is the excitation wavelength (nm). As shown in Figure 3.23, the photocurrent action spectrum of the OTE/SnO$_2$/((DMA-Fl)$_2$-DRC$_{60}$)$_n$ electrode shows a maximum IPCE value of 5.1 % at 440 nm (Figure 3.25a). Following the same experimental conditions, the observed IPCE value for OTE/SnO$_2$/(DMA-Fl-SRC$_{60}$)$_n$ electrode is relatively small, 3.0 % at 440 nm (Figure 3.25b). The maximum IPCE value of the OTE/SnO$_2$/((DMA-Fl)$_2$-DRC$_{60}$)$_n$ system in the absence of applied bias potential exhibits approximately twice as great than that of the OTE/SnO$_2$/(DMA-Fl-SRC$_{60}$)$_n$ system. This suggests that formation of uniformed nanoparticles and strong hydrogen bonds play one of important roles in the photocurrent generation rather than the networking structure.

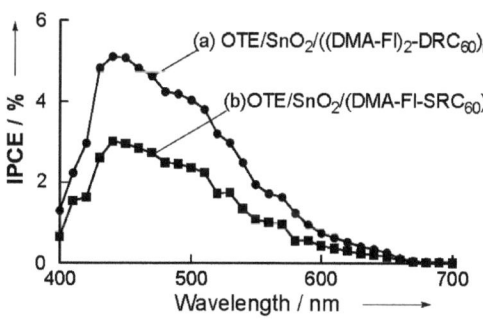

Figure 3.25: Photocurrent action spectra (IPCE vs wavelength) of (a) OTE/ SnO$_2$/ (DMA-Fl-SRC$_{00}$)$_n$ electrode prepared from cluster solutions of ([DMA-Fl] = [SRC$_{60}$] = 0.25 mM) and (b) OTE/SnO$_2$/((DMA-Fl)$_2$-DRC$_{60}$)$_n$ electrode prepared from cluster solutions of ([DMA-Fl] = 0.50 mM, [DRC$_{60}$] = 0.25 mM). Electrolyte: 0.5 M LiI, and 5 mM I$_2$ in acetonitrile.

The charge separation in the OTE/SnO$_2$/((DMA-Fl)$_2$-DRC$_{60}$)$_n$ electrode can be further modulated by the application of an electrochemical bias potential (a standard three-compartment cell as working electrode along with a Pt wire gauze counter electrode and saturated calomel reference electrode). Figure 3.26A shows I-V characteristics of the OTE/SnO$_2$/((DMA-Fl)$_2$-DRC$_{60}$)$_n$ under visible light illumination (AM 1.5). The photocurrent increases as the applied potential is scanned towards more positive potentials. Increased charge separation and the facile transport of charge carriers under a positive bias potential are responsible for enhanced photocurrent generation. At potentials greater than +0.3 V vs SCE, the direct electrochemical oxidation of iodide interferes with the photocurrent measurement. The net photocurrent generation density of OTE/SnO$_2$/((DMA-Fl)$_2$-DRC$_{60}$)$_n$ at +0.2 V vs SCE

(~0.51 mA/cm^2) was much larger than the case at 0 V vs SCE (~0.22 mA/cm^2).[257,258] Thus, by using a three-compartment cell, we can control photocurrent generation density of the OTE/SnO$_2$/((DMA-Fl)$_2$-DRC$_{60}$)$_n$ electrode. By controlling the potential of OTE/SnO$_2$ with an electrochemical bias, we can improve the charge separation and attain higher IPCE values.[258] The photocurrent action spectra of OTE/SnO$_2$/((DMA-Fl)$_2$-DRC$_{60}$)$_n$ under an applied bias potential of 0.2 V vs SCE were recorded using a standard three-compartment cell. The IPCE values of OTE/SnO$_2$/((DMA-Fl)$_2$-DRC$_{60}$)$_n$ under an applied bias potential of 0.2 V vs SCE (spectrum a in Figure 3.26 b) are much larger than those under no applied voltage condition (spectrum b) in the whole visible region. The maximum IPCE values of OTE/SnO$_2$/((DMA-Fl)$_2$-DRC$_{60}$)$_n$ at 0.2 V vs SCE attains 9.2 %, which is more than two times larger than that at 0 V vs SCE (5.1 %). This trend is in good agreement with *I-V* characteristics (Figure 3.26A).

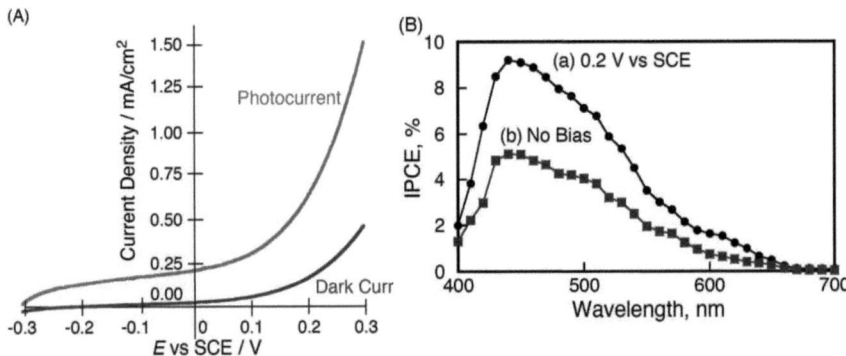

Figure 3.26: (A) *I-V* characteristics of the OTE/SnO$_2$/((DMA-Fl)$_2$-DRC$_{60}$)$_n$ electrode under white light illumination (AM 1.5 conditions). Electrolyte: LiI 0.5 M, I$_2$ 5 mM in acetonitrile; input power = 100 mW cm^{-2}. (B) Photocurrent action spectra (IPCE vs wavelength) of the OTE/SnO$_2$/((DMA-Fl)$_2$-DRC$_{60}$)$_n$ electrode (a) at an applied bias potential of 0.2 V vs SCE and (b) without applied bias potential. Electrolyte: LiI 0.5 M, I$_2$ 5 mM in acetonitrile.

3.2.7 Photoirradiation

The formation of the radical cation of DMA-Fl (DMA$^{\bullet+}$-Fl) and the radical anion of DRC$_{60}$ (DRC$_{60}$$^{\bullet-}$) clusters upon the photoexcitation of the composite clusters of ((DMA-Fl)$_2$-DRC$_{60}$)$_n$ is also confirmed by the electron spin resonance (ESR) measurements performed in frozen hexane/THF under photoirradiation. The resulting spectrum of photoirradiated ((DMA-Fl)$_2$-DRC$_{60}$)$_n$ in hexane/THF at 77 K is shown in

Figure 3.27a. The ESR spectrum consists of two signals, one of which is attributable to DRC_{60} radical anion ($DRC_{60}{}^{\bullet -}$) at a small g value (g = 2.0010), and the other is DMA-Fl radical cation ($DMA^{\bullet +}$-Fl) at a higher g value (g = 2.0038). To confirm these assignments, $DMA^{\bullet +}$-Fl and $DRC_{60}{}^{\bullet -}$ were produced independently via the chemical oxidation of DMA-Fl with $Fe(bpy)_3{}^{3+}$ and via the photoinduced electron transfer from dimeric 1-benzyl-1,4-dihydronicotinamide (($BNA)_2$) to DRC_{60} clusters,[261] respectively.

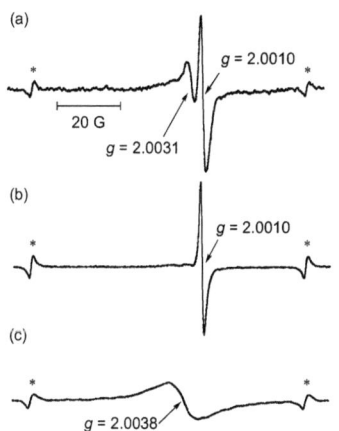

Figure 3.27: ESR spectra of a) photoirradiated ((DMA-$Fl)_2$-$C_{60})_n$ ([DMA–Fl] = 0.50 mM; [DRC_{60}] = 0.25 mM) in hexane/THF (3/1, v/v) with a high-pressure mercury lamp, b) the radical cation of DMA-Fl (0.50 mM) produced by the electron-transfer oxidation with $Fe(bpy)_3{}^{3+}$ (1.0 mM) in PhCN, and c) the radical anion of DRC_{60} clusters [($C_{60})_n$] (0.50 mM) generated in photoinduced electron transfer from dimeric 1-benzyl-1,4-dihydronicotineamide (0.50 mM) to C_{60} clusters in hexane/THF (3/1, v/v) under photoirradiation of a high-pressure mercury lamp, measured at 77 K. Asterisk denotes Mn^{2+} marker.

Figure 3.28: ESR spectra of a) photoirradiated (DMA–Fl-$SRC_{60})_n$ ([DMA–Fl] = 0.25 mM; [SRC_{60}] = 0.25 mM) in hexane/THF (3/1, v/v) with a high-pressure mercury lamp, b) the radical anion of SRC_{60} clusters [($SRC_{60})_n$] (0.50 mM) generated in photoinduced electron transfer from dimeric 1-benzyl-1,4-dihydronicotineamide (0.50 mM) to DRC_{60} clusters in hexane/THF (3/1, v/v) under photoirradiation of a high-pressure mercury lamp, and c) the radical cation of DMA–Fl (0.50 mM) produced by the electron-transfer oxidation with $Fe(bpy)_3{}^{3+}$ (1.0 mM) in PhCN, measured at 77 K. Asterisk denotes Mn^{2+} marker.

The ESR spectra of the clusters are shown in Figure 3.27a and 3.28a. A comparison of the observed spectrum in Figure 3.27a with the spectra of the authentic radical cation and radical anion in Figure 3.27b and c confirm that the observed ESR signal in Figure 3.27a is composed of two signals, one of which is due to the DMA$^{\cdot+}$-Fl and the other of which is due to the DRC$_{60}{}^{\cdot-}$ clusters. The same strategy can be extended for detection of radical species (i.e. DMA$^{\cdot+}$–Fl and SRC$_{60}{}^{\cdot-}$) of (DMA–Fl-SRC$_{60}$)$_n$ (Figure 3.28).

3.2.8 Photodynamics of the Supramolecular Complex Clusters between DMA-Fl and C$_{60}$ with Hydrogen Bond Receptors in Femtosecond Order

The occurrence of ultrafast electron transfer from the singlet excited states of DMA–Fl (DMA-^{1}Fl*) to DRC$_{60}$ in the supramolecular complex cluster was further confirmed by the femtosecond laser flash photolysis performed in the solid state on the supramolecular cluster of DMA-Fl and DRC$_{60}$ in KBr pellets.[262] As a reference test, a KBr pellet that did not contain any cluster sample did not show any absorption over the course of laser flash photolysis excited at 450 nm.

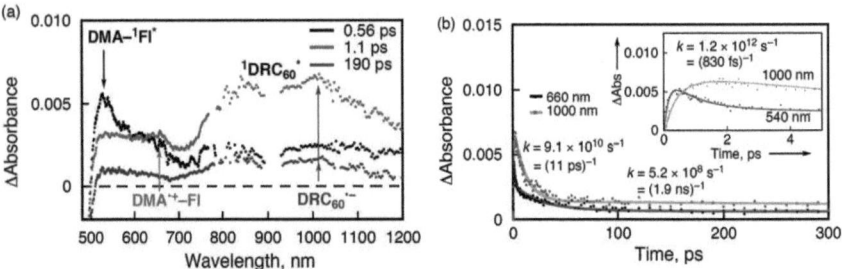

Figure 3.29: a) Transient absorption spectra of ((DMA-Fl)$_2$-DRC$_{60}$)$_n$ clusters in KBr pellets, taken by femtosecond laser excitation at 440 nm. b) Time profiles of absorbance at 660 and 1000 nm; inset: Time profiles at the shorter time range at 660 and 1000 nm. The solid curves represent the best fit to the exponential rise or decay.

In contrast, the time-resolved transient absorption spectra of supramolecular complex cluster of ((DMA-Fl)$_2$-DRC$_{60}$)$_n$ in KBr pellet are shown in Figure 3.29a. The transient absorption spectrum observed at 0.56 ps, which has an absorption maximum at 520 nm is assigned to the singlet excited state of DMA-Fl (DMA-^{1}Fl*).[244] The decay of absorption at 520 nm owing to DMA-^{1}Fl* is accompanied by a rise in the absorption at 1000 nm (inset of Figure 3.29b).

Results and Discussion

The transient absorption spectrum at 1.1 ps may be assigned to DMA-Fl radical cation (DMA$^{•+}$-Fl) and DRC$_{60}$ radical anion (DRC$_{60}^{•-}$) produced by directional electron-transfer (ET) cascade reactions from DMA to C$_{60}$ via DMA-^{1}Fl* and hydrogen bond receptor moiety as shown in figure 3.30, and DRC$_{60}$ 43 singlet excited state (^{1}DRC$_{60}^*$), which mixed with the absorption band of DRC$_{60}^{•-}$. Finally, the absorption bands of DMA$^{•+}$–Fl and DRC$_{60}^{•-}$ are still persistent at 190 ps. The rate constants of photoinduced ET, charge shifting and back ET were determined to be 1.2 × 10^{12} s^{-1}, 9.1 × 10^{10} s^{-1}, and 5.2 × 10^{8} s^{-1} by three-exponential fitting at 520 nm (DMA-^{1}Fl*), 660 nm (DMA$^{•+}$-Fl), and 1000 nm (DRC$_{60}^{•-}$), respectively.

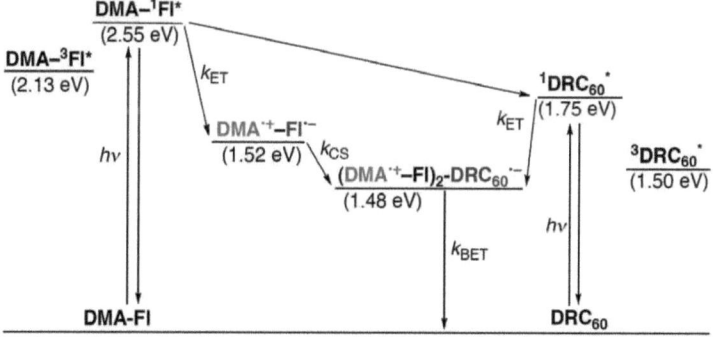

Figure 3.30: Energy diagram of DMA–Fl 8 and DRC$_{60}$ 43.

Similarly, the photodynamics of (DMA-Fl-SRC$_{60}$)$_n$ was observed and the rate constants (k_{ET}), charge shifting (k_{CS}), and back ET (k_{BET}) were determined to be 1.2 × 10^{12} s^{-1}, 9.1 × 10^{10} s^{-1}, and 5.2 × 10^{8} s^{-1} by three-exponential fitting (Figure 3.31).

3.2.9 Conclusion

Photoelectrochemical electrodes using directional electron transfer systems consisted of DMA-Fl-C$_{60}$ have been constructed for the first time. The comparison between single and double hydrogen bond supplied the difference of the formation constant of the supramolecular complexes. The formation constant of a supramolecular complex consisted of DMA-Fl with DRC$_{60}$ is bigger than that with SRC$_{60}$. The film of the ((DMA-Fl)$_2$-DRC$_{60}$)$_n$ onto the nanostructured SnO$_2$ electrode exhibited an incident photon-to-photocurrent efficiency (IPCE) of 9.2 % in a three-compartment electrochemical cell. The measured IPCE has been found greater than the (DMA-Fl-SRC$_{60}$)$_n$. In addition, photodynamics of the hydrogen bonded supramolecular assemblies composed of DMA-Fl and fullerene with single and double hydrogen bond receptors (SRC$_{60}$ and DRC$_{60}$) gave rise to the directional electron-transfer cascade reactions through the well-organized gradients. The obtained results demonstrate the potential and applied utility of a multistep electron transfer as opposite to a hole transfer along a fine-tuned and fine-balanced redox gradient in photoelectrochemical devices at first.

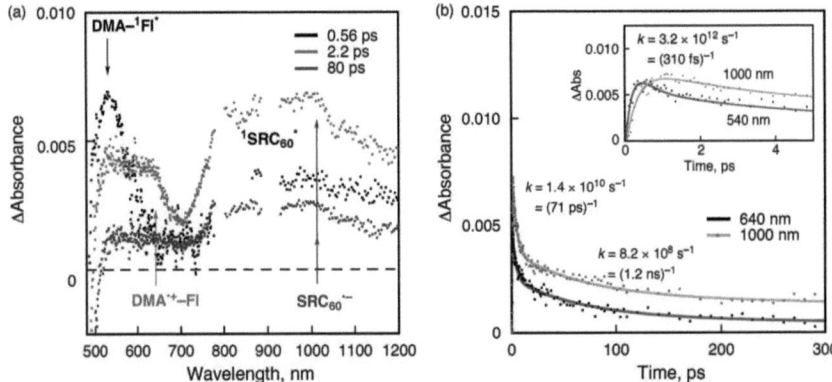

Figure 3.31: a) Transient absorption spectra of (DMA-Fl-SRC$_{60}$)$_n$ clusters in KBr pellet, taken by femtosecond laser excitation at 440 nm. b) Time profiles of absorbance at 640 and 1000 nm; inset: Time profiles at the shorter time range at 52 and 1000 nm. The solid curves represent the best fit to the exponential rise or decay.

3.3 Further Photoactive Building Blocks for Supramolecular Aggregation

Besides the above described building blocks for supramolecular nanohybrids (i.e. Hamilton receptor modified fullerenes **41-47** and **109-112**, cyanuric acid bearing porphyrin **7** and flavin **8** further compounds with one or more key or lock subunits are synthesized and characterized.

To broaden the knowledge about supramolecular fullerene-porphyrin architectures it might be interesting to change the key-lock position, i.e. to synthesize a cyanuric acid modified [60]fullerene and a Hamilton receptor bearing zinc porphyrin. Scheme 3.24 depicts the synthesis of cyanuric acid substituted fullerene **114**. Sonogashira coupling of the 4-iodophenyl isocanuric acid **63**[263] with aldehyde **20** in the presence of bis(triphenylphosphine)palladium(II) dichloride [Pd(PPh$_3$)$_2$Cl$_2$] as catalyst led to the formation of the cyanuric acid substituted aldehyde **113** in good yields. The *in situ* formed azomethine ylide of **113** and sarcosine is reacted with C$_{60}$ to afford the target compound **114** which is purified by column chromatography on SiO$_2$.

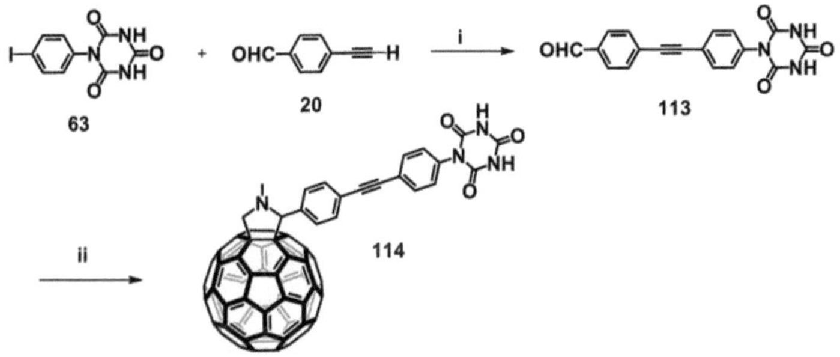

Scheme 3.24: Synthesis of the cyanuric acid substituted [60]fullerene **114**: i) Pd(PPh$_3$)$_2$Cl$_2$, CuI, THF/NEt$_3$, rt; ii) C$_{60}$, sarcosine, toluene/THF, reflux.

Hamilton receptor porphyrin **115** is designed as counter part to cyanuric acid modified [60]fullerene compounds in supramolecular donor-acceptor assemblies. The synthesis of this compound emanates from the terminal ethynylporphyrin **62** which can be coupled under modified SONOGASHIRA coupling conditions[200,201] in absence of copper(I)-catalyst with the iodo Hamilton receptor **19** (Scheme 3.25) in good yield.

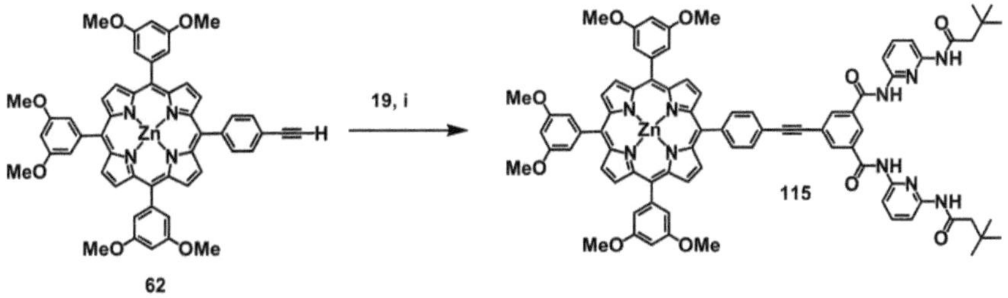

Scheme 3.25: Synthesis of porphyrin **115**. i) **19**, Pd$_2$dba$_3$, AsPh$_3$, NEt$_3$, THF, rt.

Characterization of porphyrin **115** is achieved by ^{1}H/^{13}C NMR, UV/Vis absorption and IR spectroscopy and mass spectrometry. Outstanding for the ^{1}H NMR spectrum of porphyrin derivative **115** in THF d$_8$ (Figure 3.32) are the signals of the Hamilton receptor and those from the porphyrin core. Typically the signals for the NH protons occur as singletts at 9.81 and 9.13 ppm.

Results and Discussion

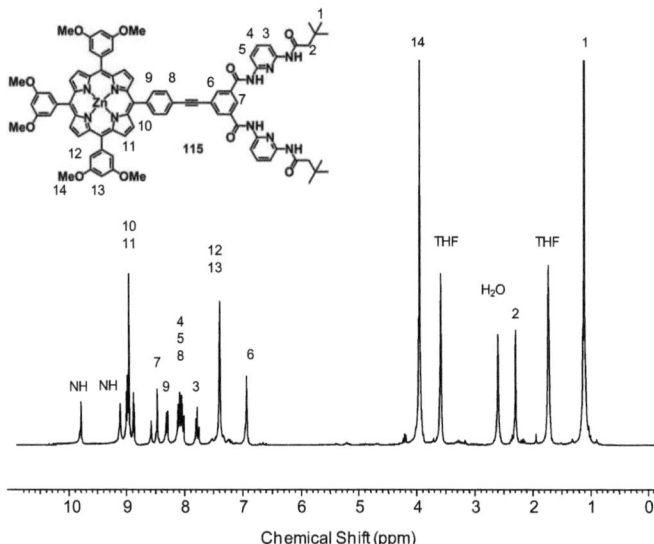

Figure 3.32: ^{1}H NMR spectrum (THF d$_8$, 400 MHz, rt) of Hamilton receptor modified porphyrin 115.

The singlet of the methylene group appears at 2.29 ppm and this of the methyl group at 1.11 ppm. Signals that can be assigned clearly to the porphyrin core are the singlet of the methylether at 3.95 ppm and the pyrrolic signals at 9.00 (d, ^3J = 4.64 Hz, 2H,), 8.96 (s, 4H) and 8.88 ppm (d, ^3J = 4.64 Hz, 2H).

The ^{13}C NMR spectrum of compound **115** in THF d$_8$ (Figure 3.33) shows characteristic signals between 20 ppm and 60 ppm which can be definitely be assigned to the methyl group (30.01 ppm), the quaternary carbon of the *tert*-butyl group (31.65 ppm), the methylene group (50.72 ppm) and the methylether (55.63 ppm). Typically the signals of the sp carbon atoms occur at 89.44 and 92.04 ppm. Specific for the six aromatic carbon atoms of the phenyl methyl ether in the porphyrin core is the signal at 159.85 ppm. Characteristic for Hamilton receptor components are furthermore the signals of the β-pyridine carbons at 110.23 and 110.23 ppm, as well as the α-pyridine carbons at 150.65 and 150.85 ppm. The carbonyl signals appear at 164.91 and 170.78 ppm.

Results and Discussion

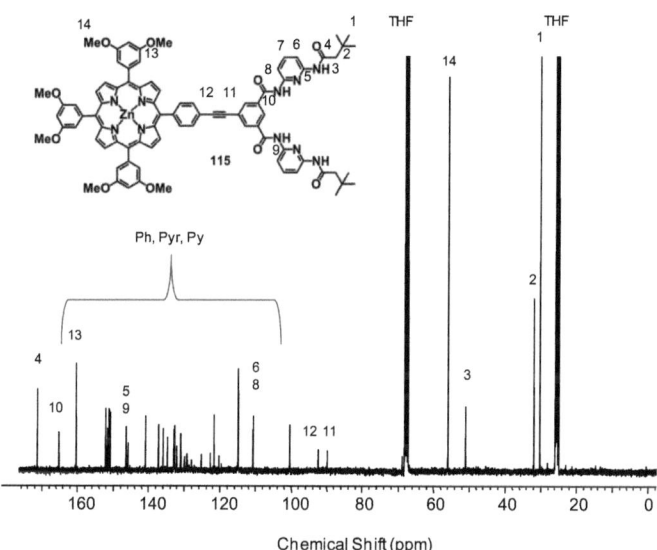

Figure 3.33: ^{13}C NMR spectrum (THF d$_8$, 100.5 MHz, rt) of Hamilton receptor modified porphyrin 115.

To expand the concept of hydrogen bond supramolecular donor-acceptor arrays chromophores that show multibinding skills are developed towards their promising photophysical properties. Compared to simple 1:1 complexes the increasing number of possible complexation positions should lead to a more efficient fluorescence quenching of the electron donor after irradiation with light. Taking this consideration into account tetra Hamilton receptor functionalized porphyrin **120** is conceived. The synthesis of a tetra Hamilton receptor functionalized zinc(II)porphyrin **120** is oriented on previous aspects and starts with the condensation of 4 eq of 4-bromobenzaldehyde **23** and 4 eq pyrrole **50** under LINDSEY conditions.[208-211] Due to the circumstance that just one single porphyrin system can be achieved by this approach the yield of 16 % is rather good. Free base porphyrin **116** can be isolated in good yield using chromatography on silica gel. Metallation with zinc(II)acetate was carried out exactly under the same conditions described above to give zinc(II)porphyrin **117**.

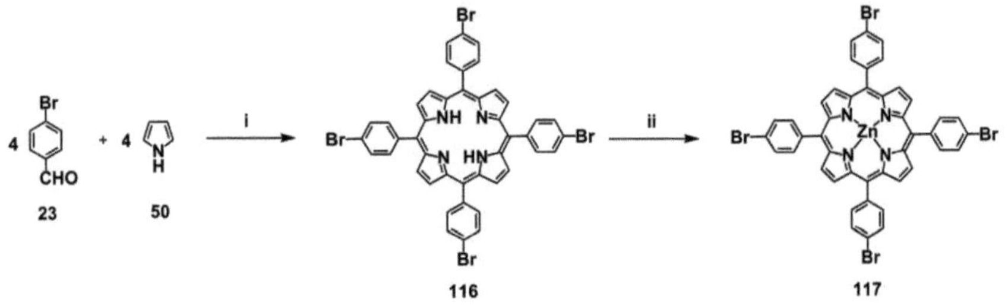

Scheme 3.26: Synthesis of symmetric TPP 117: i) 1. PPh$_4$Cl, BF$_3$*OEt$_2$, CH$_2$Cl$_2$, rt, 2. DDQ; ii) Zn(OAc)$_2$, THF, reflux.

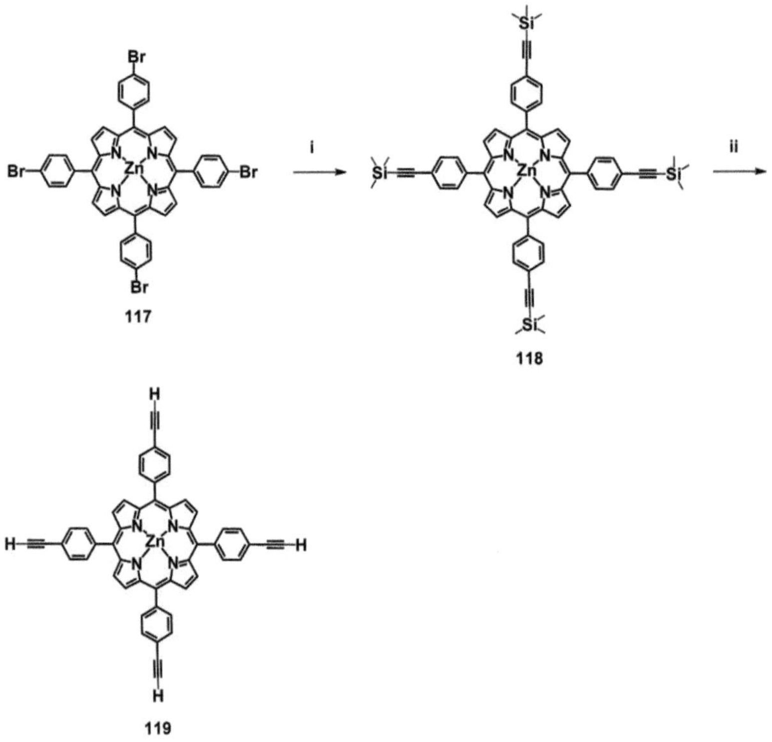

Scheme 3.27: Synthesis of tetra functionalized porphyrin 119: i) Pd(PPh$_3$)$_2$Cl$_2$, CuI, THF/HNEt$_2$, 80 °C; ii) TBAF, THF, 0 °C.

Results and Discussion

Tetrafold SONOGASHIRA coupling[200,201] of 4 eq trimethylsilylacetylene and 5,10,15,20-(p-bromo-tetraphenyl)-porphyrinato zinc (II) **117** in a mixture of THF and HNEt$_2$ under standard coupling conditions (Scheme 3.27) is then applied for further modification of the porphyrin core. Desilylation of the protecting group using TBAF as fluoride source leads to the free tetra acetylenic porphyrin compound **119** (Scheme 3.27) in a quantitative yield.

Tetra Hamilton receptor functionalized porphyrin **120** is now successfully obtainable by SONOGASHIRA coupling[200,201] of porphyrin **119** with 4 eq of the iodo Hamilton receptor **19** under palladium(0) catalysis and basic conditions (Scheme 3.28).

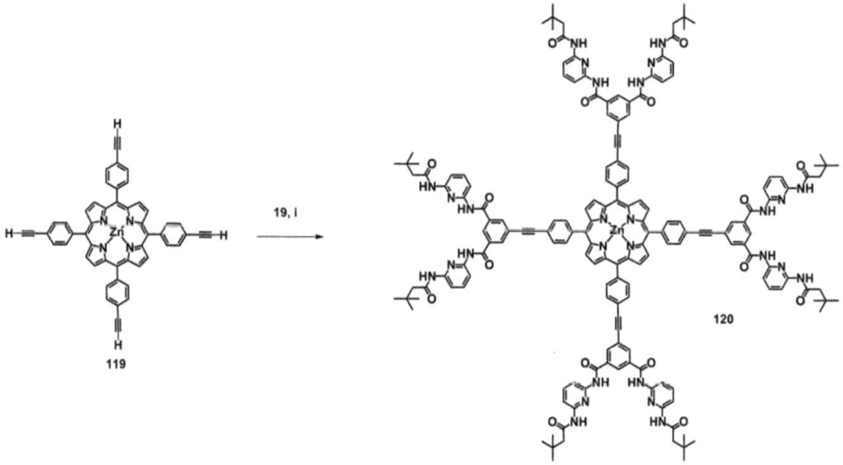

Scheme 3.28: Synthesis of tetra Hamilton receptor modified porphyrin 120: i) Pd(PPh$_3$)$_2$Cl$_2$, CuI, THF/NEt$_3$, 80 °C.

Characterization of tetra Hamilton receptor bearing porphyrin **120** is done via ^{1}H/^{13}C NMR, UV/Vis absorption and IR spectroscopy and mass spectrometry. The ^{1}H NMR spectrum of **120** in THF d$_8$ (Figure 3.34) exhibits typical features of the Hamilton receptor as well as those of the porphyrin. The signals for the NH protons appear at 9.79 and 9.09 ppm and arise as broad singlets. Characteristic as well are the signals of the methyl group at 1.09 ppm and the signals of the methylene group at 2.28 ppm. Explicitly related to the porphyrin core is the singlet of the pyrrolic protons at 8.96 ppm.

Results and Discussion

The ^{13}C NMR spectrum of porphyrin **120** in THF d$_8$ (Figure 3.35) possesses the typical signals for the Hamilton receptor subunits between 20 and 60 ppm. At 30.27 ppm occurs the signal for the methyl group, at 31.93 ppm the signal for the quartenary carbon of the *tert*-butyl group and at 50.98 ppm the signal of the methylene group. Additionally the carbonyl signals at 165.19 and 171.05 ppm are characteristic for the Hamilton receptor building block. Definitely assignable to the porphyrin ring are the signals at 89.85 and 92.16 ppm (sp carbons), 129.40 ppm (β-pyrrolic carbons) and 152.02 ppm (8C, α-pyrrolic carbons).

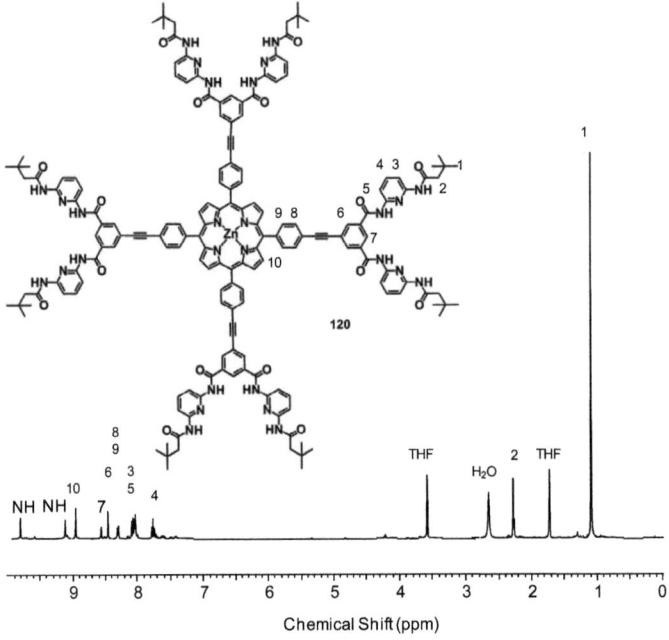

Figure 3.34: ^{1}**H NMR spectrum (THF d$_8$, 400 MHz, rt) of Hamilton receptor modified porphyrin 120.**

The purity of tetra Hamilton receptor modified zinc porphyrin **120** is proven by analytical HPLC measurements (eluent: dichloromethane/methanol 98:2). Mass spectrometry (MALDI, DCTB matrix) shows clearly a peak at 2941 m/z and no evidence for one-, two- or threefold coupled derivatives.

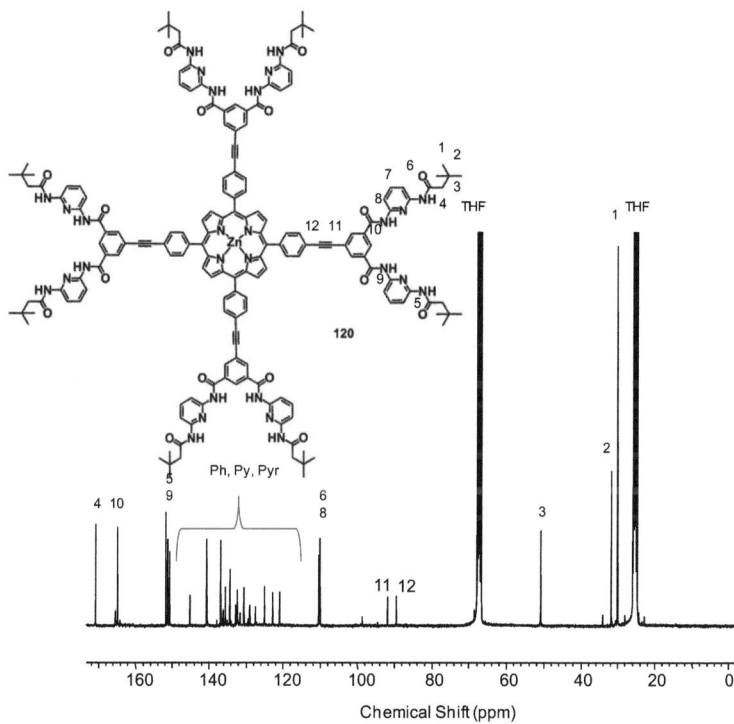

Figure 3.35: ^{13}C NMR spectrum (THF d$_8$, 100.5 MHz, rt) of Hamilton receptor modified porphyrin **120**.

Further investigations (i. e. complexation experiments) of the tetra Hamilton receptor bearing porphyrin **120** are not perfomed due to the weak solubility in apolar solvents. The complexation with cyanuric acid modified chromophores would be interesting. But the addition of THF or DMSO for dissolving the compounds induces strong interactions between solvent and key and solvent and lock respectively. This probably degrades the self-assembly process dramatically.

3.4 Supramolecular Aggregates of oligo-Phenylene-Ethynylene Wires

Poly- and oligo conjugated π-systems such as polymeric and oligomeric *p*-phenylenevinylenes (*p*-PPV/*p*-OPV), *p*-phenylenes (*p*-PP/*p*-OP), alkylthiophenes (PAT/OAT) and *p*-phenyleneethynylenes (*p*-PPE/*p*-OPE) are known for their interesting electronic, photoluminescence and electroluminescence properties.[264-271] The supramolecular arrangement of π-conjugated systems is a rather new and interesting topic in modern chemistry and nanoscience.[239,272-281] Particular linear and rigid chains in the nanometer regime consisting of π-conjugated molecules have fascinating features with respect to their semiconducting behaviour and their possible application as nanowires between electrodes.[282-288] However, to assure electron tunnelling through supramolecular arrays and to compare these with inorganic wires[289] and carbon nanotubes[290] several aspects have to be taken into account. Purity of the organic monomers is crucial due to possible trapping of holes and electrons by impurities.

Three new conjugated supramolecular π-systems consisting of four different *p*-OPEs **123, 124, 126** and **128** and a tetraphenylporphyrin **7** as chromophoric end cap are described herein. The *p*-OPEs are terminated with a Hamilton receptor at each end for the complexation with a cyanuric acid bearing tetraphenylporphyrin (TPP). Based on ^{1}H NMR titration experiments the complexation behaviour of the new *p*-OPEs has been analyzed.

3.4.1 Synthesis

The target *p*-OPEs are synthesized starting with the above described iodo Hamilton receptor **19**. This iodo compound is a suitable precursor for the synthesis of rod-like carbon chains with a Hamilton receptor as end cap. Therefore in principle various C-C cross coupling techniques are possible. Most promising in this case are SONOGASHIRA coupling techniques[200,201] with terminal acetylenes. Trimethylsilylacetylene is coupled in nearly quantitative yield at room temperature with the iodo Hamilton receptor **19** (Scheme 3.29). As Pd^0 source, Pd(PPh$_3$)$_2$Cl$_2$ is chosen and triphenylphosphine is used as additional ligand. Deprotection of the resulting Hamilton receptor **121** using TBAF (tetrabutyl ammonium fluoride) as fluoride source leads to the terminal ethynyl derivative **122**. Important for a good yield and for avoiding dimerization is to keep the temperature at or below 25 °C during the reaction process and the work-up. The obtained terminal acetylene group is then used to elongate the chain or to synthesize linear structures with two Hamilton receptor end caps for H-bonded oligomers and polymers.

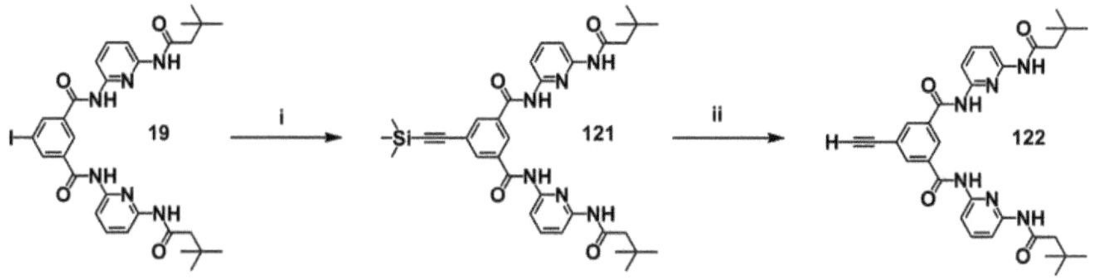

Scheme 3.29: Syntheses of Hamilton receptor derivatives **121** and **122**. i) Trimethylsilylacetylene, Pd(PPh$_3$)$_2$Cl$_2$, PPh$_3$, CuI, THF, NEt$_3$, rt; ii) TBAF, THF, rt.

Coupling of Hamilton receptor derivative **122** with iodo receptor **19** under Sonogashira conditions is used to synthesize rodlike molecule **123** (Scheme 3.30).

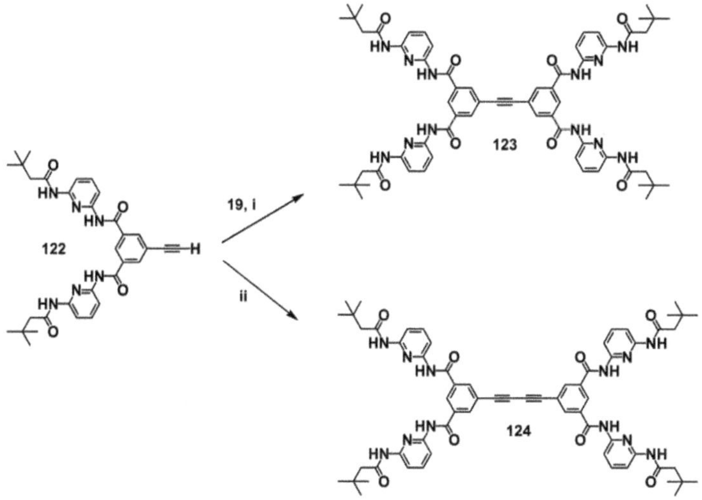

Scheme 3.30: Syntheses of target p-OPEs **123** and **124**: i) **19**, Pd(PPh$_3$)$_2$Cl$_2$, CuI, THF, HNEt$_2$, rt; ii) Pd(PPh$_3$)$_2$Cl$_2$, CuI, THF, NEt$_3$, rt.

Homo coupling of compound **122** leads to linear molecule **124** (Scheme 3.30) which contains a butadiyne linker and two Hamilton receptors as end caps. Hamilton receptor derivative **122** could also be coupled under SONOGASHIRA conditions either with 1,4-diiodobenzene **125** or with 1,4-diiodo-2,5-bis(octyloxy)benzene **127** (to improve the solubility) to give the corresponding rod-like derivatives **126** and **128** with OPE character and two Hamilton receptor end caps (Scheme 3.31).

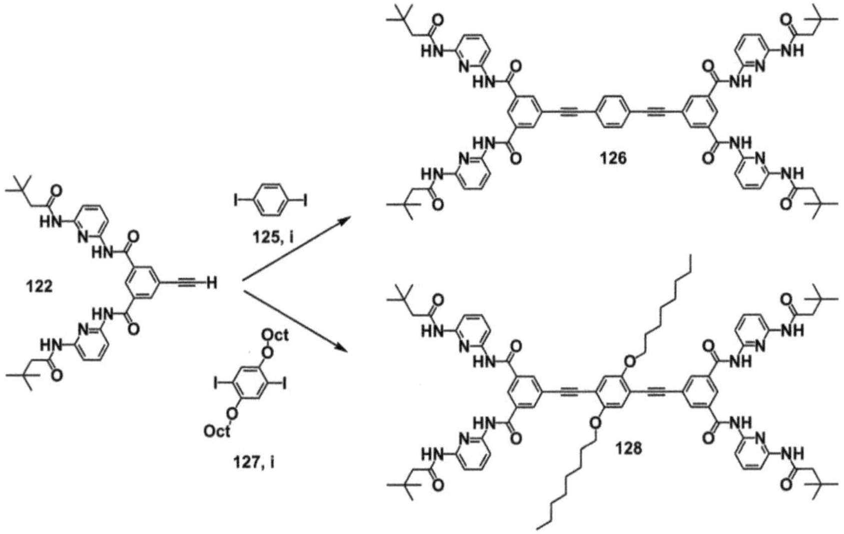

Scheme 3.31: Syntheses of target *p*-OPEs **126** and **128**. i) Pd(PPh$_3$)$_2$Cl$_2$, PPh$_3$, CuI, THF, HN*i*Pr$_2$, rt; ii) Pd(PPh$_3$)$_2$Cl$_2$, CuI, THF, NEt$_3$, 70 °C → rt.

3.4.2 Determination of Association Constants and Cooperativity of Binding

The ^{1}H NMR spectra of the Hamilton receptor bearing *p*-OPEs **124**, **126** and **128** in CDCl$_3$ show rather broad and poorly resolved signals. This is due to the presence of the large number of hydrogen-bond donors and acceptors causing the formation of intra- and intermolecular bonding networks.[96,103,121] Pronounced aggregation of **124**, **126** and **128** due to intermolecular hydrogen bonds is also the reason for the relatively low solubility of all described *p*-OPEs in chloroform. Using THF or DMSO as hydrogen bond breaking solvents leads to well resolved NMR spectra of **124, 126** and **128**. In this case, intermolecular associations are not favoured. Remarkable sharpening of the signals in CDCl$_3$ occurs also upon successive addition of the porphyrin cyanuric acid derivative **7**.

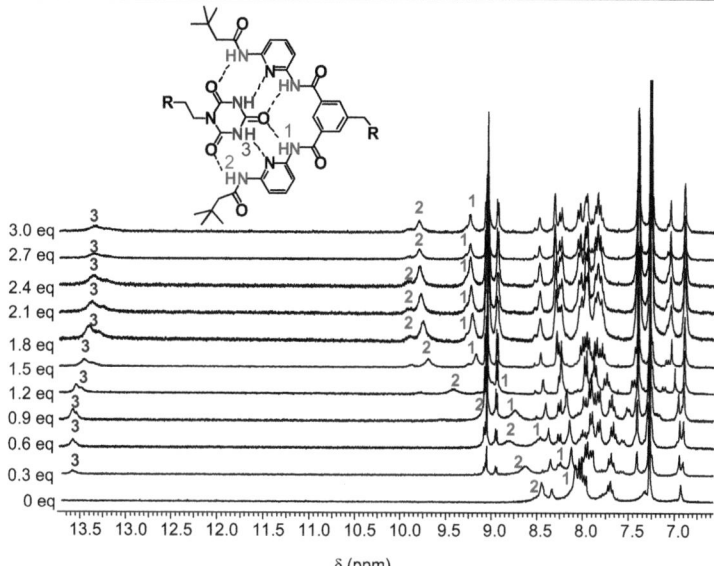

Figure 3.36: Binding motif between 124, 126 or 128 and 7 with indication of the NH protons NH^1, NH^2 and NH^3 (top) and 300 MHz ^{1}H NMR spectra of 7 at a concentration of 3.3 mM in CDCl$_3$ in the presence of various equivalents of 128 (bottom).

The sharpening of the signals is accompanied by a characteristic downfield shift of the amide proton NH^1 and NH^2 of the bis-Hamilton receptor derivatives and indicates the formation of distinct H-bonding complexes (Figure 3.36).[75,91,96,103,114,120,121,247]

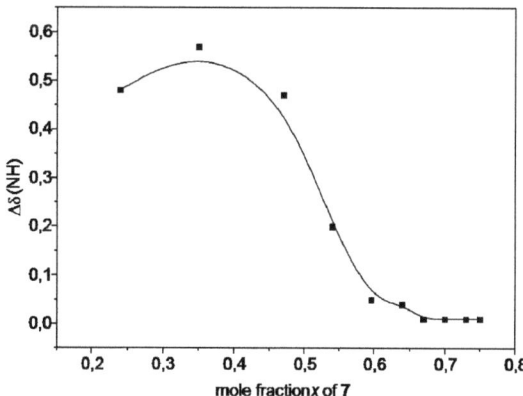

Figure 3.37: Job's plot analysis for the determination of 1:2 complex between *p*-OPE 128 and porphyrin 7.

Results and Discussion

The 1:2 binding stoichiometry of the complexes is confirmed by applying Job's method of continuous variation to the NMR experiments.[91,96,103,120] The chemical shift variation of NH^1 ($\Delta(\delta)$) as a function of the mole fraction of porphyrin **7** X(**7**) was monitored and plotted (Figure 3.37).

A series of ^{1}H NMR titration experiments in CDCl$_3$ is performed in order to determine the association constants and to analyse cooperativity phenomena. Herein, the downfield shifts of the NH^1-protons and NH^2-protons of the *p*-OPEs are determined as a function of the porphyrin concentration. In a typical titration experiment 0.5 mL of a 3.3 mM solution of a *p*-OPE is titrated with 50 µL of a 10 mM solution of porphyrin **7** for the determination of K$_n$.

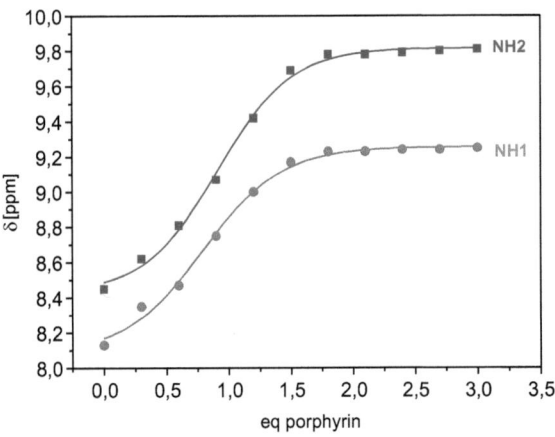

Figure 3.38: ^{1}H NMR titration plots of the chemical shifts of the NH1 and NH2 protons as a function of the amount of added porphyrin 7.

It is important to emphasize that the establishment of stable equilibria requires some time due to the intermolecular interactions between the free *p*-OPE ligands which have to be overcome before the binding of the cyanuric acid bearing porphyrin can occur. Consequently, the ^{1}H NMR spectra are taken 30 minutes after mixing of the components. Notably, a sigmoidal titration curve can be obtained by plotting the chemical shifts δ of the NH protons as function of the added porphyrin equivalents

Results and Discussion

(Figure 3.38). Such a sigmoidal behaviour is a characteristic feature for positive cooperativity associated with the subsequent binding of guest molecules.[291,292]
It is then possible to calculate the association constants for the binding of the cyanuric acid bearing porphyrin **7** by using the program Chem-Equili.[228,229] The association constants are summarized in Table 3. The calculations are based on the assumption of two equilibria (Equations (7-8)) where OPE could be **124, 126** or **128** and P is cyanuric acid bearing porphyrin **7**.

$$K_1 = \text{OPE} + P \rightleftharpoons \text{OPE:P} \quad (7)$$

$$K_2 = P + \text{OPE:P} \rightleftharpoons \text{OPE:P}_2 \quad (8)$$

For the case of statistical binding, equation (9) must be fulfilled,[291] for which t is the total number of binding sites (in this case $t=2$).

$$\frac{K_{n+1}}{K_n} = \frac{n(t-n)}{(n+1)(t-n+1)} \quad (9)$$

However, the experimental values for K_{n+1}/K_n are much higher than those obtained from equation (9), which clearly demonstrates the presence of pronounced positive cooperativity.
In other words the free binding energy of the porphyrin increases by going from the monocomplexes to the bis-complexes. This is also reflected by the sigmoidal shape of the corresponding titration curves (Figure 3.38).
The determination of the occupancy r is certainly one of the most straightforward tools for underlining cooperativity phenomena. The occupancy is the average number of ligands bound to the receptor.[96]

$$r = \frac{[1 \cdot P] + 2[1 \cdot P_2]}{[1] + [1 \cdot P] + [1 \cdot P_2]} \quad (10)$$

For statistical binding, r can be described by the SCATCHARD equation (11),[96] in which Q is the site binding constant and x the concentration of the added ligand.

$$r = \frac{t \cdot Q \cdot x}{1 + Q \cdot x} \qquad (11)$$

In the case of statistical binding, the SCATCHARD plot r/x as a function of r is a straight line. Positive cooperativity exists when the plot is no longer a straight line but a concave curve.

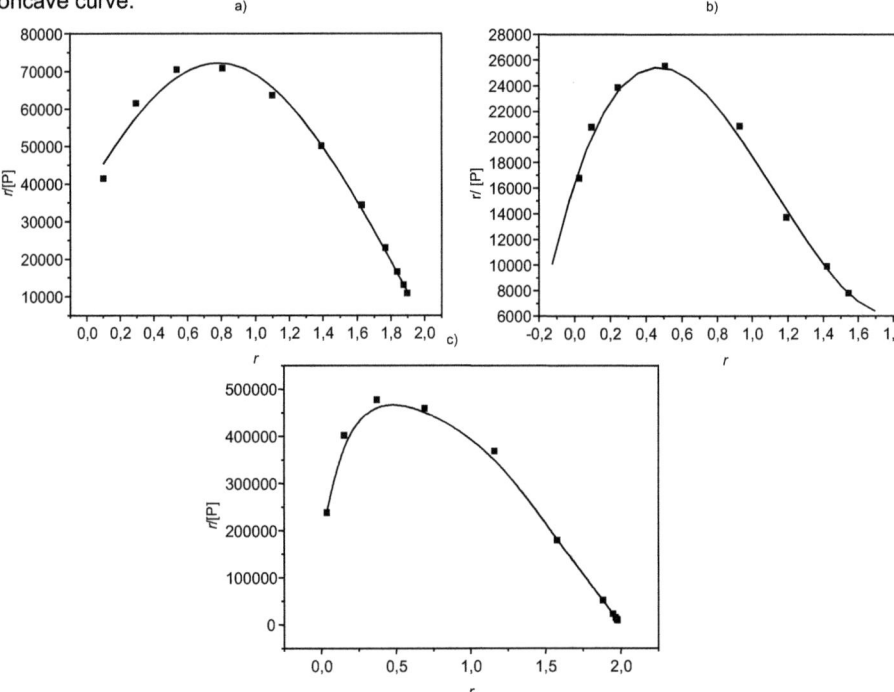

Figure 3.39: SCATCHARD plots for the *p*-OPE systems and porphyrin 7. a) 124:7, b) 126:7, c) 128:7.

All SCATCHARD plots of the titrated *p*-OPE systems display very pronounced concave behavior (Figure 3.39). The amount of cooperativity of a supramolecular system is usually quantified by the HILL coefficient n_H, which can be obtained from the maximum of the Scatchard plot according to equation 6.

$$n_H = \frac{r_{max}}{t - r_{max}} \qquad (12)$$

Results and Discussion

The higher the value of η_H, the higher is the degree of cooperativity. For infinitely high cooperativity, η_H becomes equal to t (here t = 2). However, this has never been observed in real systems. All HILL coefficients are summarized in Table 3.

	log K_1 [mol⁻¹dm³]	log K_2 [mol⁻¹dm³]	r_{max}	η_H
124:7	5.00	10.05	0.80	0.67
126:7	5.79	10.03	0.51	0.34
128:7	6.40	12.05	0.48	0.32

Table 3: Association constants, r_{max}, and Hill coefficients for the systems 124:7, 126:7 and 128:7.

The association constants K_n for the first complexations are in the expected range of 10^5 to 10^6 Lmol⁻¹. This is in good agreement with the literature-known Hamilton receptor cyanuric acid complexes.[91,96,103,121,247] Interestingly the association constants for the second complexation are extremely high (10^{10}-10^{12} Lmol⁻¹) compared to much more flexible systems that we already reported.[91,103,121,122,227] Typical association constants for further complexation of dendritic cyanurates are in the range of 10^3 to 10^6 Lmol⁻¹.[91,103,121,122,247]

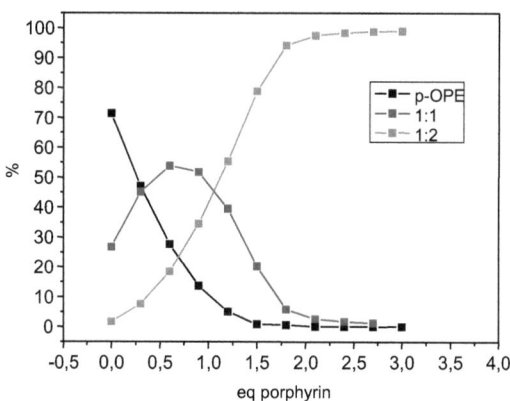

Figure 3.40: Distribution in percent of 128 as free core and within the complexes 128:7n (n = 1–2) as a function of the amount of added porphyrin obtained from analysis of the titration plots using the computer program Chem-Equili.[228,229] The total concentration of 128 within the mixtures is 3.3 mM.

Probably the stiffness of the *p*-OPEs in comparison to the more flexible alkyl chains of the literature known systems is responsible for this trend. Additionally, the significant lower solubility of the free species and the complexes compared to dendritic systems might play another important role.

The decrease of free OPE wire and the formation of the 1:1 and 1:2 complexes during the titration experiments are shown in Figure 3.40.

3.4.3 UV/Vis- and Fluorescence Titration Experiments

Electronic communication between the electroactive zinc porphyrin end-caps and the inactive bridging *p*-OPEs should be rather unlikely. To prove the absence of electronic interactions between the rod-shaped *p*-OPEs and the cyanuric acid bearing porphyrin **7** a series of UV/Vis- and steady-state fluorescence titration experiments is carried out. All steady-state fluorescence and absorption measurements are performed in dichloromethane at room temperature. Importantly, a well-defined concentration of the porphyrin derivatives is needed.

Therefore, 250 µL of a porphyrin stock solution is diluted to a total volume of 1.4 mL as a porphyrin reference. For the following titration steps 50 µL of the *p*-OPE derivative were added to 250 µl of the porphyrin stock solution and filled up to a total volume of 1.4 mL. After each addition step and waiting for 15 minutes absorption and fluorescence spectra are recorded. As expected, neither a shift of the SORET band of porphyrin **7** nor a shift of the absorption band of the *p*-OPEs could be detected (Figure 3.41).

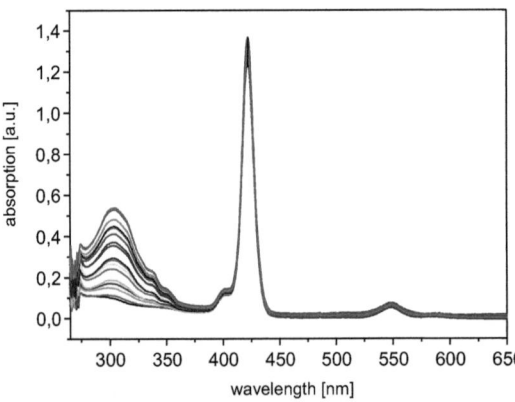

Figure 3.41: UV/Vis titration of 7 (c = 4.5 x 10^{-6} mol/L) with variable concentrations of *p*-OPE 126 starting from c_1 = 4.15 x 10^{-7} mol/L to c_{15} = 1.35 x 10^{-5} mol/L at room temperature in CH_2Cl_2.

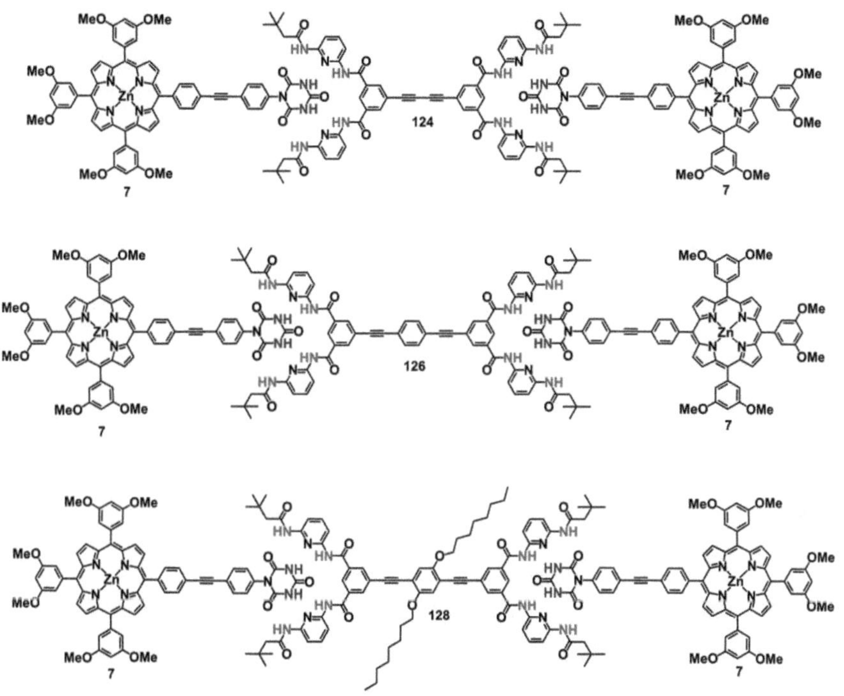

Figure 3.42: Supramolecular aggregates 124:7₂, 126:7₂ and 128:7₂.

The steady state fluorescence spectra are measured shortly after recording the absorption spectra. As excitation wavelength the SORET-absorption band (422 nm) of the porphyrin **7** is chosen. As is well known from several examples porphyrins might either activate an energy or electron transfer, if they are excited at the SORET-band.[88-91,212,293] The emission spectra of zinc porphyrin **7** show two emission bands at 598 and 642 nm. A potential electron or energy transfer can be implied, if a decrease of the fluorescence emission intensity is observed. As expected, in all analyzed systems this is not the case (Figure 3.43). The experiments prove that the electronic structure of both subunits apparently does not permit photoinduced electron or energy transfer processes.

Taking the results of the UV/Vis and fluorescence titration experiments into account p-OPEs can play an interesting and important role as inactive bridges for the supramolecular connection between electron donor-acceptor arrays like porphyrins and fullerenes.

Results and Discussion

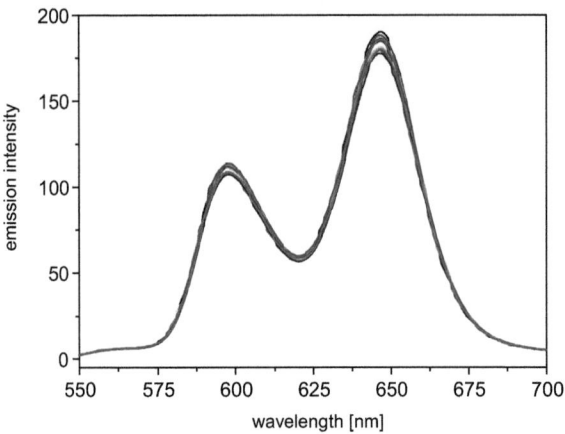

Figure 3.43: Fluorescence emission titration of porphyrin 7 with *p*-OPE 126 corresponding to the conditions of the absorption titration in Figure 3.41.

3.4.4 Conclusion

This is the first investigation of the self-assembly of supramolecular rigid *p*-oligo phenylene-ethynyl-porphyrin donor-wire nanohybrids connected *via* a Hamilton-receptor based hydrogen bonding motif. In this light three different *p*-OPEs carrying two Hamilton receptor termini and a cyanuric acid bearing zinc porphyrin derivative, are synthesized. To improve the relatively poor solubility of the *p*-OPEs in CHCl$_3$ and CH$_2$Cl$_2$ two octyloxy groups at the central benzene core in derivative **128** are introduced. The linear aggregation of these building blocks was analyzed by ^{1}H NMR titration experiments. As expected 1:2 complexes were formed which is proven by Job's plot analysis. The subsequent binding of the porphyrin to the *p*-OPE receptors shows an overall positive cooperativity for all cases. Compared to more flexible systems which contain alkyl chains and bulky groups[91,94,96,103,114,121,122,247] the association constants K$_n$ associated with twofold binding processes are rather high (log K$_1$ between 5.00 and 6.40 Lmol^{-1}, log K$_2$ between 10.03 and 12.05 Lmol^{-1}). Interestingly the most soluble *p*-OPE system **128** shows the strongest binding constants whereas the cooperativity of the less soluble *p*-OPEs **124** and **126**, expressed by the H$_{\text{ILL}}$ coefficient η_H, is higher.[294] As expected an electronic communication between both subunits can not be detected by UV/Vis and fluorescence experiments.[192,207,295,296]

Results and Discussion

3.4.5 Mass Spectrometry and Ion Beam Deposition of Hamilton Receptor bearing Rods for Surface Science in Ultrahigh Vacuum

The deposition of molecules on atomically clean surfaces in ultrahigh vacuum (UHV) is a key sample preparation technique in surface science.[297] In particular scanning tunneling microscopy (STM) allows to probe the electronic structure of a single molecule with submolecular resolution. Gas phase ions of large functional molecules, where the evaporation method often fails, can be created by electrospray ionization.[298] Herein the first results from soft landing ion beam deposition[299,300] of complex, functional molecules on surfaces in UHV analyzed with in-situ STM are presented. Homogeneously covered, otherwise clean surfaces are found after deposition, indicating the feasibility of electrospray ion beam deposition for surface science studies. Application opportunities in semiconducting devices[301] and molecular electronics[302] make the electrospray ion beam deposition to a highly interdisciplinary and interesting research topic. The STM studies give further indication on the fate of the charge carrier (H^+, Na^+, Cs^+) upon deposition.

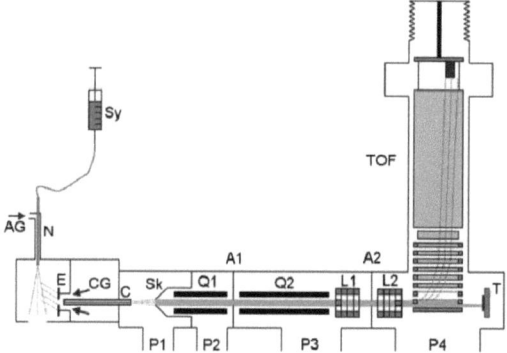

Figure 3.44: Scheme of the electrospray deposition source. The full length of the assembly is about 1 m. Sy: syringe with solution, N: spray needle, AG: assisting gas for the spray process, E: entrance plate counter electrode (-4 kV), CG: nitrogen counter gas, C: heated capillary, S: skimmer between first and second pumping stage, Q1: high-pressure quadrupole ion guide, A1: aperture between first and second ion guide, Q2: low-pressure quadrupole ion guide, L1,L2: lens system (einzel lens and steering plates), A2: aperture between third pumping stage and deposition chamber, TOF: movable orthogonal extraction time-of-flight mass spectrometer, T: deposition target. Pumping stages, P1: 1 mbar, P2: 0.1 mbar, P3: 10^{-4} mbar, P4: 10^{-6} mbar.

Results and Discussion

Method

The ion beam for deposition is created by electrospray ionization. Ion optics are used to guide the beam through the following pumping stages, where high- and low-pressure quadrupoles and electrostatic lenses are used. In the forth pumping stage a linear time-of flight mass spectrometer (TOF-MS) and a retarding grid energy detector are implemented (Figure 3.44). Before each deposition experiment, the ion beam is analyzed with respect to composition and energy by these instruments. *In-situ* deposition takes place in the 6th pumping stage at a pressure of 3×10^{-10} mbar. The incidence energy is adjusted by applying a voltage to the sample. Incidence energies as low as 2 eV per charge can be obtained. The ion beam current is monitored throughout the experiment.

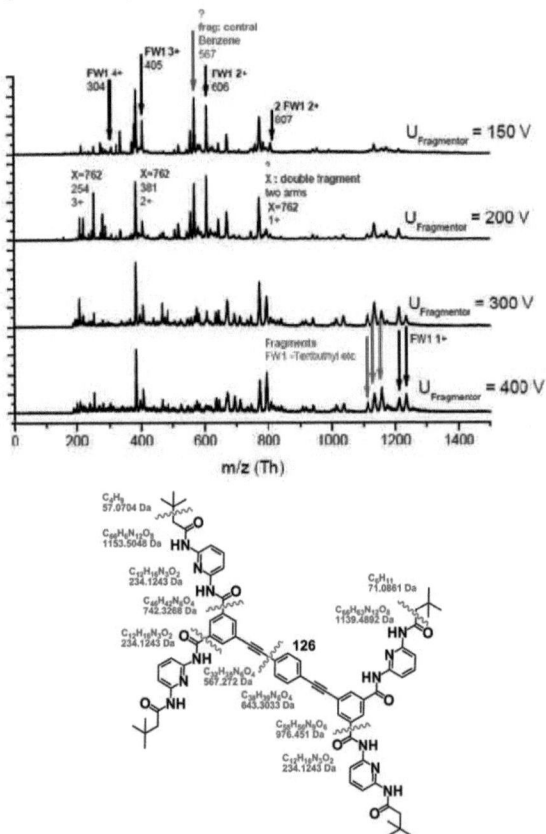

Figure 3.45: ES-TOF mass spectrograms of double Hamilton receptor 126 at different voltages and depiction of the evolving fragments.

Results and Discussion

Measurements and Data

In this particular work the double Hamilton receptor molecule **126** is used for ion beam deposition on copper surface (Cu(110)) and analyzed by STM techniques at 300 K. As a result of the ES-TOF-MS analysis (Figure 3.45), we know that the double Hamilton receptor **126** is multiply charged at four pyridin sites (m/z = 1211, 606, 405 and 304) by H^+ attachment.

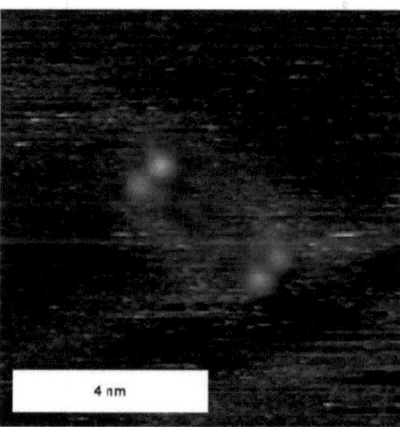

T = 300 K
U_{gap} = -0.5 V
I_{tunnel} = 50 pA
Loop Gain = 1.0 %

Figure 3.46: STM image of double Hamilton receptor 126 on Cu(110) at room temperature. The probe was annealed at 375 for 1 min.

The double Hamilton receptor (DHR) **126** has been designed and synthesized containing two distinctly separated functions for hydrogen bonding. The molecule is observed at the surface as four bright lobes, arranged in pairs, separated by 4 nm, the length of the molecule (Figure 3.46). The four bright lobes can be assigned to the four electron rich pyridine rings which generate the greatest contrast in the STM image. The central benzene ring of rod **126** and its neighboring ethynyl units can not be imaged due to their relatively low electron density. The image of double Hamilton receptor **126** demonstrates that complex molecules containing spatially separated functional groups can be landed softly on a clean metal surface in a way that the shape of the molecule persists.

3.5 Porpyhinato Phosphonium Salts as Precursor for Wittig-Horner Olefinations

OPV porphyrins are very interesting compounds for electron transfer investigations. Due to their rigidity and electron tunneling possibility OPV subunits are powerful building blocks for electron transfer model systems.[192,198,296]

In the first approach it is attempted to synthesize a tetraphenyl porphyrin core with three *tert*-butyl groups for increasing the solubility and one methylenebromo group for further derivatization. The idea is to create a mono-substituted porphyrin which can be easily transferred into further derivatives. 4-Methoxy-methyl-benzaldehyde **129**[303] is chosen as educt which is easily synthesizable in two steps and easy to handle. Another reason is that it provides certain rigidity in the desired final product. The synthesis is performed with a statistical batch under modified LINDSAY-conditions.[208-211] Four equivalents pyrrole **50**, one equivalent 4-methoxy-methylbenzaldeyhde **129** and three equivalents 4-*tert*-butylbenzaldehyde **48** are dissolved in dichloromethane under addition oft ethanol and tetraphenyl phosphonium chloride (Scheme 3.32).

Tetraphenyl phosphonium chloride is an additive which allows working in much more concentrated solutions. 4-*tert*-Butylbenzaldehyde iss been chosen to increase the solubility in the final product. Boron trifluoride etherate seems to be the more prosperous catalyst in comparison to trifluoro acetic acid to build up the porphyrin indicated by higher yields. The final oxidation is performed with DDQ. There is no difference observed between light exclusion or not. The purification on silica gel with dichloromethane as eluent allows the separation of the different products. The TPP eluated first and then the desired mono substituted porphyrin. Both products were collected separately and the higher substituted porphyrins are collected together. The yield of 13 % is acceptable.

Using HBr (33 % in AcOH) as bromide source bromo-methyl porphyrin **131** is easily achievable (Scheme 3.32). The bromo group is important for the transformation into phosphorous ylides in the later synthetic route. Metallation with zinc is quite easy in the case of tetraphenylporphyrins. Zinc porphyrins are well analyzed towards their photophysical properties. Therefore metalloporphyrin systems with zinc as metal are very attractive for electron transfer investigations. The metallation is carried out in refluxing THF using zinc(II)acetate as zinc(II) source (Scheme 3.32).

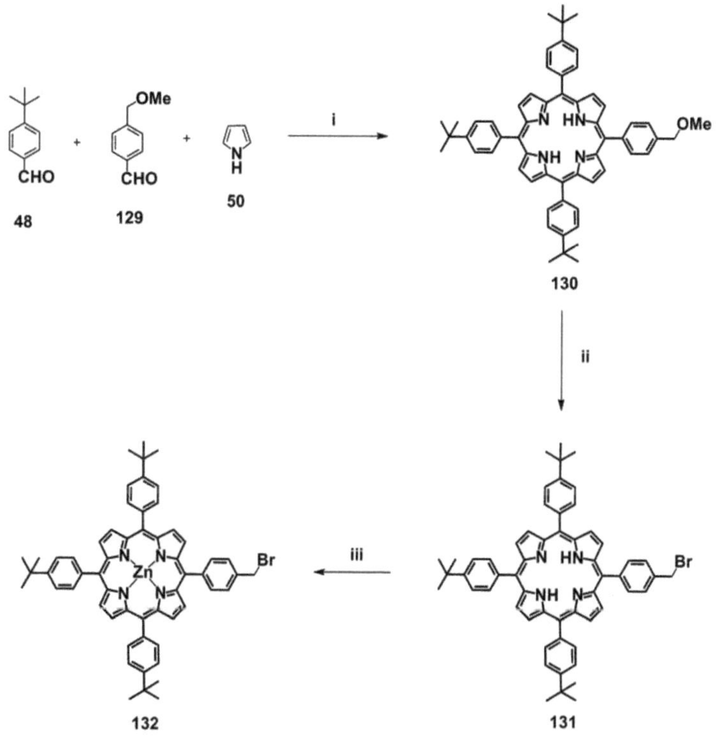

Scheme 3.32: Synthesis of porphyrin 132. i) 1. BF$_3$*OEt$_2$, PPh$_4$Cl, CH$_2$Cl$_2$, ethanol, rt; 2. DDQ, rt; ii) HBr (33 % in AcOH), CH$_2$Cl$_2$, rt; iii) Zn(OAc)$_2$, THF, reflux.

As precursors for the phosphor ylids in Wittig type reactions, phosphonates and phosphonium salts are used in a wide range of different substrates. In this particular case the first attempt is to choose the phosphonate strategy. Therefore bromo-methyl-porphyrin **131** is heated up to 160 °C in triethyl phosphonate for 1.5 h (Scheme 3.33). The reaction can be carried out nearly quantitatively. Porphyrin phosphonate **132** can be easily isolated by distillation of the excessive triethyl phosphonate. Unfortunately the phosphonate **133** is poorly soluble in common organic solvents. Due to this circumstance it is not possible to detect sharp peaks in the ^{1}H NMR spectrum. For a well resolved ^{13}C NMR spectrum the solubility of phosphonate **133** is too low. A tested Wittig-Horner olefination reaction with terephthalaldehyde monodiethylacetal, potassium *tert*-butoxide and 18-crown-6 in THF and CH_2Cl_2 failed which is expectable due to the poor solubility.

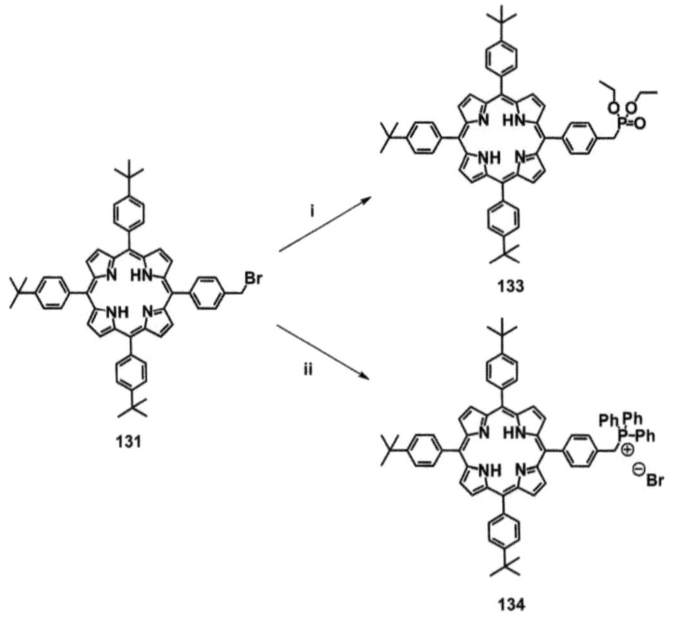

Scheme 3.33: Synthesis of porphyrinato phosphonate 133 and porphyrinato phosphonium salt 134. i) P(OEt)$_3$, reflux, ii) PPh$_3$, THF, reflux.

To solve the solubility problem the phosphonate is replaced by a phosphonium salt which is quite easy to synthesize starting from bromo-methyl porphyrin by nucleophilic substitution with triphenyl phosphine in refluxing THF (Scheme 3.33). A tested olefination reaction using phosphonium salt **134** as ylide precursor and 3,4-dimethoxy benzaldehyde as carbonyl compound leads to the desired OPV-porphyrin in acceptable yields.

Conclusion

The successful synthesis of the porphyrinato phosphonium salt **134** for Wittig Horner olefination reactions is reported herein for the first time. Starting from relatively cheap and well available precursor compounds target porphyrin **134** is achievable in five reaction steps. First olefination experiments with ylide precursor **134** show promising results.

3.6 Syntheses and Characterization of oligo-Phenylene-Vinylene Wires

Oligomers of unsaturated conjugated π-systems show interesting electron tunneling properties.[264,265,267-271] Most common in this particular light are oligo-phenylene and oligo-thiophene systems. Using vinylene or acetylene subunits between the phenylene rings is a promising strategy for the fine tuning of optical, photophysical and electronic properties of possible wirelike devices. First photochemical investigations have shown that oligo-phenylene-vinylene (OPV) systems are often more promising for the tunneling of electrons than oligo-phenylene-acetylene (OPA) systems.[192,198,296]

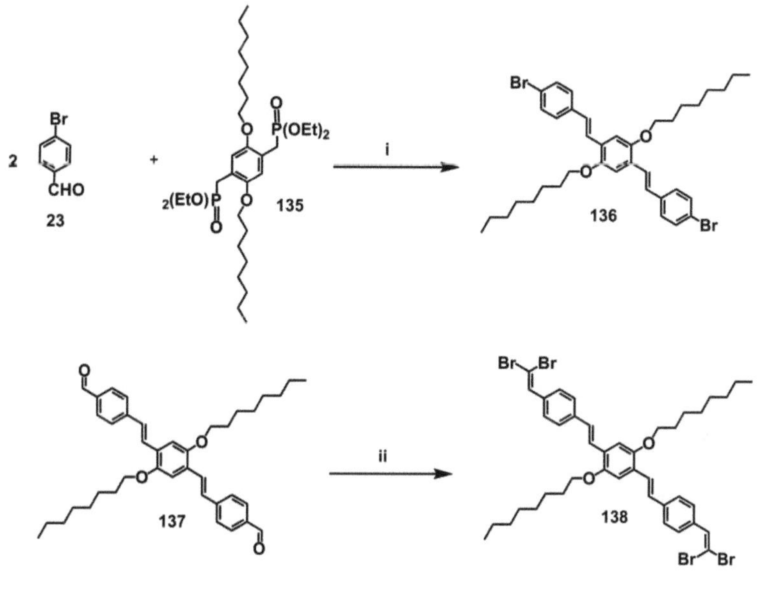

Scheme 3.34: Syntheses of the OPV wires **136** and **138**. i) KO^{t}Bu, 18-Crown-6, THF, rt; ii) PPh$_3$, CBr$_4$, CH$_2$Cl$_2$, 0 °C.

As reference components OPV wirelike systems with bromo endgroups for further modifications are designed. Key precursor for all synthesized *p*-oligo-phenylene vinylene wires is the bisphosphonate **135**[304,305] shown in Scheme 3.34. OPV wires are known for their rather low solubility in organic solvents. Therefore the alkyl chains are designed which provide a good solubility in most organic solvents (i.e. CHCl$_3$,

CH_2Cl_2, toluene, THF, etc.). WITTIG-HORNER olefination[306,307] with 4-bromo benzaldehyde **23** and potassium *tert*-butoxide as base leads to *p*-bis-bromo-OPV **136** in good yields. Olefination of bisformyl OPV wire **137**[308] applying the COREY-FUCHS protocol[309] with PPh_3 and CBr_4 accesses *p*-OPV **138** with two terminal bis-bromoolefines (Scheme 3.34).

Crucial for the applications of organic wirelike molecules especially when molecular electronics are concerned is their connection to metallic contacts. Thiols are proven to be excellent anchoring groups for contacting organic molecules to gold surfaces.[283,310,311] For this purpose p-OPV wire **141** is synthesized combining the contactation possibility of the thiol groups, the OPV character and the alkyl chains for better solubility. Application of a modified WITTIG-HORNER olefination[306,307] using the above mentioned bisphosphonate **135** and 4-thiomethyl benzaldehyde **139** gives bisthiomethyl wire **140** in a moderate yield. After basic deprotection with NaSMe and hydrolysis *p*-bisthiol OPV **141** can be obtained in good yields (Scheme 3.35).

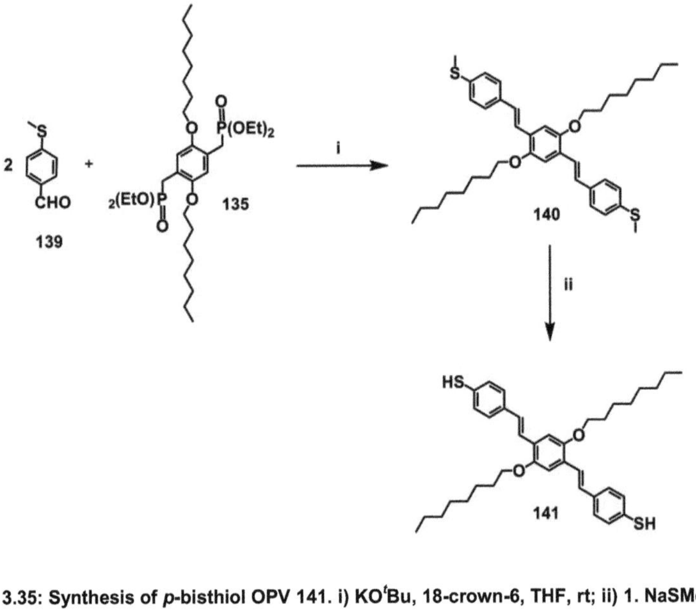

Scheme 3.35: Synthesis of *p*-bisthiol OPV **141**. i) KO^{t}Bu, 18-crown-6, THF, rt; ii) 1. NaSMe, DMA, 140 °C, 2. H_2O, 0 °C.

The characterization of compound **141** is achieved by NMR and IR spectroscopy, mass spectrometry and elemental analysis. The ^{1}H NMR spectrum of **141** in $CDCl_3$ at rt (Figure 3.47) shows several typical features. The triplet of the methyl group occurs at 1.08 ppm (^3J = 7.94 Hz, 6H), whereas the CH$_2$ chain protons split into three multiplets which have a chemical shift of 1.49, 1.54 and 2.04 ppm. The methylene-ether groups produce a triplet at 4.21 ppm (^3J = 6.47 Hz, 4H). Typically the singlet of the thiol protons arises at 3.66 ppm. Terminally the sp^2 protons can be assigned to the signals between 7 and 8 ppm.

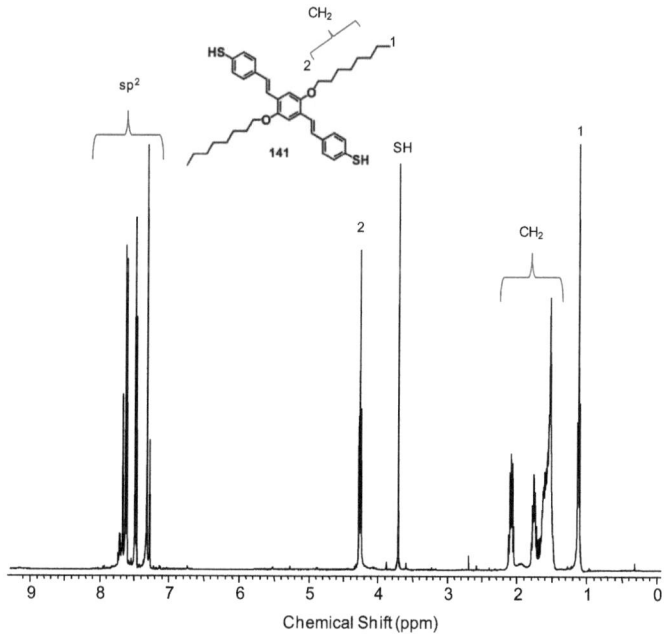

Figure 3.47: ^{1}H NMR spectrum (CDCl$_3$, 400 MHz, rt) of bis-thiol wire 141.

The ^{13}C NMR spectrum of bis-thiol wire **141** is depicted in Figure 3.48. The signals of the methyl group and the CH$_2$ chain are found in the region between 10 and 35 ppm. The methylene-ether groups cause a signal at 69.51 ppm. Characteristic for methylene-phenylene ether is the signal at 151.04 ppm which can be assigned to the sp^2 carbon of the phenylene-methylene group. The other sp^2-carbon atoms produce signals in the region of 110-140 ppm.

To elongate the sp^2 chain phosphonium salt **143** was synthesized for further WITTIG-HORNER olefinations. Therefore the benzylbromo compound **142** was converted by the addition of triphenylphosphine (Scheme 3.36) to the desired phosphorus ylide precursor **143**. Extended OPV-thiol **144** can then be synthesized condensating the corresponding bisformyl OPV rod **137** with 4-methylthiobenzyl-triphenylphosphonium bromide **143** under WITTIG-HORNER conditions[306,307] (Scheme 3.36).

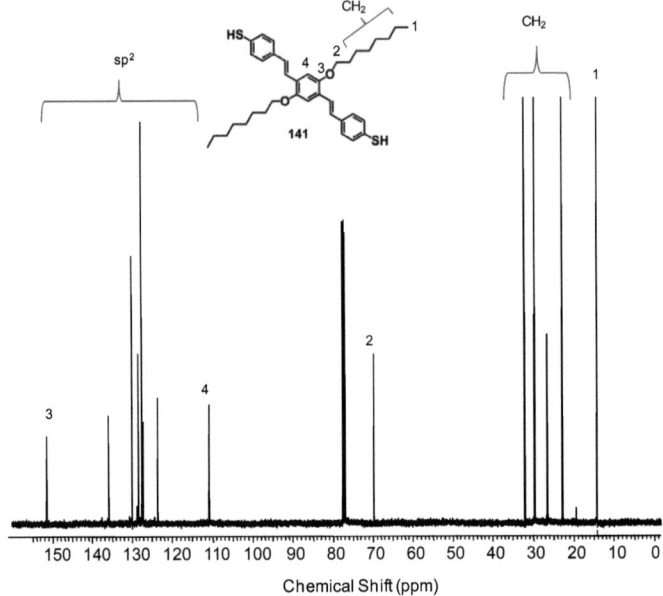

Figure 3.48: ^{13}C NMR spectrum (CDCl$_3$, 100.5 MHz, rt) of bis-thiol wire **141**.

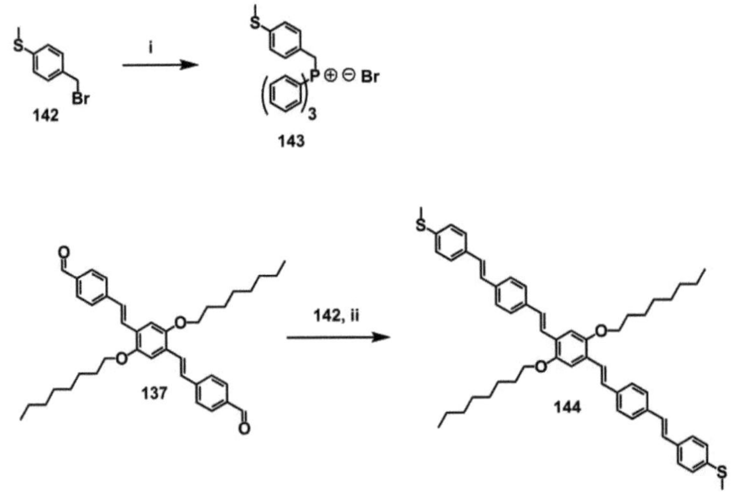

Scheme 3.36: Synthesis of OPV-thioether **144**. i) PPh$_3$, Et$_2$O/CH$_2$Cl$_2$, rt; ii) 4-methylthiobenzyl-triphenylphosphonium bromide, KO^{t}Bu, 18-Crown-6, THF/CH$_2$Cl$_2$, rt.

Results and Discussion

Conclusion

The syntheses of diverse oligo-phenylene-vinylene rods (**136**, **138**, **140**, **141** and **143**) are successful using WITTIG-HORNER olefination reactions under modified conditions. Additional the development of the novel phosphonium salt **143** is described which is a powerful precusor for the syntheses of extended OPV thiol rods. In terms of molecular electronics OPV bisthiol **141** is propbably the most interesting compound towards the contactation possibilty of thiol groups and its extraordinary good dissolution behavior in common organic solvents. Further functionalization options of OPV **136** and **140** demonstrate the formation chance of mixed oligo-phenylene (OPV)/oligo-phenylene-acetylene (OPA) wires.

3.7 Fullerene-Porphyrin Containing Triads for the Fine Tuning of Solar Energy Conversion

Combining the rapidly evolving fields of nanostructured materials and organic chemistry is an attractive strategy for constructing large and complex, yet highly ordered, molecular and supramolecular entities with defined functions. Especially in the field of electron transfer systems a number of interesting donor-acceptor systems have been established in the last years.[38-51] All of them include the assembly of an electron donor (e.g. porphyrins, ferrocenes, fluorenes, etc.) and an electron acceptor (e.g. [60]fullerene). Whereas porphyrins and fluorenes show electron donating behavior only after irradiation with light ferrocenes possess electron donating properties already in the ground state.[312-314] Most of the described solar energy converting arrays are arranged via covalent bridges, only a few are assembled via supramolecular approaches.[91-95]

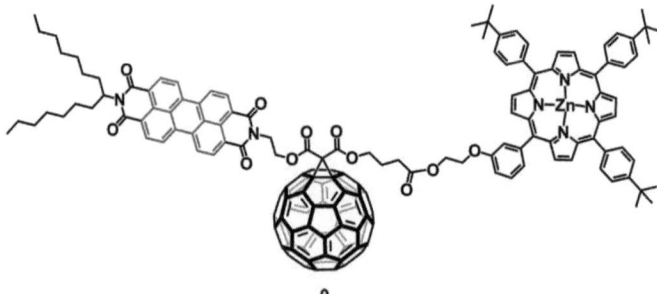

Figure 3.49: Example for a triad containing two light harvesting building blocks and C_{60} as electron acceptor.

In this context, porphyrins due to their biological relevancy[315] and remarkable photoelectronic properties[316] have stimulated a great interest in devising, designing, synthesizing and testing porphyrin containing electron donor acceptor architectures. The photophysical and (opto)electronic properties of metalloporphyrins[317,318] are truly unique, especially when they are combined with fullerenes. What renders fullerenes exceptional is their capability to accept reversibly up to six electrons,[319,320] while exhibiting a low reorganization energy that is associated with each of this electron transfer steps.[18,245,321] As a matter of fact, a considerable number of electron donor acceptor conjugates – based on the combination of porphyrins and fullerenes –

emerged in recent years that combine all the key functions of the photosynthetic reaction center, that is, light harvesting and charge transfer.[16,322-328]

To expand the relatively short light absorption area of porphyrins (the main absorption band is around 420 nm) additional chromophores/fluorophores (i.e. perylene, fluorene or pyrene) are introduced to develope triads consisting of two light harvesting subunits and C_{60} (Figure 3.49).

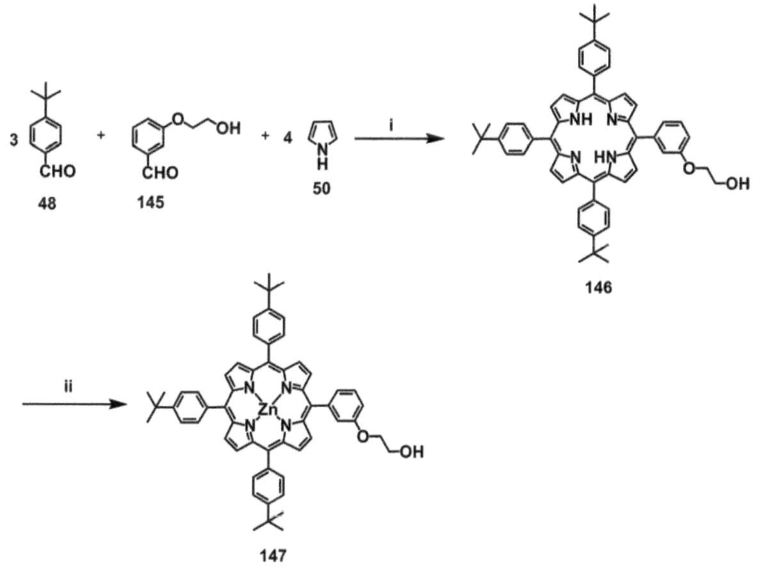

Scheme 3.37: Synthesis of porphyrin 147. i) 1. TFA, CH_2Cl_2, rt; 2. NEt_3, DDQ ii) $Zn(OAc)_2$, THF, reflux.

The synthesis of these complex structures starts with the creation of the basic tetraphenyl zinc porphyrin **147** which exhibits a hydroxyl endgroup at its periphery. The hydroxyl functionality is chosen due to the great accessibility of the corresponding aldehyde **145** and for further derivatization possibilities. To increase the solubility three steric hindering *tert*-butyl groups at the phenyl rings are introduced in this approach. The synthesis is performed with a statistical batch under modified LINDSAY-conditions[208-211] (Scheme 3.37). Four equivalents pyrrole **50**, one equivalent 3-ethoxy-hydroxy-benzaldeyhde **145**[329] and three equivalents 4-*tert*-butylbenzaldehyde **48** are dissolved in dichloromethane. Then is added to the rapidly stirring solution. After stirring for one hour triethylamine is added for neutralization.

Finally DDQ is spend to oxidize the upcoming porphyrinogene. Metallation of porphyrin **146** occurred by adding zinc(II)acetate and under reflux in THF (Scheme 3.37).

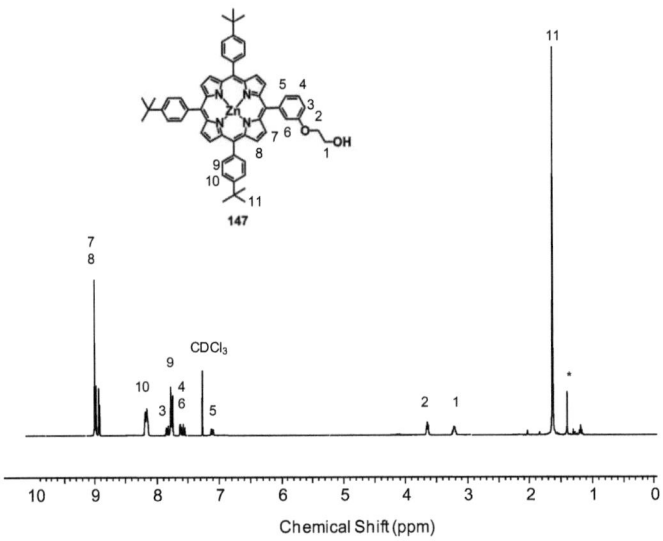

Figure 3.50: ¹H NMR spectrum (CDCl₃, 400 MHz, rt) of porphyrin **147**.

Characterization of porphyrin **147** is achieved by ¹H/¹³C NMR, IR and UV/Vis spectroscopy, mass spectrometry and elementary analysis. The ¹H NMR spectrum of porphyrin compound **147** (Figure 3.50) shows typical features associated with the substitution pattern of the phenyl rings. The singlet for the *tert*-butyl group arises at 1.63 ppm (s, 27H), whereas the aromatic protons of the 4-*tert*-butyl phenyl rings split into doublets and are detected at 8.16 ppm (d, ³J = 7.58 Hz, 6H) and 7.84 ppm (d, ³J = 7.54 Hz). The ethoxyhydroxy substituted phenyl ring causes a set of two doublets at 7.57 ppm (d, ³J = 7.06 Hz, 1H) and 7.13 ppm (d, ³J = 7.15 Hz , 1H), one triplet at 7.52 (t, ³J = 8.29 Hz , 1H) and one singlet at 7.62 ppm (s, 1H). The methylene groups of the C₂ spacer cause triplets that occur at 3.65 ppm (t, ³J = 7.20 Hz , 2H, PhOCH_2) and 3.22 ppm (m, 2H, HOCH_2).

Characteristic for the ¹³C NMR spectrum of porphyrin derivative **147** (Figure 3.51) are signals of the *tert*-butyl group at 33.92 ppm (3C, *C*(CH₃)₃) and 30.76 (9C, *C*H₃). Furthermore signals of the methylene groups at 67.50 ppm (1C, HO*C*H₂) and 59.78 (1C, PhO*C*H₂) can be assigned as typical for compound **147**. The signals of the aromatic carbon atoms occur as expected between 110 and 160 ppm. The signal of the phenylene ether carbon can be found at 155.53 ppm which is in good agreement with estimated values.

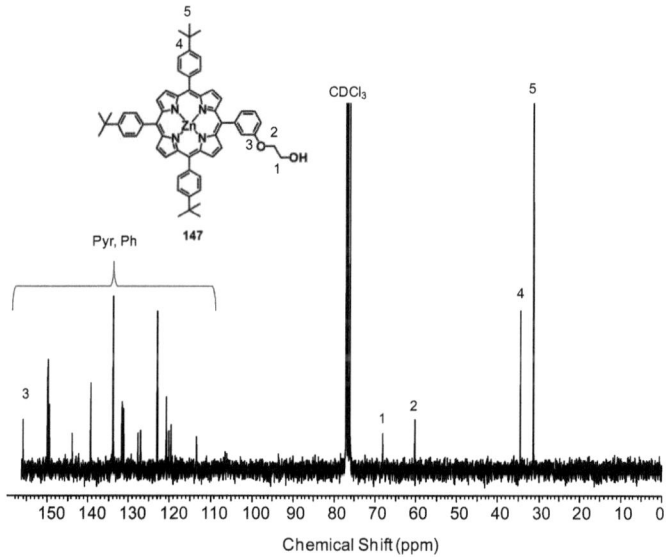

Figure 3.51: ^{13}C NMR spectrum (CDCl$_3$, 100.5 MHz, rt) of porphyrin 147

For connecting three different chromophores an unsymmetric building block with different derivatization possibilities is crucial for the further procedure. Unsymmetric malonate **148**[330] (Scheme 3.38) is the ideal choice for this purpose. The carboxylic group can be easily transferred into esters or amides, the *tert*-butyl ester can be cleaved under mild acidic conditions and gives therefore another chance to build up esters or amides and the acidic malonyl protons can be removed simply by adding non-nucleophilic bases for cyclopropanation reactions.

3.7.1 Synthesis and Characterization of a Perylene-Porphyrin-Fullerene Triad

Perylenes are bright colored due to their extended conjugated π-system. They absorb blue and green light and therefore appear red or purple.[331] Most eye-catching besides their color is their enormous fluorescence. Some compounds achieve quantum yields up to 99 %.[332] The electronic properties of perylenes have come to the centre of interest in various fields of science. Especially the attribute as "archetype molecular semiconductor"[333] makes perylenes interesting for electronic applications. Soluble perylene dyes are used for example in laser dyes.[334-336] Additional applications as organic electronic devices like organic field effect transistors are very auspicious.[337-341]

Perylene alcohol **149**[342] is a distinguished building block for the outlined triads. It combines the famous photophysical properties of perylene dyes with a high solubility in organic solvents. Whereas the C_{15} swallow tail at one imide position assures solubility the hydroxyl group at the other imide position affords further functionalization of perylene bisimide **149** (Scheme 3.38). Malonate **148** can then be modified using perylene alcohol **149**[342] under modified STEGLICH conditions[219-221] (Scheme 3.38).

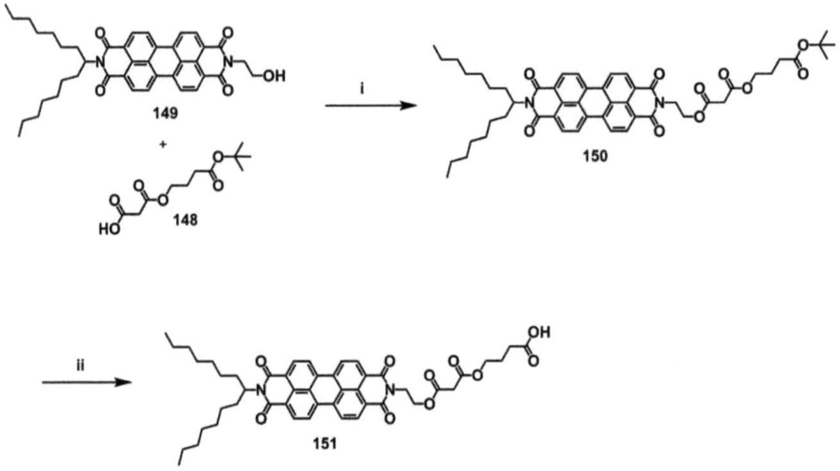

Scheme 3.38: Synthesis of perylene bisimide 151. i) DCC, DMAP, 1-HOBT, CH$_2$Cl$_2$, 0 °C→rt; ii) TFA, CH$_2$Cl$_2$, rt.

Acidic ester hydrolysis using TFA in dichloromethane is then applied to cleave the *tert*-butyl ester **150** and to obtain carboxylic acid derivative **151** (Scheme 3.38).

Results and Discussion

At this point of the synthetic route perylene dye **151** and zinc porphyrin **147** can be connected via an ester bridge using the well known STEGLICH reaction protocol[219-221] (Scheme 3.39). Dyad **152** can be isolated in a good yield (75 % related to used perylene derivative **151**) and shows interesting features towards its absorption behavior (Figure 3.52). Beside the most intensive absorption band of the zinc porpyhrin at 421.5 nm (log ε = 5.73) the typical absorption bands of the perylene dye are detected at 529.5 nm (log ε = 4.99) and 493.0 (log ε = 4.76). As estimated the extinction coefficients of the perylene bands are one order of magnitude smaller than the value of the SORET band.

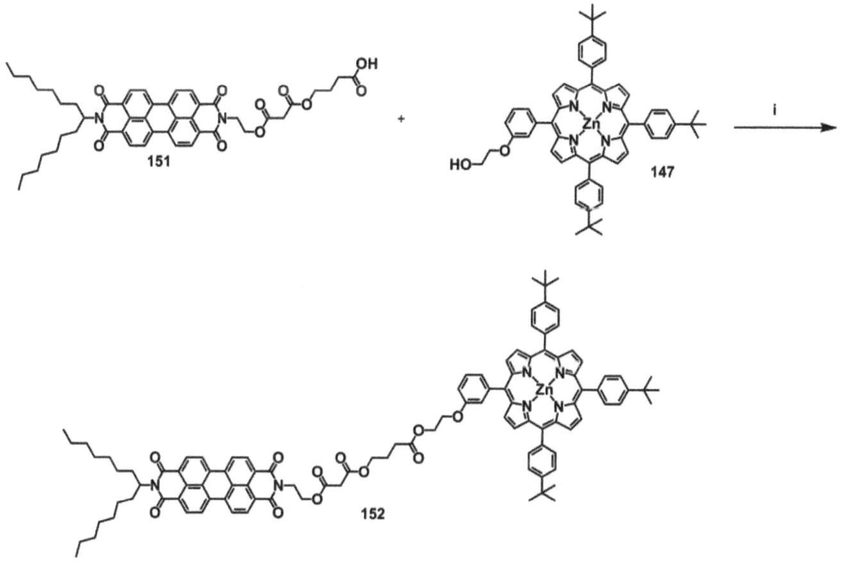

Scheme 3.39: Synthesis of perylene-porphyin dyad 152. i) DCC, DMAP, 1-HOBT, CH$_2$Cl$_2$, 0 °C→rt.

The ^{1}H NMR spectrum of dyad **152** in CDCl$_3$ at room temperature (Figure 3.53) shows features which can be assigned to either the perylene building block or the porphyrin unit. Typical for the swallow tail of the perylene bisimide is the triplet of the methyl groups which occurs at 0.83 ppm (t, ^3J = 8.36 Hz, 6H) and the multiplet of the branching group (C*H*) which arises at 5.12 ppm. The methylene groups between perylene bisimide and malonate split into multiplets and are detected at 4.58 (m, 2H, NCH₂) and 4.50 (m, 2H, OCH₂).

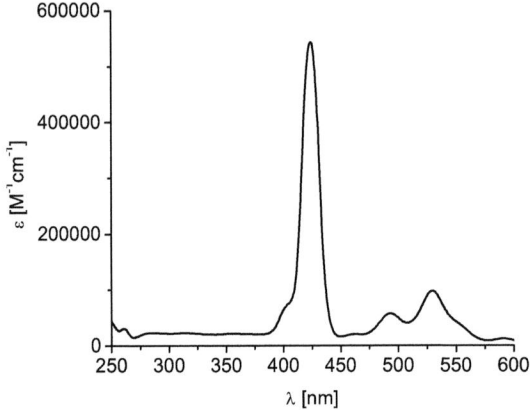

Figure 3.52: UV/Vis absorption spectrum of dyad 152 in CH$_2$Cl$_2$ at rt.

Typical for malonates is the singlet at 3.02 ppm. The other clearly assignable signals of the methylene groups occur at 4.09 ppm (t, ^3J = 7.20 Hz, 2H, PhOCH$_2$), 3.65 ppm (t, ^3J = 7.20 Hz, 2H, PhOCH$_2$) and 3.43 ppm (m, 2H, OCH$_2$).

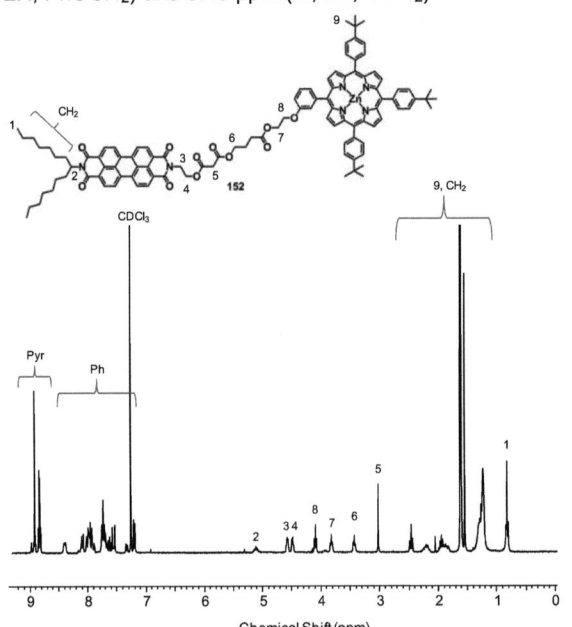

Figure 3.53: ^{1}H NMR spectrum (CDCl$_3$, 400 MHz, rt) of perylene-porphyrin dyad 152.

Results and Discussion

Perylene-porphyrin dyad **152** can then be further functionalized by applying modified BINGEL conditions.[138-140,343] Therefore a slight excess of [60]fullerene is dissolved in toluene before the malonate and CBr$_4$ as halogenation reagent is added. DBU dissolved in toluene is then dropped to the rapid stirring solution during at least 1 h. The slow adding of the base and the right dilution (1 mg C$_{60}$ per 1 mL solvent) are crucial to enhance the chance to get a high yield of the wanted monoadduct **9** (Scheme 3.40). The isolation of the desired monoadduct **9** can be achieved relatively easy using column chromatography on silica gel. After eluation of excessive C$_{60}$ with toluene a gradient of ethyl acetate in toluene is applied to separate the monoadduct from bisadducts and unreacted malonate.

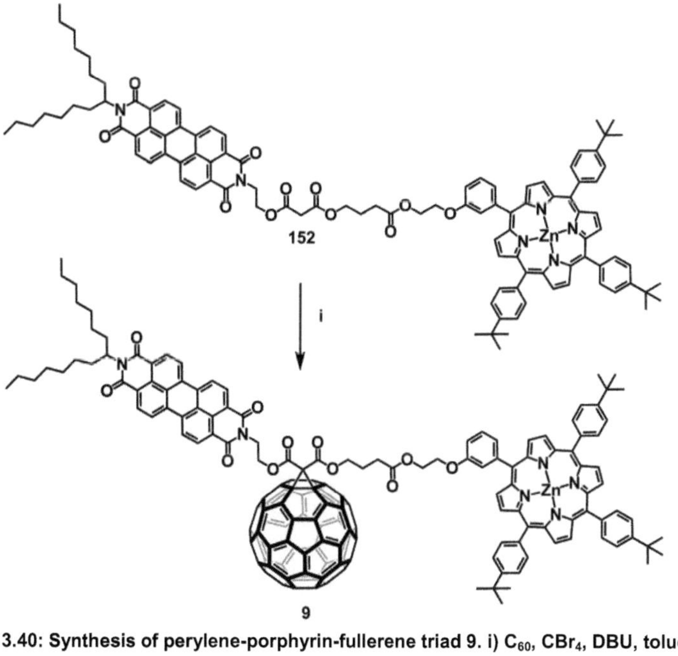

Scheme 3.40: Synthesis of perylene-porphyrin-fullerene triad 9. i) C$_{60}$, CBr$_4$, DBU, toluene, rt.

Characterization of triad **9** is achieved by ^{1}H/^{13}C NMR, UV/Vis and IR spectroscopy as well as mass spectrometry. Probably the most interesting properties of this triad are found in the UV/Vis absorption spectrum in CH$_2$Cl$_2$ (Figure 3.54). Additionally to the main absorption band of the porphyrin (SORET band) at 425.5 nm (log ε = 4.90) two bands assigned to the perylene dye at 530.0 nm (log ε = 4.21), 491.0 nm (log ε = 4.10)

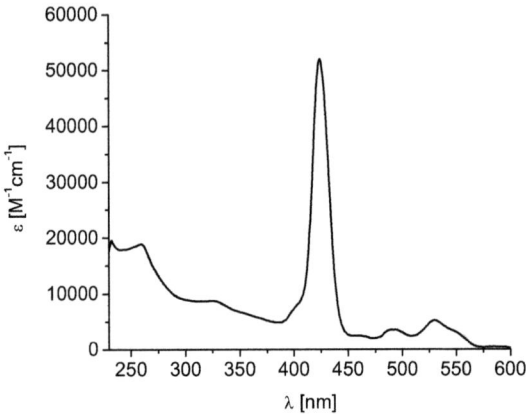

Figure 3.54: UV/Vis absorption spectrum of triad 9 in CH$_2$Cl$_2$ at rt.

and two bands allocated to the [60]fullerene at 328.0 (log ε = 3.94) and 260.0 nm (log ε = 4.47) are detected. As expected the extinction coefficients of the perylene and fullerene bands are one order of magnitude lower than the SORET band of the porphyrin.

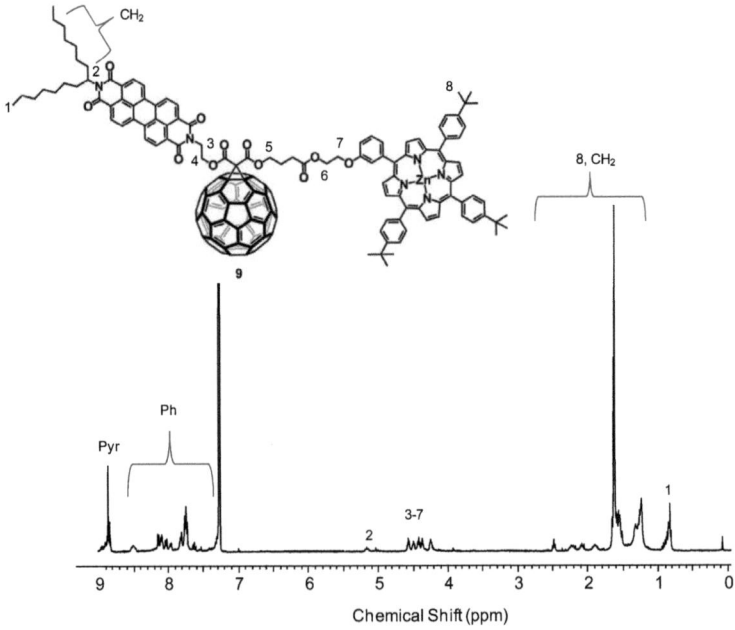

Figure 3.55: ^{1}H NMR spectrum (CDCl$_3$, 400 MHz, rt) of perylene-porphyrin-fullerene triad 9.

Taking the absorption spectrum of zinc(II)porphyrin-fullerene entities[38,45,48,63,83,91,344] as reference a clear broadening of the absorption spectrum in the visible region of the triad **9** can be postulated. Combining the features of the perylene dye with those of the zinc porphyrin makes triad **9** particular to an interesting alternative for the conversion of solar light.

In comparison to the ^{1}H NMR spectrum of dyad **152** only minor changes can be observed in the ^{1}H NMR of triad **9** (Figure 3.55). Besides the loss of singlet of the malonyl methylene a slightly shift of the methylene groups' signals can be monitored.

3.7.2 Synthesis and Characterization of a Fluorene-Porphyrin Dyad

Fluorenes have gained intense attention concerning their excellent electronic and photophysical properties. Especially in the focus of developing organic light emitting diods (OLEDs) oligo- and polyfluorenes play an important role.[345-347] Low gap-band copolymers consisting of fluorenes are even applicated in plastic solar cells with efficiencies up to 2.2 %.[348-351] For these reasons triads containing fluorenes should show interesting features especially concerning their photophysical properties.

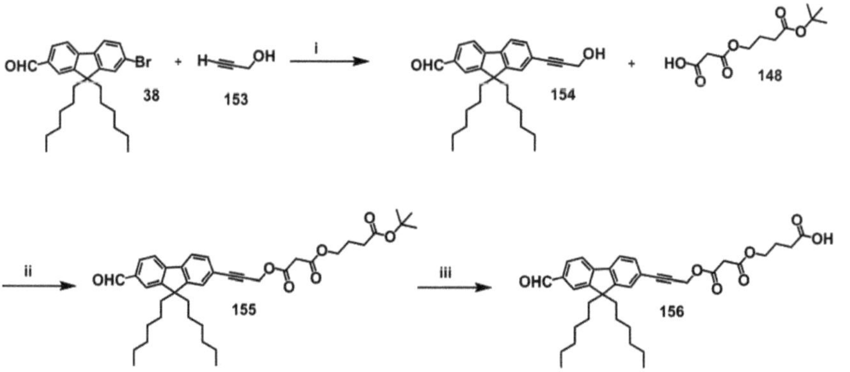

Scheme 3.41: Synthesis of fluorene compound 156. i) Pd(PPh$_3$)$_2$Cl$_2$, CuI, THF/NEt$_3$, reflux; ii) DCC, DMAP, 1-HOBT, CH$_2$Cl$_2$, 0 °C→rt; iii) TFA, CH$_2$Cl$_2$, rt.

The synthesis of the fluorene containing triad starts from fluorene compound **38** which can be obtained in good yields under conditions described in the literature.[207] Additionally to the bromo and the formyl group of fluorene fluorophore **38** which are important for chemical functionalization possibilities the two hexyl chains are crucial to advance the solubility in common organic solvents. Applying a modified SONOGASHIRA coupling protocol[200,201] fluorene compound **38** is then reacted with propargyl alcohol **153** in refluxing THF/NEt$_3$ and Pd(PPh$_3$)$_2$Cl$_2$/CuI as catalytic system (Scheme 3.41). Resulting fluorene derivative **154** can then be coupled with unsymmetric malonate **148** under STEGLICH conditions[219,352,353] (Scheme 3.41). Acidic cleavage of *tert*-butyl ester **155** with TFA leads to carboxylic acid derivative **156** in quantitative yield (Scheme 3.41).

Fluorene fluorophore **156** is connected with zinc porphyrin **147** in the next step of the synthetic plan *via* esterification using the well known STEGLICH reaction protocol[219,352,353] (Scheme 3.42). Dyad **157** can be isolated in an exceptable yield (55 % related to used fluorene derivative **156**) and shows interresting features towards its absorption behavior (Figure 3.56).

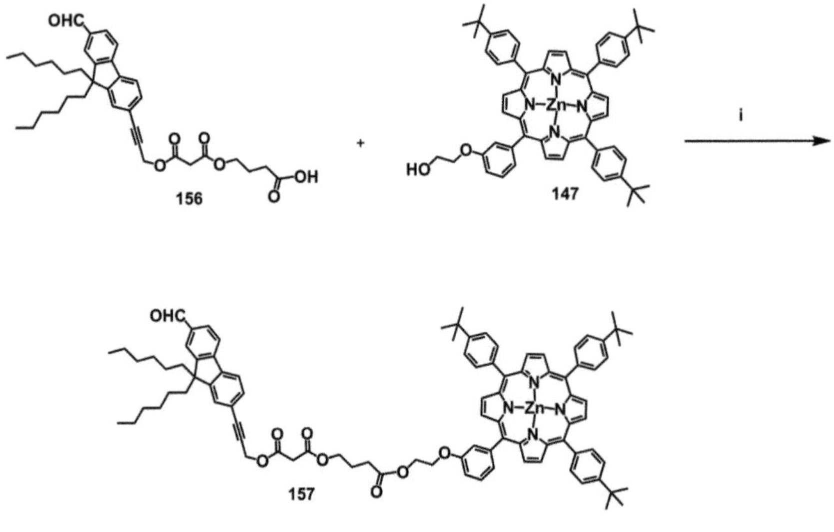

Scheme 3.42: Synthesis of diad 157. i) DCC, DMAP, 1-HOBT, CH$_2$Cl$_2$, 0°C→rt.

Characterization of dyad **157** occurred by ^{1}H/^{13}C NMR, UV/Vis and IR spectroscopy as well as mass spectrometry. Certainly of most interest are the absorption probabilities of fluorene-porphyrin compound **157**. The UV/Vis absorption spectrum of **157** in CH$_2$Cl$_2$ at room temperature is depicted in figure 3.56. The most intense absorption band (SORET-band) is detected at 421.0 nm (log ε = 5.50), the Q-band of the porphyrin system is found at 548.5 nm (log ε = 4.23) in the visible region.

Results and Discussion

Typically for fluorene systems are the three absorption bands at 341.5 nm (log ε = 4.61), 260 nm (log ε = 4.58) and 216 nm (log ε = 4.50) in the UV region. Remarkable in this context is that the extinction coefficient of the fluorene bands are at least one order of magnitude less intense compared to the SORET band, but slightly higher than the Q-band. The loss of one Q-band in the absorption spectrum of this diad is mentionable as well.

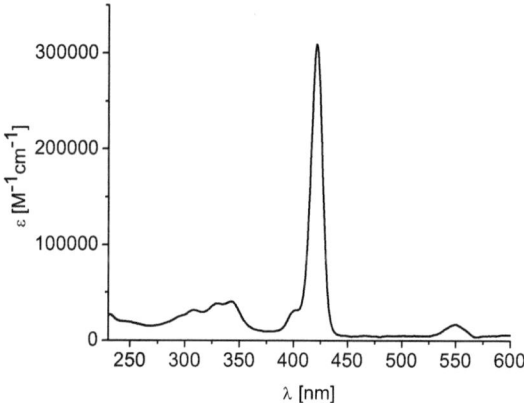

Figure 3.56: UV/Vis absorption spectrum of dyad **157** in CH_2Cl_2 at rt.

The analysis of the ^{1}H NMR spectrum of compound **157** shows characteristic signals that can be associated either with the porphyrin core or with the fluorene entity (Figure 3.57). Typically the singlet of the formyl group occurred at 9.96 ppm, whereas the singlet of the methylene group adjacent to the alkyne group arises at 4.91 ppm. The triplet of the side chain methyl groups can bet detected at 0.75 ppm (t, ^3J = 6.94 Hz, 6H). Characteristic for the malonate is the singlet at 3.32 ppm. The signals effected by the ester or ether methylene groups can be found at 4.52 ppm (t, ^3J = 6.42 Hz, 2H), 4.16 (t, ^3J = 7.25 Hz, 2H.) and 3.37 ppm (t, ^3J = 7.22 Hz, 2H). Characteristic for the porphyrin *tert*-butyl phenyl rings are the two doublets in the aromatic region at 8.15 ppm (d, ^3J = 8.30 Hz, 6H) and 7.76 ppm (d, ^3J = 8.30 Hz, 6H) and the singlet of the *tert*-butyl group at 1.63 ppm.

The ^{13}C NMR spectrum of dyad **157** in $CDCl_3$ at room temperature is shown in Figure 3.58. The formyl carbon is detected at 192.24 ppm, the ester carbonyls at 172.63, 165.94 and 165.74 ppm. The signals of the pyrrolic and aromatic carbons occur as expected in the region of 115-160 ppm. Typically of alkynyl compounds the signals for the sp-carbons are found at 87.48 and 82.95 ppm.

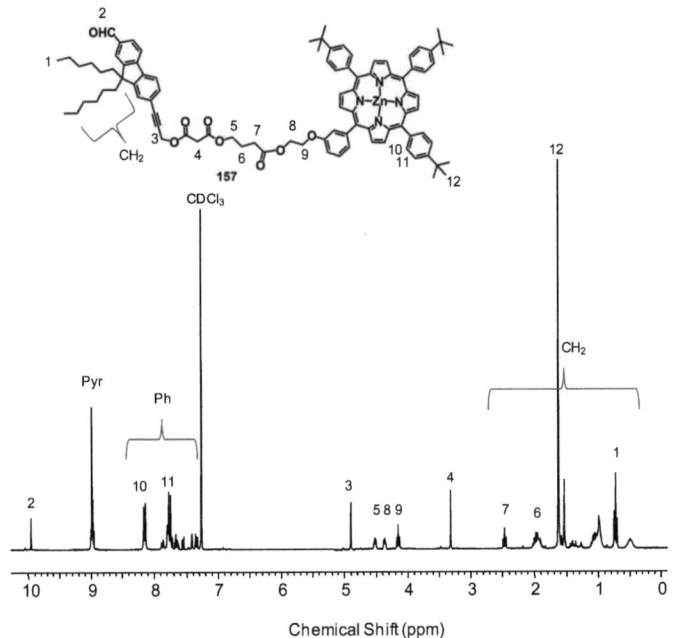

Figure 3.57: ^{1}H NMR spectrum (400 MHz) of dyad 157 in CDCl$_3$ at rt.

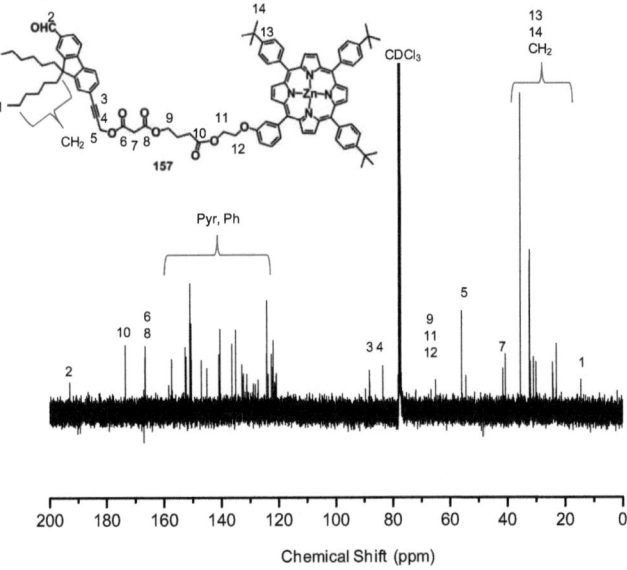

Figure 3.58: ^{13}C NMR spectrum (100.5 MHz) of dyad 157 in CDCl$_3$ at rt.

The ester and ether methylene carbons cause signals at 66.10, 64.49, 63.06 and 53.77 ppm. Whereas the signal at 41.03 ppm can be assigned to the malonyl methylene group. The *tert*-butyl groups and the alkyl chain methylene groups provoke signals in the region between 20 and 40 ppm. The signal of the side chain methylene groups can be found at 13.92 ppm.

3.7.3 Synthesis and Characterization of a Pyrene-Porphyrin-Fullerene Triad

Pyrenes are excellent fluorophores that show extremely long fluorescence lifetime (τ is up to 400 ns in deaerated solution, compared to $\tau > 10$ ns for most organic fluorophores).[230,354-361] Due to their powerful photophysical properties pyrene systems are often used in intermolecular charge transfer processes.[362]

Starting from commercial available pyrene butanol **158** DCC coupling under STEGLICH conditions[219-221] with unsymmetric malonate **148** is applied to synthesize pyrene malonate **159** (Scheme 3.43) In an acceptable yield (60 %). Purification of the desired pyrene derivative **159** can be effected easily with column chromatography on silica gel.

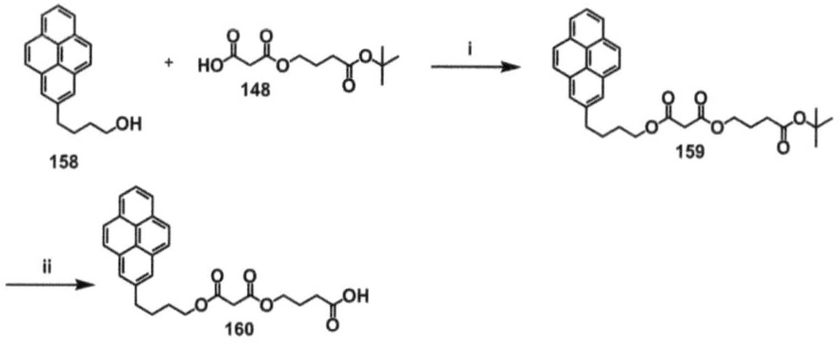

Scheme 3.43: Synthesis of pyrene carboxylic acid 160. i) DCC, DMAP, 1-HOBT, CH$_2$Cl$_2$, 0°C→rt; ii) TFA, CH$_2$Cl$_2$, rt.

Acidic cleavage of *tert*-butyl ester **159** with TFA in dichloromethane at room temperature leads to the corresponding carboxylic acid pyrene derivative in quantitative yield after evaporation of the solvent (Scheme 3.43). Further purification of the crude product is not necessary.

Pyrene compound **160** is then coupled with zinc porphyrin **147** under STEGLICH conditions[219-221] (Scheme 3.44). The yield of 65 % can be evaluated as relatively good for such complex systems. Careful column chromatography is used for the isolation of the arising pyrene-porphyrin fluorophore-chromophore hybrid.

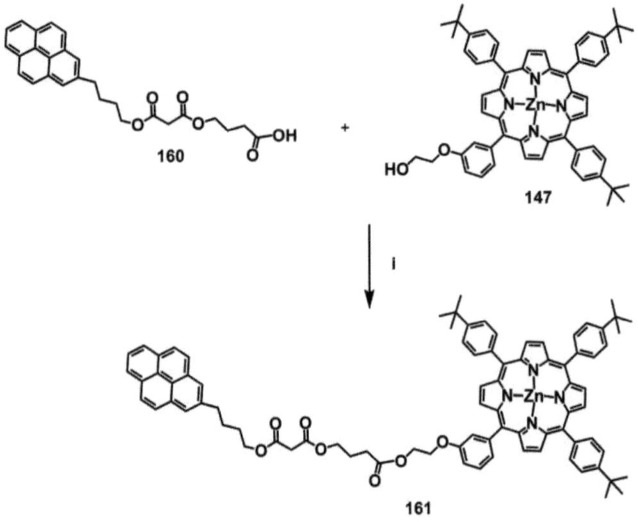

Scheme 3.44: Synthesis of pyrene-porphyrin dyad **161**. i) DCC, DMAP, 1-HOBT, CH_2Cl_2, 0 °C→rt.

The isolated pyrene-porphyrin component **161** can then be further functionalized at its malonyl methylene group by applying modified BINGEL conditions as described above.[138-140,343] The synthesis of triad **162** is shown Scheme 3.45. The isolation of the synthesized pyrene-porphyrin-fullerene triad **162** can be achieved easily using careful column chromatography on silica gel. After eluation of excessive C_{60} with toluene a gradient of ethyl acetate in toluene is used to separate the monoadduct from bisadducts and unreacted malonate. In comparison to other BINGEL modification reactions at [60]fullerene the yield of 40 % for this relatively complex systems is rather good.

Results and Discussion

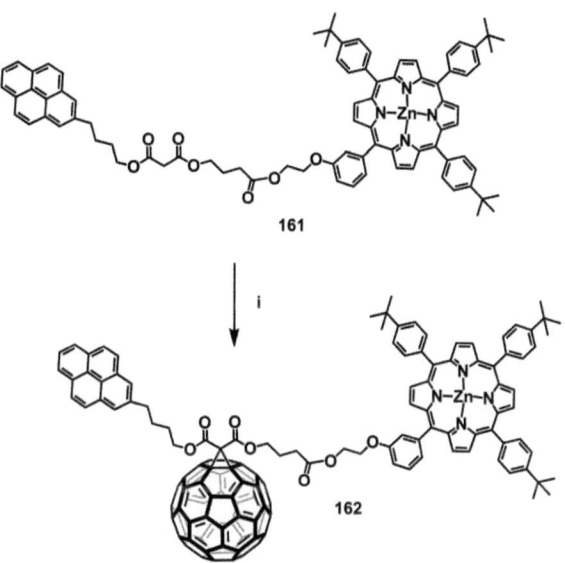

Scheme 3.45: Synthesis of pyrene-porphyrin-fullerene triad 162. i) C_{60}, CBr_4, DBU, toluene, rt.

Characterization of triad **162** is achieved by using $^1H/^{13}C$ NMR, UV/Vis and IR spectroscopy as well as mass spectrometry. The probably most interesting features for applications in photoactive devices can be derived from pyrene-porphyrin-fullerene's absorption behavior. The correlated UV/Vis spectrum of **162** in CH_2Cl_2 at room temperature is shown Figure 3.59.

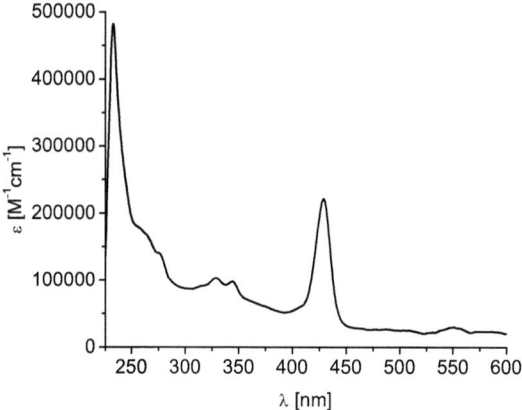

Figure 3.59: UV/Vis absorption spectrum of pyrene-porphyrin-fullerene triad 162 in CH_2Cl_2 at rt.

As expected the most intense absorption band can be found at 429.0 nm (log ε = 5.35) and is typical for the SORET band of zinc(II)porphyrin. The Q-band at 548.0 nm (log ε = 4.49) also belongs to the porphyrin entity, whereas the absorption structure in the region of 350 – 300 nm (two bands at 344.5 (log ε = 5.00) and 328.0 nm (log ε = 5.02) are characteristic for pyrene structures. Absorption bands associated with fullerene monoadducts are detected at 277.0 (log ε = 5.15), 264.0 (log ε = 5.22) and 231.5 nm (log ε = 5.70). Interestingly the second Q-band of the zinc(II)porphyrin is missing like it is in the case of the perylene-porphyrin-fullerene triad.

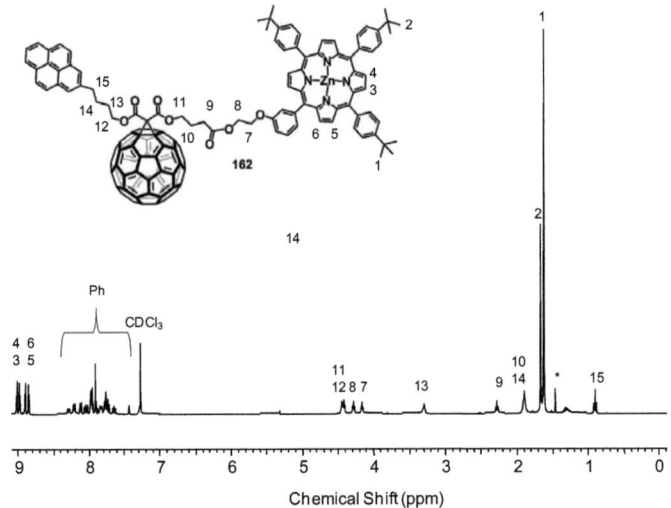

Figure 3.60: ^{1}H NMR spectrum (CDCl$_3$, 400 MHz, rt) of pyrene-porphyrin-fullerene triad 162.

Characteristic for the ^{1}H NMR spectrum of triad **162** (Figure 3.60) in CDCl$_3$ at room temperature are the signals of the different methylene groups, the *tert*-butyl signals and the signals caused by the pyrrolic protons. The methylene group in direct neighborhood to the pyrene is detected at 0.90 ppm and splits into a triplet (t, ^3J = 7.07 Hz, 2H), the other methylene groups without adjacent heteroatoms arise at 1.89 (m, 4H), 2. 28 (t, ^3J = 7.07 Hz, 2H, CH$_2$) and 3.30 ppm (m, 2H). The signal of the *tert*-butyl methylene group splits interstingly into two singlets at 1.67 (s, 9H) and 1.63 ppm (s, 18 H). This has propably to do with inter- and intramolecular interactions between the extended π-systems. The ether and ester methylene groups provoke signals at 4.16 (t, ^3J = 5.31 Hz, 2H), 4.28 (t, ^3J = 5.68 Hz, 2H) and 4.43 ppm (m, 4H).

The pyrrolic protons of the porphyrin core cause signals at 9.00 (m, 4 H) and 8.89 ppm (m, 4H).

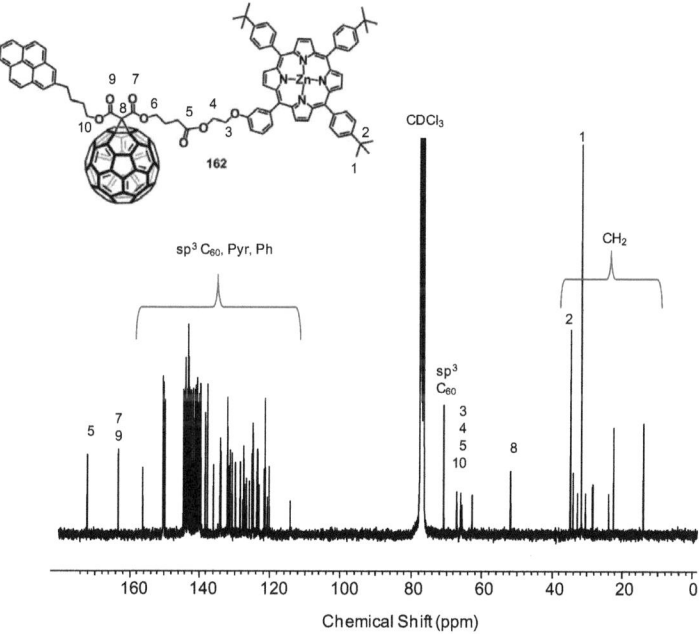

Figure 3.61: ^{13}C NMR spectrum (100.5 MHZ) of triad 162 in CDCl$_3$ at rt.

Characteristic chemical shifts of the carbon atoms of nanohybrid **162** are detected in its ^{13}C NMR spectrum (Figure 3.61). Typical for the *tert*-butyl groups of the porphyrin subunit are the signals at 34.11 (3C, C(CH$_3$)$_3$) and 31.72 ppm (9C, CH$_3$). The signals for the methylene groups occur between 14 and 35 ppm. The quaternary malonyl carbon produces a signal at 51.83 ppm which is a significant shift in comparison to the methylene malonate signal in the precursor spectrum (41.73 ppm). The signals for the ether and ester methylene groups arise at 67.11 (1C, OCH$_2$), 65.95 (1C, OCH$_2$), 65.61 (1C, OCH$_2$) and 62.81 ppm (1C, PhOCH$_2$). The sp^3-carbon atoms of the fullerene adduct can be found at 70.75 ppm, whereas the sp^2-carbon atoms of the C$_{60}$ entity as well as the aromatic carbon atoms of the pyrene and the porphyrin are detected in the region between 110 and 160 ppm as expected. Characteristic for phenyl-alkyl ethers is the signal at 156.44 ppm which can be assigned to the aromatic carbon of the ether bridge. Typical for ester groups in this triad are the three signals at 172.28, 163.29 and 163.24 ppm.

3.7.4 Conclusion

This is the first attempt to synthesize triads that contain besides porphyrin and [60]fullerene building blocks an additional chromophore (perylene) **9** or fluorophore (pyrene) **162**. Supplementary the synthesis and features of a fluorene-porphyrin fluorophore-chromophore hybrid is reported. All these complex systems can be obtained in exceptable yields with justifiable costs using standard or modified reaction protocols. Taking the absorption behavior of the described dyads **152, 157** and **161** and this of the triads **9** and **162** into account a significant expansion in the absorption of natural occuring light is achieved. In case of the additional fluorophores the absorption ability is expended into the UV region (compounds **157, 161** and **162**) whereas the perylene systems **152** and **9** show further absorption bands in the visible region. In all cases the main absorption band persists the SORET band of the zinc(II)porphyrin building block. In the particular case of light harvesting devices the shown nano systems are promising candidates for the fine tuning of electron transfer processes and for more efficient sunlight converting systems. Applications in solar cells or similar devices should be probed in the near future.

Results and Discussion

3.8 Porphyrinato Phosphonic Acids for Applications in Printable Electronics

Organic substrates are interesting materials in the hot research topic of nanostructured devices. This probably has to do with their relatively cheap preparation in comparison to expensive crystalline and amorphous inorganic materials (e.g. silicon, GaAs, etc.). HALIK et al. have reported intensively the fabrication of organic devices and the relationship between molecular structure and electrical performance in the light of organic thin film transistors (organic TFT or OFET).[363]

The design and engineering of organic semiconductors, conductors and insulators play the key role for a future technology beyond silicon. In focus for the first applications are organic non-volatile memory cells, charge/energy storage devices and rectifiers.[364,365] Fundamental steps are the hierarchically self-assembly and the characterization of organic monolayers as molecular dielectrics in capacitors or organic transistor devices.[366] Therefore the organic target molecule has to be linked *via* a suitable anchoring group on the surface (e.g. electrode material). Besides thiols on gold surfaces,[283,310,311] catecholes[367-369] and carboxylic acids on zinc oxide surfaces,[370,371] or phosphonic acids[372] have the property to bond rather strong on alumina surfaces.

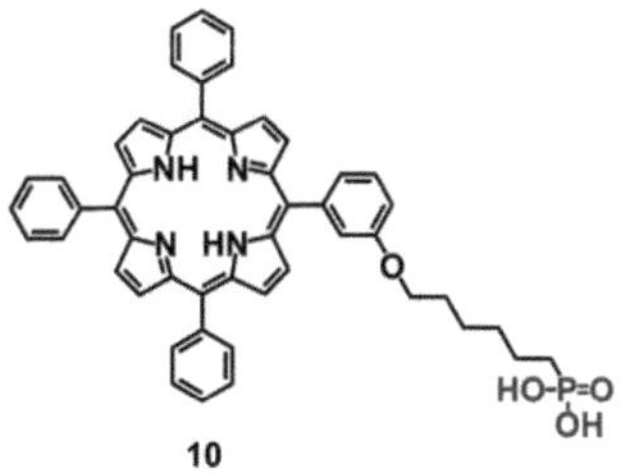

Figure 3.62: Porphyrinato phosphonic acid 10 for application in electronic devices.

In this particular work a tetraphenylporphyrin (TPP) as electroactive organic subunit is connected *via* a C_6 spacer to the phosponic acid (Figure 3.62) for the application in hierachically ordered electronic devices. The spacer unit is needed as insulating structure to prevent short circuits.

Phosphonic acid bearing porphyrins are expected to be model compounds for binding on alumina surfaces. The basic concept in this context is to synthesize a tetraphenylporphyrin (TPP) which has just one phosphonic acid functionality and three additional unsubstituted phenyl rings to provide a sufficient solubility and to avoid bulky groups that eventually hinder the self-assembly process simultaneously. Therefore 3-(6-hydroxyhexyloxy)-benzaldehyde and benzaldehyde are chosen as starting materials. The synthesis is performed under statistical LINDSEY-conditions (Scheme 3.46).[208-211]

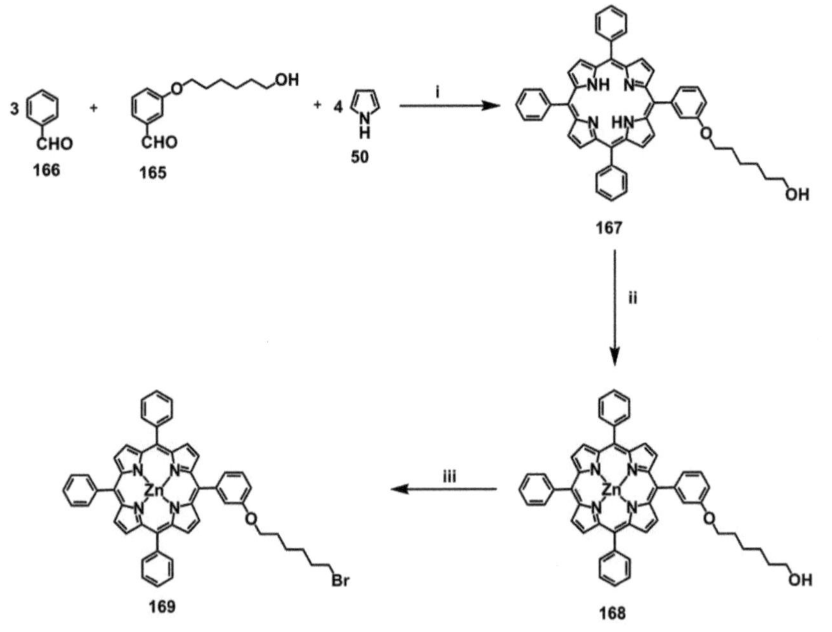

Scheme 3.46: Syntheses of the precursor porpyhrins 167-169. i) 1. TFA, CH$_2$Cl$_2$, 2. TEA, 3. DDQ; ii) Zn(OAc)$_2$, CHCl$_3$/MeOH 3:1, rt; iii) PPh$_3$, CBr$_4$, CH$_2$Cl$_2$, rt.

Four equivalents of pyrrole **50**, one equivalent of 3-(6-hydroxyhexloxy)-benzaldehyde **165** and three equivalents of benzaldehyde **166** are dissolved in dichloromethane. Then TFA is added and the solution is stirred for 1 h before NEt$_3$ is added to neutralize the reaction mixture. The final oxidation is performed with DDQ. Separation of all statistically formed porphyrin isomers is possible by careful column chromatography on silica gel with CH$_2$Cl$_2$. Mono substituted hydroxyl TPP **167** was isolated in 7 % yield after precipitating with *n*-pentane. The yield is typical for this type of reaction. Metallation with zinc is straight foward in the case of tetraphenylporphyrins. Zinc porphyrins have been intensively investigated with respect to their photophysical properties and are very attractive building blocks for architectures enabling electronic devices.[91,212-214] Bromination of the hydroxy group can be easily achieved by APPEL's halogenation protocol[373] using PPh_3 and CBr_4 (Scheme 3.46).

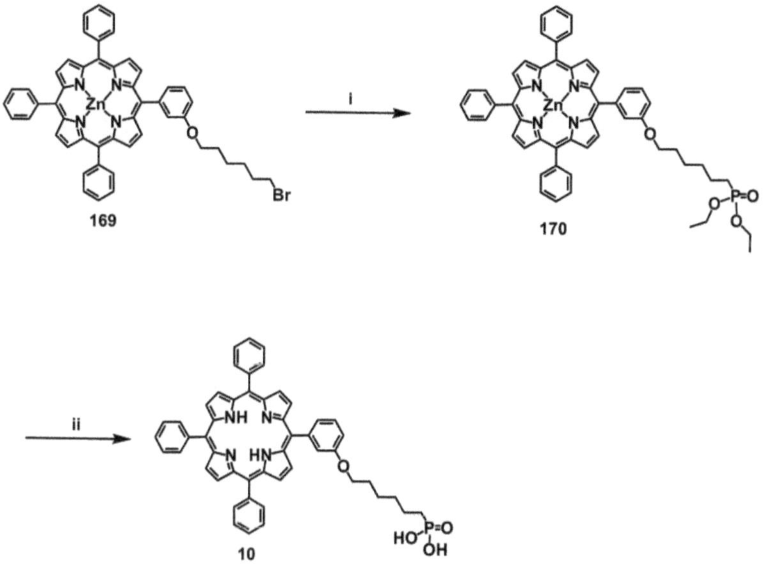

Scheme 3.47: Synthesis of the target porphyrin 10. i) $P(OEt)_3$, reflux ii) 1. $BrSiMe_3$, CH_2Cl_2, 0 °C→rt, 2. MeOH.

To introduce the phosphonate to the porphyrin the well knwon ARBUSOW reaction protocol[374,375] is applied. Refluxing bromoporphyrin **169** in triethylphosphite gives porphyrin phosphonate **170** under loosing bromethane (Scheme 3.47). Cleavage of the ethyl esters is achieved under mild conditions using $BrSiMe_3$ at 0°C in CH_2Cl_2 (Scheme 3.47). Additionally this LEWIS acidic conditions leads to demetallation and the isolation of the free base porphyrin **10**.

Characterization of porphyrinato phosphonic acid **10** is done by the classical analyzing tools of organic chemistry ($^1H/^{13}C$ NMR, UV/Vis and IR spectroscopy, mass spectrometry and elementary analysis). A closer look on the ^1H NMR spectrum suggests a strong aggregation behavior of the free base porphyrin. Probably the interaction between the nitrogen atoms in the porphyrin core and the phosphonic acid anchoring group leads to partially protonation of the free base nitrogens. The rather broad signals of the aliphatic and aromatic protons in the ^{1}H NMR spectrum (Figure 3.63) and the week and broad signal for the NH protons are clear hints for this hypothesis.

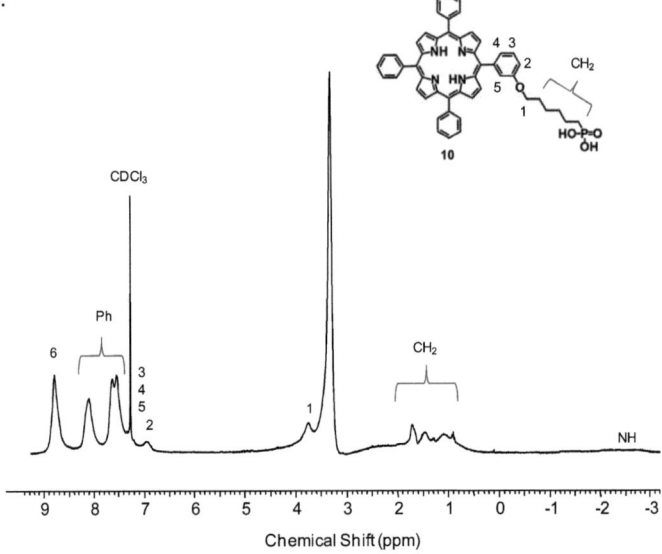

Figure 3.63: ^{1}H NMR spectrum (CDCl$_3$/MeOH, 400 MHz, rt) of porphyrinato phosphonic acid 10.

Adding a few drops of methanol to break up the aggregates is not successful as it is shown in Figure 3.63.

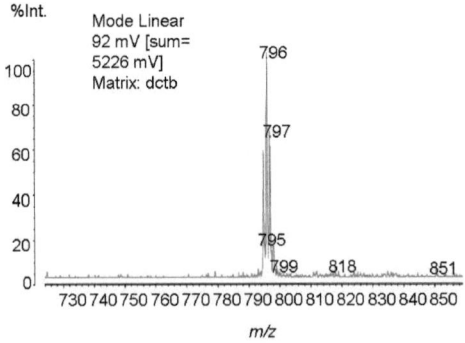

Figure 3.64: The mass spectrum of porphyrin 10 shows the product peak at 796 m/z very clearly. No evidence for educt, partially cleaved ester or metallated species is found.

The mass spectrogram of porphyrin **10** (Figure 3.64) on the other hand shows very clearly the product peak at 796 m/z and lacks any hind of educt or only partially cleaved ester. Additionally no metallated species can be found in the spectrum.

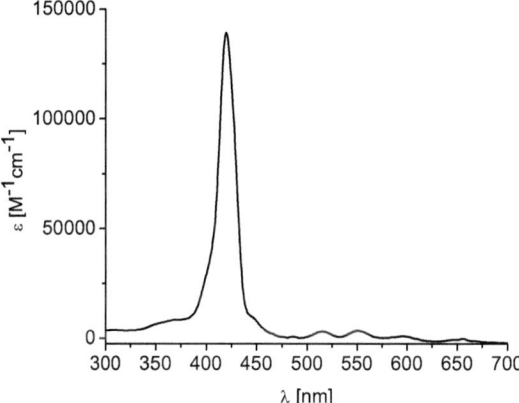

Figure 3.65: The UV/Vis absorption spectrum in CH_2Cl_2 of compound **10** shows the typical features of a free base porphyrin.

The UV/Vis spectrum of porphyrin **10** in CH_2Cl_2 (Figure 3.65) exhibits the characteristic features of a free base porphyrin. The main absorption band (SORET band) occurs at 419.0 nm, the four Q-bands appear at 515.5, 550.0, 593.5 and 655 nm.

First deposition experiments on alumina surfaces in the Department of Physics show promising results. The suspected short circuits due to the relative short C_6 alklyl chain can not be detected.

4 Summary

Within this thesis the synthesis and characterization of novel supramolecular wire-like donor-acceptor nanohybrids were investigated. In this particular light the focus was on fullerene-porphyrin architectures that are constructed *via* the six-folded hydrogen bonding Hamilton receptor cyanuric acid motif. Besides the synthesis and characterization of the fullerene and porphyrin building blocks their complexation behavior and the photophysical properties of the assembled nanosystems were of tremendous interest. In great difference to former examinations rigid spacers instead of flexible alkyl chains between chromophore and key/lock were introduced (i.e. *p*-phenylene-ethynylene, *p*-phenylene-vinylene, *p*-ethynylene, *p*-phenylene, *p*-thiophene and fluorene). To increase the poor solubility of Hamilton receptor bearing fulleropyrrolidines in nonpolar solvents alkyl chains in **47**, **94** and **95** or dendrimers in **91** were introduced at the pyrrolidine ring.

The association constants K_{ass} of the assembled hydrogen bonding complexes were analyzed using ^{1}H NMR titration techniques as well as steady state fluorescence quenching approaches. Additionally time-resolved measurements (i.e. time-resolved fluorescence and transient absorption) were applied to illuminate the electron transfer behavior of the self-assembled donor-acceptor systems. These techniques gave clear evidence for spacer mediated electron transfer processes. For instance, on one hand complex **94:7** shows clear electron transfer, on the other hand elongation with a second *p*-phenylene-ethynylene repetition unit causes preclusion of the photoinduced electron transfer.

To enlarge the concept of supramolecular arranged donor-acceptor systems the cyanuric acid bearing porphyrin **7** was replaced with dimethylamino flavin (DMA-Fl) **8**. Directional electron transfer systems containing DMA-Fl **8** and Hamilton receptor modified [60]fullerenes DRC_{60} **43** and SRC_{60} **109** were then used for the construction of photoelectrochemical electrodes. A film of the nanohybrids onto nanostructured SnO_2 electrode exhibited an incident photon-to-photocurrent efficiency (IPCE) of 9.2 % in a three-compartment electrochemical cell, which is significant higher than the measured IPCE of a $(DMA-Fl-SRC_{60})_n$ film. In addition, photodynamics of the hydrogen bonded supramolecular assemblies composed of DMA-Fl and fullerene with single and double hydrogen bond receptors (SRC_{60} and DRC_{60}) gave rise to the directional electron-transfer cascade reactions through the well-organized gradients. The results obtained demonstrate the potentiality and applied utility of a multistep electron transfer as opposite to a hole transfer along a fine-tuned and fine-balanced redox gradient in photoelectrochemical devices at first.

The investigations of the self-assembled supramolecular rigid p-oligo phenylene-ethynyl-porphyrin nanohybrids connected *via* a Hamilton-receptor based hydrogen bonding motif underlines the importance of wire-like systems. In this light, three different p-OPEs carrying two Hamilton receptor termini were synthesized. To improve the relatively poor solubility of the p-OPEs in chloroform and dichloromethane two octyloxy groups at the central benzene core were introduced in derivative **128**. The linear aggregation of these building blocks with porphyrin **7** was then analyzed by ^{1}H NMR titration experiments. As expected 1:2 complexes were formed which is proven by Job`s plot analysis. The subsequent binding of the porphyrin to the p-OPE receptors shows an overall positive cooperativity in all cases. Compared to more flexible systems which contain alkyl chains and bulky groups the binding constants K_n associated with twofold binding processes are rather high. Interestingly the most soluble p-OPE system **128** shows the strongest binding constants whereas the cooperativity of the less soluble p-OPEs **124** and **126**, expressed by the HILL coefficient η_H, is higher. As expected an electronic communication between both subunits could not be detected by UV/Vis and fluorescence experiments.[192,207,295,296]

Besides supramolecular approaches molecular electron rich systems play a certain role in this thesis. The successful synthesis of the porphyrinato phosphonium salt **134** for WITTIG HORNER olefination reactions is reported herein. Additionally the syntheses and characterization of diverse oligo-phenylene-vinylene rods (**136**, **138**, **140**, **141** and **143**) is emphasized. In terms of molecular electronics OPV bisthiol **141** is probably the most interesting one towards the contactation possibility of its thiol groups and its extraordinary good dissolution behavior in common organic solvents. Further functionalization options of OPV **136** and **140** demonstrate the formation chance of mixed oligo-phenylene-vinylene (OPV)/oligo-phenylene-acetylene (OPA) wires.

Summary

To extend the absorption features of donor-acceptor systems triads that contain besides porphyrin and C_{60} building blocks an additional chromophore (perylene) **9** or fluorophore (pyrene) **162** were developed and probed towards their photophysical properties in this work. Taking the absorption behavior of the described dyads **152, 157** and **161** and this of the triads **9** and **162** into account a significant expansion in the absorption of natural occuring light is achieved. In case of the additional fluorophores the absorption ability is expended into the UV region (compounds **157, 161** and **162**) whereas the perylene systems **152** and **9** show further absorption bands in the visible region. In all cases the main absorption band persists the SORET band of the zinc(II)porphyrin building block.

Conclusively the successful synthesis of the phosphonic acid bearing porphyrin **10** which should find its application in organic non-volatile memory cells, charge/energy storage devices or rectifiers is described in this thesis. First deposition experiments on alumina surfaces in the Department of Physics show promising results. The suspected short circuits due to the relative short C_6 alkyl chain can not be detected.

5 Zusammenfassung

Im Rahmen der vorliegenden Arbeit wurde die Synthese und Charakterisierung neuartiger supramolekular-angeordneter stäbchenförmiger Donor-Akzeptor-Nanohybride untersucht. Ein besonderes Augenmerk wurde dabei auf Fulleren-Porphyrin-Architekturen gelegt, die über sechsfache Wasserstoffbrückenbindung des Hamiltonrezeptor-Cyanursäure-Bindungsmotivs aufgebaut sind. Neben der Synthese und der Charakterisierung der einzelnen Fulleren- und Porphyrinbausteine waren die Untersuchung ihres Komplexierungsverhaltens sowie die photopyhsikalischen Eigenschaften der selbstorganisierten Nanosysteme von herausragendem Interesse. Im Unterschied zu vorherigen Forschungsansätzen, bei denen flexible Alkylketten zwischen Chromophor und Schlüssel/Schloss installiert wurden, sind in der aktuellen Arbeit bewusst starre Zwischenstücke (p-Phenyl-Ethinyl, p-Phenyl-Vinyl, p-Ethinyl, p-Phenyl, p-Thiophen und Fluoren) eingebaut worden. Um die Löslichkeit der eher schlechtlöslichen hamiltonrezeptor-modifizierten Fullerenpyrrolidine in unpolaren Lösungsmitteln zu erhöhen, wurden zusätzliche Alkylketten in den Verbindungen **47**, **94** und **95** oder Dendrimere in Verbindung **91** an den Pyrrolidinring eingeführt.

Die Assoziationskonstanten K_{ass} der sich selbstorganisierenden Wasserstoffbrückenkomplexe wurden sowohl mittels ^{1}H NMR Titrationsexperimenten als auch mit statischer Fluoreszenzlöschungsuntersuchungen bestimmt. Zusätzlich wurden zeitaufgelöste Experimente (zeitaufgelöste Fluoreszenz- und Transientenabsorptionsspektroskopie) angewand, um das Elektronentransferverhalten der selbstorganisierten Donor-Akzeptorsysteme zu beleuchten. Durch diese Methoden konnten klare Hinweise auf spacervermittelten Elektronentransferprozesse erbracht werden.

Um das vorliegende Konzept der supramolekular angeordneten Donor-Akzeptorsysteme zu erweitern wurde das Dimethylaminoflavin (DMA-Fl) **8** anstelle des cyanursäuretragenden Porphyrins **7** für weitere Untersuchungen verwendet. Durch die Komplexierung mit den hamiltonrezeptormodifizierten Fullerenkomponenten DRC_{60} **43** und SRC_{60} **109** war es möglich photoelektrochemische Elektroden mit gerichtetem Elektronentransferpotential zu konstruieren. Dabei hat sich gezeigt, dass ein $((DMA-Fl)_2-DRC_{60})_n$-Film auf nanostrukturiertem SnO_2 eine deutlich höhere Effizienz der Lichtumwandlung (9,2 %) in einer dreiteiligen

Zusammenfassung

elektrochemischen Zelle zeigt als ein entsprechender ((DMA-Fl)-SRC$_{60}$)$_n$ Film. Photodynamische Untersuchungen der über Wasserstoffbrückenbindungen organisierten Hybride haben darüberhinaus den gerichten kaskadenartigen Elektronentransfercharakter der hier zugrundeliegenden Prozesse bestätigt. Diese Ergebnisse zeigen deutlich das Potential und die Anwendungsorientiertheit von Multielektronentransferprozessen entlang feineingestellter Redoxgradienten gegenüber Lochtransferprozessen in photoelektrochemischen Bausteinen auf.

Die Untersuchung selbstorganisierter supramolekularer *p*-Oligo-Phenyl-Ethinyl-Porphyrinstäbchen, die ebenfalls über das Hamiltonrezeptorbindungsmotiv angeordnet sind, unterstreicht die Bedeutung stäbchenartiger Systeme. In diesem Zusammenhang wurden mehrere *p*-OPEs mit zwei terminalen Hamiltonrezeptoren synthetisiert. Um die bescheidene Löslichkeit der *p*-OPEs in Chloroform und Dichlormethan signifikant zu erhöhen wurden beim Beispiel **128** zwei Oktylketten am zentralen Benzolring eingeführt. Das Komplexierungsverhalten der genannten *p*-OPEs mit dem cyanursäuremodifizierten Poprhyrin **7** wurde daraufhin mittels ^{1}H-NMR Titrationsexperimenten untersucht. Die zu erwartenden 1:2-Komplexe konnten durch Job's Plot Methoden bewiesen werden. Desweiteren belegen die Experimente in allen Fällen eine positive Kooperativität der zweiten Komplexierung des Porphyrins. Im Vergleich zu flexiblen Systemen konnten deutlich höhere Bindungskonstanten für die zweite Komplexierung bestimmt werden. Interessanterweise ist zwar die Assoziationskonstante im Fall des optimal löslichen *p*-OPEs Derivats **128** am höchsten, doch scheint die Kooperativität im Falle der schlechter löslichen Komponenten **124** und **126**, ausgedrückt durch den Hillkoeffizienten η_H, stärker ausgebildet zu sein. Eine elektronische Kommunikation zwischen den beiden Untereinheiten kann durch UV/Vis Absorptions und Fluoreszenzlöschungsexperimenten ausgeschlossen werden.

Neben supramolekularen Ansätzen waren auch molekulare, elektronenreiche Verbindungen Gegenstand dieser Arbeit. Neben der erfolgreichen Synthese des phosphoniumhaltigen Porphyrins **134** für WITTIG HORNER Olefinierung wurden auch die Synthese und Charakterisierung diverser oligo-Phenyl-Vinyl-Stäbchen (**136, 138, 140, 141** und **143**) durchgeführt. Besonders hervorzuheben ist dabei oPV Bisthiol **141**, das aufgrund seiner Goldkontaktierungsmöglichkeit sicherlich interessante

Eigenschaften für die molekulare Elektronik bietet. Weitere Funktionalisierungsmöglichkeiten der oPVs **136** und **140** deuten auf mögliche gemischte oPV/oPE Stäbchen hin.

Zusätzlich wurden Triaden, die neben Porphyrin- und Fullereneinheiten ein weiteres Chromophor (Perylen) **9** oder Fluorophor (Pyren) **162** beinhalten, entwickelt und auf ihre photophysikalischen Eigenschaften untersucht. Untersuchungen der Absorptionseigenschaften der beschriebenen Diaden **152**, **157** und **161**, bzw. der Triaden **9** und **162** zeigen eine signifikante Erweiterung des Apsorptionsspektrums auf. Im Falle des zusätzlichen Chromophors (**152** und **9**) steigt die Absoprtionsfähigkeit im sichtbaren Bereich deutlich an, während im Falle des zusätzlichen Fluorophors (**157, 161** und **162**) die Absorption im UV-Bereich erheblich erweitert wird. In allen Fällen bleibt jedoch die SORET-Bande des Zink(II)porphyrins die Hauptabsorptionsbande.

Abschließend wurde eine erfolgreiche Synthese der Porphyrinphosphonsäure **10** konzipiert und durchgeführt. Porphyrinposphonsäure **10** soll in organischen Speicherzellen, für die Ladungs- bzw. Energiespeicherung und in flexiblen Displays seine Anwendung finden. Erste Depositionsversuche auf Aluminiumoxid im Department für Physik haben vielversprechende Ergebnisse erbracht. So konnten Kurzschlüsse, die aus den relativ kurzen C_6 Alkylketten hätten resultieren können, nicht beobachtet werden.

6 Experimental Part

6.1 Utilized Chemicals and Instruments

Chemicals: The utilized chemicals were ordered from Acros®, Aldrich® or Merck® and were not purified except it is annotated. All solvents were freshly distilled before using them. CH_2Cl_2 and EtOAc were distilled over K_2CO_3. Dry solvents were fabricated by the standard methods.[376] CH_2Cl_2 was dried with $LiAlH_4$, DMF with phosphorus pentoxide and THF with sodium under inert atmosphere. After drying the solvents were distilled and stored using standard Schlenk techniques. In some cases HPLC grade solvents (toluene, THF, CH_2Cl_2 and methanol) were used. These were ordered from VWR® and were not further purified or dried. C_{60} was extracted from a C_{60}/C_{70} blend ordered from Hoechst® by absorption at silica gel/activated charcoal and further column chromatography at silica gel and toluene as eluent.

In the case of analytical and preparative work the following instruments were used:

Thin Layer Chromatography (TLC):
TLC plates silica gel 60 F_{254}, layer thickness 0.2 mm, Merck KGaA, Darmstadt.
The detection of transparent compounds was done by using UV-light (254 nm and 366 nm) or by developing with 1 % aqueous $KMnO_4$ solution and sulphuric acid $H_3(P(Mo_3O_{10})_4)/Ce(SO_4)_2$ solution respectively.

Flash Chromatography (FC):
Silica gel 60 (230-400 mesh, 0.04-0.063 nmn), Macherey-Nagel.

Infrared (IR) Spectroscopy:
ASI React IR™ 1000, Analytical Services, Inc., Huntsville, AL.
The detection of IR spectra results from a diamond specimen holder. Absorption is mentioned in wavenumbers $\tilde{\nu}$ [cm^{-1}].

UV/Vis Spectroscopy:
Shimadzu UV 3102 PC UV/Vis-NIR Scanning Spectrophotometer, Shimadzu Corporation, Analytical Instruments Division, Kyoto Japan.

The detection of the UV/Vis spectra happened in the named solvents.

Mass Spectrometry:
FAB: Micromass Zabspec, NBA matrix
MALDI-TOF: Shimadzu Axima Confidence, DCTB, DHB or SINmatrix

NMR Spectroscopy:
Jeol JNM EX 400 and Jeol JNM GX (^{1}H: 400 MHz, ^{13}C: 100.5 MHz), Bruker Avance 300 (^{1}H: 300 MHz, ^{13}C: 75.4 MHz), Bruker.
Every NMR spectrum was detected in $CDCl_3$ or other denoted solvents. The chemical shifts in ppm attended relative to tetramethyl silane (TMS). Data were processed with Mestrec or ACD Labs.

Elemental Analysis:
CE Instruments EA 1110 CHNS

Fluorescence Spectroscopy:
Flouromax3 Jobin Yvon Horiba, HORIBA Jobin Yvon GmbH, München.

Centrifuge:
Sigma 204 with rotor 11032 (4 x 100 mL), Sigma Laborzentrifugen, Osterode.

All compounds were named systematically using the IUPAC nomenclature when it was practically and possible.

6.2 Syntheses and Spectroscopic Data

General Operation Procedure for xylene oxidation (GOP I)
$KMnO_4$ (2.1 eq) was added to a suspension of the xylene derivative (1 eq) in 120 ml *tert*-butanol/H_2O (1:1). After heating the suspension for 1 h at 100 °C another amount of $KMnO_4$ (2.1 eq) was added. The suspension was heated at 100 °C for another 20 h and then cooled down to room temperature. After filtration over celite 500 and washing with water the filtrate was reduced to one-third and acidified with conc. HCl. The appearing white precipitate was dissolved in conc. $NaHCO_3$ solution und washed

three times with 100 ml diisopropylether. After further acidification with conc. HCl the white precipitate was collected and dried at 80 °C over night.

General Operation Procedure for chlorination of carboxylic acids (GOP II)

The carboxylic acids were suspended in $SOCl_2$ under dry conditions. After adding one droplet of DMF the reaction mixture was refluxed for four hours and the excessive $SOCl_2$ was removed by distillation.

General Operation Procedure for esterification/amidation *via* carboxylic acid chlorides (GOP III)

A solution of the chloride (1 eq per reaction possibility) in dry THF or dry CH_2Cl_2 (40 mL) was added dropwise to a solution of the amine or alcohol (1 eq per reaction possibility) and triethylamine or pyridine (1 eq per reaction possibility) in dry THF or dry CH_2Cl_2 (100 mL) at 0 °C. The solution was stirred at rt for 12 h, the residue was filtered off and the solvent was removed under reduced pressure. Purification was achieved by column chromatography on silica gel.

General Operation Procedure for SONOGASHIRA coupling reactions (GOP IV)

The reaction was performed under inert conditions in the presence of nitrogen in dry solvents and an excess of amine (i.e. triethylamine, diethylamine or diisopropylamine). The iodo compound (1 eq), CuI (0.04 eq) and the palladium(0) catalyst (0.02 eq) were completely dissolved before the acetylene (1.2 eq) was added at once. The reaction mixture was then stirred at room temperature for at least 12 h. After filtration and removing the solvent the crude product was purified by column chromatography on silica gel.

General Operation Procedure for cleavage of TMS protecting groups (GOP V)

The reaction was carried out under inert conditions and an atmosphere of nitrogen. The TMS protected acetylene compound (1 eq) was dissolved in dry THF (20 mL) and the reaction mixture was cooled down to 0 °C. Then TBAF (1 M in THF, 1.1 eq) was added dropwise and the reaction mixture was stirred at 0°C for 2 h before some drops of water were added. After distillation of the solvent the crude product was isolated by column chromatography on silica gel.

General Operation Procedure for cleavage of diethyl acetals, *tert*-butyl esters and Boc protected amines (GOP VI)

The diethyl acetal, *tert*-butyl ester or Boc protected amine was dissolved in CH_2Cl_2 (20 mL) and trifluoroacetic acid (2 mL) was added at room temperature. The solution was stirred for 3 h.

a In the case of the diethyl acetal cleavage saturated $NaHCO_3$ solution (20 mL) was added and the organic phase was separated. After drying over $MgSO_4$ the solvent was removed and the desired aldehyde was dried in high vacuum.

b In the case of *tert*-butyl ester or Boc protected amine cleavage the solvent was removed in vacuum and the obtained acid or ammonium salt was dried in high vacuum.

General Operation Procedure for SUZUKI coupling reactions (GOP VII)

The reaction was performed under an atmosphere of nitrogen in dry DMF. The iodo component (1 eq), the boronic acid/ester (1 eq) and $Pd(PPh_3)_4$ (0.04 eq) were dissolved completely before dry potassium carbonate (2 eq) was added at once. The reaction mixture was then stirred at room temperature for at least two days. After filtration ethyl acetate (50 mL) was added and the mixture was washed three times with brine (100 mL). The solvent was removed in vacuum and the remainder was purified by column chromatography on silica gel.

General Operation Procedure for preparation of mono fulleropyrrolidines (GOP VIII)

The aldehyde (1 eq) was dissolved in THF (5 mL) under inert conditions before toluene/ODCB (1 mL/1 mg C_{60}) and C_{60} (1.2 eq) were added. The mixture was stirred for 30 minutes under exclusion of light until C_{60} was dissolved completely. Then the glycine derivative (1.2 eq) was added and the mixture was refluxed over night. After cooling down to room temperature the mixture was poured onto a silica gel column and the excessive C_{60} was separated by elution with toluene. The desired monoadducts could be obtained by applying a gradient of EtOAc in toluene. The occurring brown solids were dissolved in CS_2 and precipitated with *n*-pentane to exclude solvent inclusions.

Experimental Part

General Operation Procedure for preparation of 1:3 tetraphenyl porphyrins (GOP IX)

a A solution of pyrrole **50** (4 eq), benzaldehyde I (1 eq), benzaldehyde II (3 eq), ethanol (10 mL) and tetraphenyl phosphonium chloride (0.1 eq) in dichloromethane (1 L per 0.03 mol pyrrole) was stirred for 30 minutes. Then boron trifluoride etherate (3 eq) was added changing the color of the solution to purple red. The mixture was stirred for at least 35 minutes and no longer than 40 minutes. Now 2,3-dichloro-5,6-dicyano-1,4-benzoquinone (3 eq) were added under atmospheric conditions. The reaction mixture was stirred for 15 hours. The solvent was reduced under evaporation and filtered over silica in dichloromethane. The product was obtained by precipitation with *n*-pentane.

b Pyrrole **50** (4 eq), benzaldehyde I (1eq) and benzaldehyde II (3 eq) were dissolved in CH_2Cl_2 (1 L per 0.03 mol pyrrole). TFA (3 eq) was added and the reaction mixture was stirred for 1 h before TEA (3.1 eq) was added for neutralization. After stirring for another 8 min DDQ (3 eq) was added and the mixture was stirred over night. The mixture was filtered over silica in dichloromethane and the solvent was removed in vacuum. The product was obtained by precipitation with *n*-pentane.

General Operation Procedure for preparation of tetraphenyl porphyrinato zinc(II) (GOP X)

Free base porphyrin (1 eq) was dissolved in THF (20 mL) and zinc-acetate (3 eq) was added. The mixture was heated to reflux over a period of 4 hours. The solvent was removed under reduced pressure and the residue was purified with column chromatography on silica gel. Final crystallization in *n*-pentane gave purple red solids.

General Operation Procedure for STEGLICH esterification/amidation (GOP XI)

The reaction was carried out under dry conditions under an atmosphere of nitrogen. The carboxylic acid (1 eq), 1-HOBT (1.1 eq) and DMAP (1.1 eq) were dissolved totally in dry CH_2Cl_2 at room temperature. After cooling down to 0 °C DCC (2 eq) were added and the mixture was stirred at 0 °C for half an hour. Then the corresponding alcohol/amine (1.2 eq) was added at once and the mixture was stirred for at least 18h allowing the mixture to reach room temperature. After filtration of DCU the solvent was evaporated and the remainder was dissolved in cold EtOAc

(-18 °C) to separate the resulting DCU completely. After distillation of the EtOAc the crude product was purified by column chromatography on silica gel.

General Operation Procedure for preparation of *N*-benzyl glycine esters (GOP XII)

The reaction was processed under inert conditions and an atmosphere of nitrogen. The aldehyde (1 eq) and the glycine ester hydrochloride (10 eq) were suspended in a mixture of dry THF (50 mL) and dry methanol (6 mL). Then triethylamine (10 eq) was added and the reaction mixture was stirred for one hour at room temperature. $NaBH_3CN$ (2 eq) was now added and the mixture was stirred over night. After adding water (50 mL) carefully the organic compounds were extracted with CH_2Cl_2 (50 mL) for three times. The combined organic layers were dried over $MgSO_4$ and the solvent removed in vacuum. Purification of the crude product occurred by column chromatography on silica gel ($CH_2Cl_2 \rightarrow CH_2Cl_2$/MeOH 97:3).

General Operation Procedure for WITTIG-HORNER olefinations (GOP XIII)

Potassium *tert*-butoxide (2.1 eq) and 18-crown-6 (0.03 eq) were dissolved under inert conditions in dry THF (100 mL). Then the bisaldehyde (1 eq) and the phosphonate/phosphonium salt (2.1 eq) dissolved in dry THF (50 mL) were added dropwise at room temperature over a period of 1 h. After stirring for additional two hours water (50 mL) was added and the occurring precipitate was filtered. The collected crude product was then purified using column chromatography on silica gel.

General Operation Procedure for preparation of mono fullerocyclopropanes (GOP XIV)

C_{60} (1.2 eq) was added to rapid stirring toluene (1 mL/1 mg C_{60}) under inert conditions and exclusion of light at room temperature. After stirring for 30 minutes to dissolve C_{60} totally, CBr_4 (1.1 eq) and the corresponding malonate (1 eq) were added. Then DBU (2.2 eq) dissolved in toluene was added dropwise over a period of at least 30 min. The solution was stirred for further 3 h, subsequently solvent was reduced to the half and poured on a silica gel column. Excessive C_{60} was separated by eluation with toluene. The desired monoadduct could be isolated by applying a gradient of EtOAc to toluene.

6.2.1 Supramolecular Fullerene Building Blocks and Their Precursors

5-Bromo-isophthalic acid (13)

5-Bromo-isophthalic acid was prepared according to GOP I using KMnO$_4$ (43.34 g, 0.27 mol) and 5-bromo-xylene (12 g, 8.82 mL, 0.06 mol).

Yield: 12.7 g (80 %).

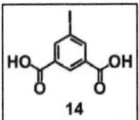

^{1}H NMR (DMSO d$_6$, 400 MHz, RT): δ [ppm] = 13.52 (broad s, 2H, COOH) 8.38 (s, 1H), 8.25 (s, 2H).

^{13}C NMR (DMSO d$_6$, 100.5 MHz, RT): δ [ppm] = 165.35 (2C, COOH), 135.74 (2C, PhC), 133.49 (2C, PhC), 128.96 (1C, PhC), 122.03 (1C, PhC).

IR (diamond, RT): $\tilde{\nu}$ [cm^{-1}] = 2990, 21360, 1698, 1449, 1399, 1275, 1266, 903, 753, 712.

EA: calculated for C$_8$H$_5$O$_4$Br (245.03): C 39.21, H 2.06; found: C 40.21, H 2.29.

5-Iodo-isopthalic acid (14)

5-iodo-isophthalic acid was prepared according to GOP I using KMnO$_4$ (43.34 g, 0.27 mol) and 5-bromo-xylene (15 g, 9.40 mL, 0.06 mol).

Yield: 16.1 g (85 %).

^{1}H NMR (DMSO d$_6$, 400 MHz, RT): δ [ppm] = 13.54 (broad s, 2H, COO*H*), 8.40 (s, 3H, Ph*H*).

^{13}C NMR (DMSO d$_6$, 100.5 MHz, RT): δ [ppm] = 165.2 (2C, C=O), 141.4 (2C, PhC), 133.1 (2C, PhC), 129.1 (1C, PhC), 94.8 (1C, PhC).

IR (diamond, RT): $\tilde{\nu}$ [cm^{-1}] = 3535, 3440, 3085, 2645, 2542, 1710, 1624, 1571, 1461, 1292, 1212.32, 1159, 1107, 799, 751.

EA: calculated for C$_8$H$_5$O$_4$I (292.03): C 32.90, H 1.73; found: C 33.05, H 1.86.

5-Bromo-isophthaloyl dichloride (15)

5-bromo-isophthaloyl dichloride was synthesized according to GOP II starting with 5-bromo-isophthalic acid (6.0 g, 24.0 mmol).
Yield: 6.63 g (100%).

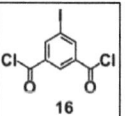

¹H NMR (DMSO d₆, 400 MHz, RT): δ [ppm] = 8.36 (s, 1H), 8.19 (s, 2H).

¹³C NMR (DMSO d₆, 100.5 MHz, RT): δ [ppm] = 165.38 (2C, COCl), 135.78 (2C, PhC), 133.59 (2C, PhC), 128.92 (1C, PhC), 122.08 (1C, PhC).

5-Iodo-isophthaloyl dichloride (16)

5-iodo-isophthaloyl dichloride was prepared according to GOP II starting with 5-iodo-isophthalic acid (1.5 g, 5.51 mmol).
Yield: 1.68 g (100%).

¹H NMR (DMSO d₆, 400 MHz, RT): δ [ppm] = 8.38 (s, 3H, Ph*H*).

¹³C NMR (DMSO d₆, 100.5 MHz, RT): δ [ppm] = 165.21 (2C, C=O), 141.41 (2C, Ph*C*), 133.0 (2C, Ph*C*), 129.12 (1C, Ph*C*), 94.8 (1C, Ph*C*).

N,N'-Bis[6-(3,3-dimethylbutyrylamino)pyridin-2-yl]-5-bromo-isophthalamide (18)

Synthesis was accomplished by GOP III with dichloride **15** (2.71 g, 9.6 mmol), *N*-(6-aminopyridine-2-yl)-3,3-dimethylbutyramide **17** (4.35 g, 21 mmol) and triethylamine (2.7 mL, 19.3 mmol) in THF as solvent. Purification by column chromatography on silica gel (CH₂Cl₂/ethyl acetate (2:1) as eluent) gave a yellowish solid.
Yield: 5.53 g (92 %).

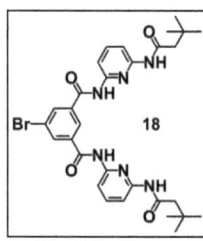

¹H NMR (THF d₈, 400 MHz, RT):

δ [ppm] = 9.76 (bs, 2H, N*H*), 9.10 (bs, 2H, N*H*), 8.46 (s, 1H, Ph*H*), 8.29 (s, 2H, Ph*H*), 8.01 (m, 4 H, py*H*), 7.73 (d, ³*J* = 8.05 Hz, 2H, py*H*), 2.27 (s, 4H, C*H*₂), 1.08 (s, 18 H, C*H*₃).

¹³C NMR (THF d₈, 100.5 MHz, RT):

δ [ppm] = 170.80 (2C, C=O), 164.25 (2C, C=O), 151.70 (2C, α-pyC), 151.05 (2C, α-pyC), 140.56 (2C, PhC), 138.07 (2C, PhC), 134.36 (1C, PhC), 126.63 (2C, pyC), 123.09 (1C, PhC), 110.49 (2C, β-pyC), 110.20 (2C, β-pyC), 50.68 (2C, CH_2), 31.63 (2C, $C(CH_3)_3$), 29.98 (6C, CH_3).

MS (FAB, NBA): m/z = 625 [M]⁺.

IR (diamond, RT): ṽ [cm⁻¹] = 2955, 2361, 1684, 1671, 1558, 1541, 1521, 1508, 1446, 1298, 1277, 1241, 764, 750.

EA: calculated for $C_{30}H_{37}O_5N_6Br$*EtOAc (711.65): C 57.38, H 6.09, N 11.81; found: C 56.97, H 5.79, N 12.22.

N,N'-Bis[6-(3,3-dimethylbutyryl-amino)pyridin-2-yl]-5-iodo-isophthalamide (19)

The iodo Hamilton Receptor was synthesized by GOP III with dichloride **16** (1.68 g, 5.1 mmol), *N*-(6-aminopyridine-2-yl)-3,3-dimethylbutyramide **17** (2.12 g, 10.2 mmol) and triethylamine (1.44 mL, 10.2 mmol) in THF as solvent. Purification by column chromatography on silica gel (CH_2Cl_2/ethyl acetate (2:1) as eluent) gave a yellowish solid.
Yield: 3.23 g (94 %).

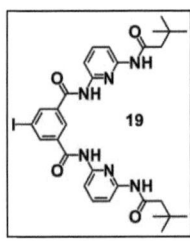

¹H NMR (THF d₈, 400 MHz, RT): 9.73 (bs, 2H, N*H*), 9.09 (bs, 2H, N*H*), 8.46 (s, 1H, Ph*H*), 8.45 (s, 2H, Ph*H*), 8.00 (m, 4 H, py*H*), 7.73 (d, ³*J* = 8.1 Hz, 2H, py*H*), 2.26 (s, 4H, C*H*₂), 1.08 (s, 18H, C*H*₃).

¹³C NMR (THF d₈, 100.5 MHz, RT):
δ [ppm] = 170.81 (2C, C=O), 164.24 (2C, C=O), 151.73 (2C, α-pyC), 151.09 (2C, α-pyC), 140.52 (2C, PhC), 140.27 (2C, PhC), 137.91 (1C, PhC), 127.13 (1C, pyC), 110.42 (2C, β-pyC), 110.11 (2C, β-pyC), 94.50 (2C, PhC), 50.68 (2C, CH_2), 31.62 (2C, $C(CH_3)_3$), 30.03 (6C, CH_3).

MS (FAB, NBA): m/z = 671 [M]⁺.

IR (diamond, RT): $\tilde{\nu}$ [cm^{-1}] = 3429, 3303, 2957, 2929, 1690, 1584, 1516, 1447, 1365, 1298, 1156.

EA: calculated for $C_{30}H_{37}O_5N_6I \cdot H_2O$ (688.54): C 52.33, H 5.42, N 12.21; found: C 52.65, H 5.15, N 12.28.

N,N'-Bis[6-(3,3-dimethylbutyrylamino)pyridin-2-yl]-5-ethynyl-phenyl-4-formyl-isophthalamide (21)

Hamilton Receptor **21** could be obtained *via* GOP IV using iodo Hamilton Receptor **19** (500 mg, 0.75 mmol), Pd(PPh$_3$)$_2$Cl$_2$ (6 mg, 0.01 mmol), CuI (4 mg, (0.02 mmol) and 4-ethynyl-benzaldehyde **20** (126 mg, 0.97 mmol) in dry THF (12 mL) and NEt$_3$ (8 mL). The reaction mixture was stirred for 20 h at rt. Column chromatography of the residue (toluene/EtOH 95:5) was used to isolate the desired product.
Yield: 524 mg (99 %)

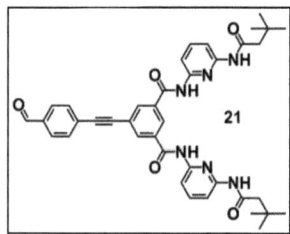

^{1}H NMR (DMSO d$_6$, 400 MHz, RT):
δ [ppm] = 10.50 (bs, 2H, N*H*), 10.05 (s, 1H, *H*C=O), 9.81 (bs, 2H, N*H*), 8.55 (s, 1H, Ph*H*), 8.35 (d, ^3J = 6.61, 2H, Ph*H*), 7.98 (m, 5H, Ph*H*, py*H*), 7.94 (m, 4H, py*H*), 7.83 (d, ^3J = 7.97 Hz, 2H, Ph*H*), 2.32 (s, 4H, C*H$_2$*), 1.03 (s, 18 H, C*H$_3$*).

^{13}C NMR (DMSO d$_6$, 100.5 MHz, RT):
δ [ppm] = 192.00 (1C, *H*C=O), 170.60 (2C, *C*=O), 164.15 (2C, *C*=O), 150.33 (2C, α-py*C*), 149.79 (2C, α-py*C*), 139.67 (2C, Ph*C*), 138.83 (2C, Ph*C*), 136.33 (2C, Ph*C*), 132.93 (2C, Ph*C*), 131.94 (2C, Ph*C*), 129.32 (2C, py*C*), 123.99 (2C, Ph*C*), 110.36 (2C, β-py*C*), 109.95 (2C, β-py*C*), 82.18 (1C, ethynyl *C*), 75.93 (1C, ethynyl *C*), 54.52 (2C, *C*H$_2$), 30.22 (2C, *C*(CH$_3$)$_3$), 29.33 (6C, *C*H$_3$).

MS (FAB, NBA): m/z = 673 [M]$^+$.

FT-IR (diamond, RT): $\tilde{\nu}$ [cm^{-1}] = 3282, 2954, 2361, 1683, 1585, 1540, 1509, 1446, 1300, 1237, 1155, 1126, 802, 763, 746, 728.

4-(2-(4-(2-(Triisopropylsilyl)ethynyl)phenyl)benzaldehyde (24)

Benzaldehyde **24** was synthesized *via* GOP IV with 4-bromobenzaldehyde **23** (90 mg, 0.48 mmol), Pd(PPh$_3$)$_2$Cl$_2$ (3.5 mg, 0.01 mmol), CuI (3 mg, 0.01 mmol) and (2-(4-ethynylphenyl)ethynyl)triisopropylsilane **22** (150 mg, 0.53 mmol) in toluene (20 mL) and HNEt$_2$ (10 mL). The reaction mixture was stirred for 24 h at 75 °C. Column chromatography (toluene) of the residue was used to purify the product.
Yield: 130 mg (70 %).

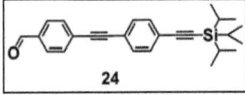

^{1}H NMR (CDCl$_3$, 400 MHz, RT):
δ [ppm] = 10.00 (s, 1H, *HC*=O), 7.85 (d, ^3J = 6.71 Hz, 2H, Ph*H*), 7.66 (d, ^3J = 8.18 Hz, 2H, Ph*H*), 7.46 (m, 4H, Ph*H*), 1.12 (s, 18H, C*H$_3$*), 1.11 (s, 3H, C*H*).

^{13}C NMR (CDCl$_3$, 100.5 MHz, RT):
δ [ppm] = 191.34 (1C, *H*C=O), 135.52 (1C, Ph*C*), 132.10 (2C, Ph*C*), 132.05 (2C, Ph*C*), 131.54 (1C, Ph*C*), 129.58 (1C, Ph*C*), 129.32 (2C, Ph*C*), 124.10 (2C, Ph*C*), 122.25 (1C, Ph*C*), 106.42 (1C, ethynyl*C*), 93.41 (1C, ethynyl*C*), 93.06 (1C, ethynyl*C*), 90.20 (1C, ethynyl*C*), 18.64 (3C, *C*H), 11.29 (6C, *C*H$_3$).

MS (FAB, NBA): m/z = 387 [M]$^+$.

IR (diamond, RT): $\tilde{v}$ [cm^{-1}] = 2361, 2342, 2153, 1702, 1601, 1561, 1514, 1494, 1386, 1300, 1276, 1248, 1223, 1207, 1162, 1108, 1013, 841, 827, 792, 755.

4-(2-(4-Ethynylphenyl)ethynyl) benzaldehyde) (25)

Aldehyde **25** was accessible *via* GOP V using compound **24** (130 mg, 0.34 mmol) and TBAF (400 µL, 0.4 mmol, 1 M solution in THF). The residue was purified by column chromatography on silica gel (toluene).
Yield: 77 mg (99 %).

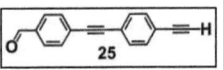

^{1}H NMR (CDCl$_3$, 400 MHz, RT):
δ [ppm] = 10.01 (s, 1H, *HC*=O), 7.86 (d, ^3J = 6.59 Hz, 2H, Ph*H*), 7.67 (d, ^3J = 8.06 Hz, 2H, Ph*H*), 7.48 (m, 4H, Ph*H*), 3.18 (s, 1H, ethynyl*H*).

¹³C NMR (CDCl₃, 100.5 MHz, RT):

δ [ppm] = 191.34 (1C, HC=O), 135.52 (1C, PhC), 132.10 (2C, PhC), 132.05 (2C, PhC), 131.54 (1C, PhC), 129.58 (1C, PhC), 129.32 (2C, PhC), 124.10 (2C, PhC), 122.25 (1C, PhC), 106.42 (1C, ethynylC), 93.41 (1C, ethynylC), 82.58 (1C, ethynylC), 81.05 (1C, ethynylC).

MS (FAB, NBA): m/z = 230 [M]⁺.

IR (diamond, RT): $\tilde{v}$ [cm⁻¹] = 2362, 2345, 2150, 1793, 1691, 1601, 1561, 1508, 1493, 1428, 1382, 1322, 1301, 1290, 1210, 1184, 1167, 1132, 1101, 1019, 957, 859, 841, 830, 758.

Elongated Ethynyl-formyl Hamilton receptor (26)

Hamilton receptor **26** was prepared according to GOP IV using iodo Hamilton Receptor **19** (106 mg, 0.16 mmol), Pd(PPh₃)₂Cl₂ (2 mg, 0.01 mmol), CuI (9 mg, 0.05 mmol) and 2-(4-ethynylphenyl)ethynyl)benzaldehyde **25** (40 mg, 0.18 mmol) in dry THF (10 mL) and NEt₃ (5 mL). The reaction mixture was stirred for 48 h at room temperature. The residue was purified by column chromatography (CH₂Cl₂/EtOAc/MeOH 2:1:0.01).
Yield: 117 mg (97 %).

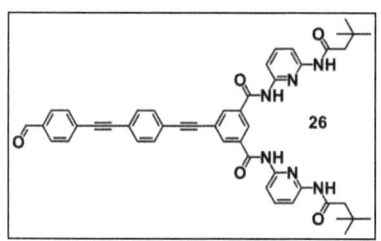

¹H NMR (THF d8, 400 MHz, RT):

δ [ppm] = 10.05 (s, 1H, HC=O), 9.45 (bs, 2H, NH), 8.82 (bs, 2H, NH), 8.40 (s, 3H, PhH), 8.37 (d, ³J = 6.68 Hz, 4H, PhH), 7.96 (d, ³J = 8.45 Hz, 2H, PhH), 7.94 (d, ³J = 7.98 Hz, 2H, PhH), 7.91 (d, ³J = 7.97 Hz, 2H, pyH), 7.63 (d, ³J = 8.12 Hz, 2H, pyH), 7.53 (m, 2H, pyH), 2.20 (s, 4H, CH₂), 1.04 (s, 18 H, CH₃).

¹³C NMR (THF d₈, 100.5 MHz, RT):

δ [ppm] = 192.45 (1C, HC=O), 170.60 (2C, C=O), 164.15 (2C, C=O), 150.33 (2C, α-pyC), 149.79 (1C, PhC), 139.67 (2C, α-pyC), 138.83 (2C, PhC), 137.12 (2C, PhC), 136.33 (4C, PhC), 134.44 (2C, PhC), 132.93 (4C, PhC), 131.94 (2C, PhC), 129.32

(2C, PyC), 123.99 (1C, PhC), 110.36 (2C, β-pyC), 109.95 (2C, β-pyC), 82.18 (2C, ethynyl C), 75.93 (2C, ethynyl C), 54.52 (2C, CH_2), 30.22 (2C, $C(CH_3)_3$), 29.33 (6C, CH_3).

MS (FAB, NBA): m/z = 772 [M]$^+$.

FT-IR (diamond, RT): ṽ [cm^{-1}] = 3282, 2955, 2362, 1685, 1584, 1541, 1512, 1445, 1301, 1238, 1153, 1122, 802, 765, 748, 729.

5-(3,3-Diethoxyprop-1-ynyl)-N^1,N^3-bis(6-(3,3-dimethylbutyrylamino)pyridin-2-yl)isophthalamide (28)

The synthesis of Hamilton receptor component **28** was achieved by GOP IV starting from iodo Hamilton receptor **19** (200 mg, 0.3 mmol), Pd(PPh$_3$)$_2$Cl$_2$ (2 mg, 0.01 mmol), CuI (2 mg, 0.01 mmol) and 3,3-diethoxyprop-1-yne **27** (51 µL, 0.36 mmol) in THF (20 mL) and HN^{i}Pr$_2$ (5mL). The reaction mixture was stirred for 20 h and the residue was cleaned up by column chromatography on silica gel (CH$_2$Cl$_2$/EtOAc 4:1). Yield: 197 mg (98 %).

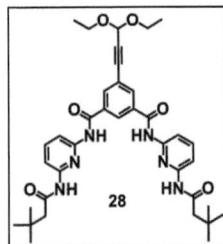

^{1}H NMR (DMSO d$_6$, 400 MHz, RT):

δ [ppm] = 10.67 (bs, 2H, NH), 10.04 (bs, 2H, NH), 8.52 (s, 1H, PhH), 8.26 (s, 2H, PhH), 7.87 (d, ^3J = 6.61 Hz, 2H, pyH), 7.82 (m, 2H, pyH), 7.79 (d, ^3J = 6.63 Hz, 2H, pyH), 5.60 (s, 1H, CH), 3.71 (m, 4H, OCH$_2$), 2.31 (s, 4H, CH$_2$), 1.20 (t, ^3J = 7.70 Hz, 2H, CH$_2$CH$_3$), 1.02 (s, 18H, CH$_3$).

^{13}C NMR (DMSO d$_6$, 100.5 MHz, RT):

δ [ppm] = 170.68 (2C, C=O), 164.11 (2C, C=O), 149.35 (2C, α-pyC), 148.79 (2C, α-pyC), 138.79 (2C, pyC), 133.66 (2C, PhC), 132.64 (2C, PhC), 127.08 (1C, PhC), 120.35 (1C, PhC), 109.40 (2C, β-pyC), 108.84 (2C, β-pyC), 89.80 (1C, CH), 85.35 (1C, ethynylC), 81.72 (1C, ethynylC), 59.25 (2C, OCH$_2$), 47.87 (2C, CH$_2$), 29.68 (2C, C(CH$_3$)$_3$), 28.35 (6C, CH$_3$), 13.83 (2C, CH$_2$CH$_3$).

MS (FAB, NBA): m/z = 672 [M]$^+$.

Experimental Part

IR (diamond, RT): $\tilde{v}$ [cm^{-1}] = 2956, 2360, 2342, 1672, 1585, 1508, 1446, 1297, 1277, 1260, 1241, 800, 764, 750.

N^1,N^3-Bis(6-(3,3-dimethylbutyrylamino)pyridin-2-yl)-5-(3-oxopro-1-ynyl)isophthalamide (29)

This derivative was synthesized via GOP VIa using Hamilton receptor derivative **28** (100 mg, 0.145 mmol).
Yield: 85 mg (98 %).

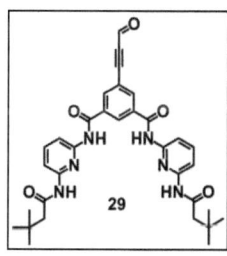

^{1}H NMR (DMSO d$_6$, 400 MHz, RT):

δ [ppm] = 10.74 (s, 2H, N*H*), 10.03 (s, 2H, N*H*), 9.50 (s, 1H, *H*C=O), 8.64 (s, 1H, Ph*H*), 8.48 (s, 2H, Ph*H*), 7.89 (d, ^3J = 6.61 Hz, 2H py*H*), 7.83 (m, 2H, py*H*), 7.79 (d, ^3J = 6.63 Hz, 2H, py*H*), 2.31 (s, 4H, C*H$_2$*), 1.02 (s, 18H, C*H$_3$*).

^{13}C NMR (DMSO d$_6$, 100.5 MHz, RT):

δ [ppm] = 178.66 (1C, *H*C=O), 170.96 (2C, C=O), 164.11 (2C, C=O), 151.49 (2C, α-py*C*), 150.88 (2C, α-py*C*), 140.96 (2C, py*C*), 136.93 (2C, Ph*C*), 136.12 (2C, Ph*C*), 131.30 (1C, Ph*C*), 120.13 (1C, Ph*C*), 111.53 (2C, β-py*C*), 111.11 (2C, β-py*C*), 92.32 (1C, ethynyl*C*), 89.55 (1C, ethynyl*C*), 50.00 (2C, *C*H$_2$), 31.83 (2C, *C*(CH$_3$)$_3$), 30.50 (6C, *C*H$_3$).

MS (FAB, NBA): m/z = 596 [M]$^+$.

IR (diamond, RT): $\tilde{v}$ [cm^{-1}] = 2959, 1678, 1605, 1585, 1508, 1493, 1445, 1302, 1241, 803, 754.

4-Ethenyl benzaldehyde (31)

4-Bromo styrene **30** (1 g, 0.71 mL, 5.4 mmol) was dissolved in THF (20 mL) under inert conditions and cooled down to -78 °C. Then *n*-BuLi (3.75 mL of 1.6 mol solution in hexanes), 6 mmol) was added dropwise over a period of 10 min. Afterwards the reaction mixture was stirred for 2 h at -78 °C before DMF (462 μL, 6 mmol) was added at once. After stirring at low temperature for another 30 minutes the cooling

bath was removed and the solution was allowed to reach room temperature. Saturated NH$_4$Cl$_{aq}$ solution (10 mL) and CH$_2$Cl$_2$ (20 mL) were added and the water layer was extracted with CH$_2$Cl$_2$ (50 mL) three times. After drying over MgSO$_4$ the solvent was evaporated and the yellow residue was purified by column chromatography on silica (hexanes/Et$_2$O 9:1).
Yield: 667 mg (87 %).

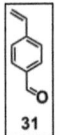

¹H NMR (CDCl$_3$, 400 MHz, RT): δ [ppm] = 10.00 (s, 1H, HC=O), 7.80 (d, ^3J = 8.2 Hz, 2H, PhH), 7.50 (d, ^3J = 8.2 Hz, 2H, PhH), 6.7 (dd, ^3J = 17.6 Hz, 10.9 Hz, 1H, CHCH$_2$), 5.90 (d, ^3J = 17.6 Hz, 1H, CHCH$_2$), 5.40 (d, ^3J = 10.9 Hz, 1H, CHCH$_2$).

¹³C-NMR (CDCl$_3$, 100.5 MHz, RT):
δ [ppm] = 192.15 (1C, HC=O), 143.83 (1C, PhC), 136.26 (1C, CHCH$_2$), 136.03 (1C, CHCH$_2$), 130.49 (2C, PhC), 127.13 (2C, PhC), 117.88 (1C, PhC).

EA: calculated for C$_9$H$_8$O (138.47): C 80.89, H 6.50; found: C 80.98, H 6.86.

N,N'-Bis[6-(3,3-dimethylbutyrylamino)pyridin-2-yl]-5-ethenyl-phenyl-4-formyl-isophthalamide (32)
Iodo Hamilton receptor **19** (250 mg, 0.37 mmol) was dissolved in dry THF (10 mL) and NEt$_3$ (5 mL) before Pd$_2$dba$_3$ (4 mg, 0.01 mmol) and AsPh$_3$ (1 mg, 0.01 mmol) were added. 4-Formyl-styrene **31** (64 mg, 0.49 mmol) was added and the reaction mixture stirred at 70 °C for 3 d. After cooling down to room temperature the solvent was evaporated and the crude product purified by column chromatography (CH$_2$Cl$_2$/THF 4:1).
Yield: 100 mg (40%).

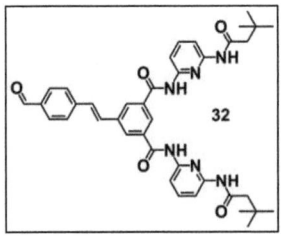

¹H NMR (THF d$_8$, 400 MHz, RT):
δ [ppm] = 10.87 (s, 1H, HC=O), 9.98 (bs, 2H, NH), 9.66 (bs, 2H, NH), 9.02 (s, 1H, PhH), 8.39 (s, 2H, PhH), 8.03 (d, ^3J = 8.07 Hz, 2H, PhH), 8.01 (d, ^3J = 8.17 Hz, 2H, PhH), 7.93 (d, ^3J = 8.13 Hz, 1H, CH), 7.82 (d, ^3J = 8.12 Hz, 1H, CH), 7.73 (m, 4H, pyH), 7.55 (m, 2H, pyH), 2.61 (s,

4H, CH_2), 1.09 (s, 18H, CH_3).

^{13}C NMR (THF d$_8$, 100.5 MHz, RT):

δ [ppm] = 191.29 (1C, HC=O), 170.80 (2C, C=O), 165.42 (2C, C=O), 151.71 (2C, α-pyC), 151.28 (2C, α-pyC), 143.51 (1C, CH), 140.54 (1C, CH), 138.76 (1C, PhC), 136.78 (2C, pyC), 130.60 (1C, PhC), 129.75 (4C, PhC), 127.90 (2C, PhC), 126.77 (1C, PhC), 129.32 (1C, PhC), 123.99 (2C, PhC), 110.30 (2C, β-pyC), 110.05 (2C, β-pyC), 50.72 (2C, CH$_2$), 29.98 (2C, C(CH$_3$)$_3$), 25.20 (6C, CH$_3$).

MS (FAB, NBA): m/z = 675 [M]$^+$.

FT-IR (diamond, RT): $\tilde{\nu}$ [cm^{-1}] = 3284, 2955, 2357, 1685, 1580, 1543, 1501, 1452, 1306, 1238, 1155, 1127, 808, 765, 742, 725.

EA: calculated for C$_{39}$H$_{42}$O$_5$N$_6$*0.5 CH$_2$Cl$_2$ (717.26): C 66.14, H 6.04, N 11.72; found: C 65.77, H 6.50, N 11.70.

N,N-Bis(6-(3,3-dimethylbutyrylamino)pyridin-2-yl)-4'-formyl-biphenyl-3,5-dicarboxamide (34)

Compound **34** was accessible via GOP VII with iodo Hamilton receptor **19** (100 mg, 0.15 mmol), 4-formylphenylboronic acid **33** (27 mg, 0.18 mmol), Pd(PPh$_3$)$_4$ (9 mg, 0.01 mmol) and K$_2$CO$_3$ (79 mg, 0.35 mmol) in DMF (15 mL). Column chromatography on silica gel (CH$_2$Cl$_2$/EtOAc 4:1) of the residue leads to a white powder.
Yield: 51 mg (71 %).

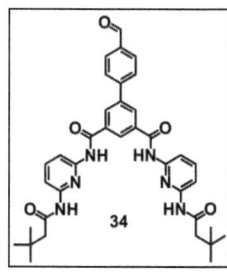

^{1}H NMR (THF d$_8$, 400 MHz, RT):

δ [ppm] = 10.06 (s, 1H, HC=O), 9.80 (s, 2H, NH), 9.14 (s, 2H, NH), 8.52 (s, 1H, PhH), 8.45 (s, 2H, PhH), 8.02 (d, ^3J = 6.61 Hz, 2H, PhH), 8.00 (d, ^3J = 6.61 Hz, 2H, PhH), 7.97 (m, 4H, pyH), 7.71 (m, 2H, pyH), 2.27 (s, 4H, CH$_2$), 1.08 (s, 18H, CH$_3$).

Experimental Part

^{13}C NMR (THF d$_8$, 100.5 MHz, RT):
δ [ppm] = 192.09 (1C, HC=O), 171.14 (2C, C=O), 165.67 (2C, C=O), 151.92 (2C, α-pyC), 151.45 (2C, α-pyC), 145.75 (1C, PhC), 141.24 (2C, PhC), 140.83 (2C, pyC), 137.17 (2C, PhC), 131.07 (1C, PhC), 130.77 (2C, PhC), 128.71 (2C, PhC), 127.78 (1C, PhC), 120.13 (1C, PhC), 111.53 (2C, β-pyC), 111.11 (2C, β-pyC), 50.97 (2C, CH$_2$), 31.92 (2C, C(CH$_3$)$_3$), 30.29 (6C, CH$_3$).

MS (FAB, NBA): m/z = 649 [M]$^+$.

IR (diamond, RT): $\tilde{\nu}$ [cm^{-1}] = 2959, 1680, 1606, 1585, 1512, 1446, 1298, 1243, 1157, 1134, 799, 750.

EA: calculated for C$_{37}$H$_{40}$N$_6$O$_5$ *(2H$_2$O) (684.75): C 64.93; H 6.48; N 12.27; found: C 65.45, H 6.45, N 12.09.

N,N-Bis(6-(3,3-dimethylbutyrylamino)pyridine-2-yl)-5-(5-formylthiophen-2-yl)isophthalamide (36)

Thiophene compound **36** was synthesized *via* GOP VI starting with iodo Hamilton receptor **19** (250 mg, 0.37 mmol), 4-formylthiophenboronic acid **35** (70 mg, 0.48 mmol), Pd(PPh$_3$)$_4$ (22 mg, 0.02 mmol) and K$_2$CO$_3$ (124 mg, 0.89 mmol) in DMF (10 mL). The reaction mixture was stirred for 24 h at room temperature. The residue was purified by column chromatography on silica gel (CH$_2$Cl$_2$/EtOAc 4:1).
Yield: 135 mg (56 %).

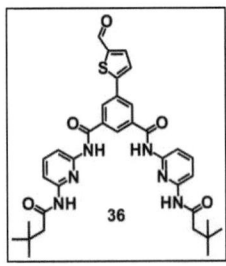

^{1}H NMR (DMSO d$_6$, 400 MHz, RT): δ [ppm] = 10.75 (s, 2H, N*H*), 10.05 (s, 2H, N*H*), 9.96 (s, 1H, *H*C=O), 8.54 (s, 1H, Ph*H*), 8.52 (s, 2H, Ph*H*), 8.12 (d, ^3J = 4.03 Hz, 1H, thio*H*), 7.99 (d, ^3J = 4.03 Hz, 1H, thio*H*), 7.86 (m, 4H, py*H*), 7.81 (m, 2H, py*H*), 2.30 (s, 4H, C*H*$_2$), 1.01 (s, 18H, C*H*$_3$).

^{13}C NMR (DMSO d$_6$, 100.5 MHz, RT):
δ [ppm] = 184.32 (1C, HC=O), 170.94 (2C, C=O), 164.58 (2C, C=O), 150.56 (2C, α-pyC), 150.02 (2C, α-pyC), 142.97 (1C, thioC), 140.02 (2C, pyC), 138.97 (1C, thioC),

135.35 (2C, Ph*C*), 132.89 (1C, thio*C*), 128.51 (2C, Ph*C*), 128.43 (2C, Ph*C*, thio*C*), 126.82 (1C, Ph*C*), 110.70 (2C, β-py*C*), 110.07 (2C, β-py*C*), 49.07 (2C, *C*H$_2$), 30.89 (2C, *C*(CH$_3$)$_3$), 29.57 (6C, *C*H$_3$).

MS (FAB, NBA): m/z = 654 [M]$^+$.

IR (diamond, RT): $\tilde{\nu}$ [cm^{-1}] = 2959, 1683, 1656, 1609, 1584, 1512, 1443, 1398, 1298, 1277, 1261, 1157, 804, 791, 764, 750.

EA: calculated for C$_{35}$H$_{38}$N$_6$O$_5$ * (0.5 DMSO) (678.78): C 61.88, H 5.60, N 12.38, S 6.90; found: C 61.68, H 5.79, N 11.92, S 6.18.

9,9-Dihexyl-2-formyl-7-Hamilton receptor-fluorene (40)

Fluorene bearing Hamilton receptor **40** was accessible *via* GOP VI using Iodo Hamilton receptor **19** (138 mg, 0.21 mmol), 9,9-dihexyl-2-formyl-7-pinacolyl-borano-fluorene **39** (100 mg, 0.21 mmol), Pd(PPh$_3$)$_4$ (8 mg, 0.01 mmol) and K$_2$CO$_3$ (57 mg, 0.41 mmol) in DMF (10 mL). The remainder was purified by column chromatography on silica gel (CH$_2$Cl$_2$/EtOAc 6:1).
Yield: 167 mg (88 %)

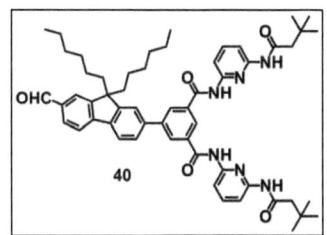

^{1}H NMR (THF d$_8$, 300 MHz, RT):

δ [ppm] = 10.06 (s, 1H, *H*C=O), 9.74 (s, 2H, N*H*), 8.99 (s, 2H, N*H*), 8.48 (s, 2H, Ph*H*), 8.46 (s, 1H, Ph*H*), 8.01 (m, 10H, py*H*, Ph*H*), 7.74 (m, 2H, py*H*), 2.27 (s, 4H, CH$_2$), 1.09 (m, 38H, hexyl*CH$_2$*, CH$_3$), 0.74 (t, ^{3}J = 7.15 Hz, 6H, hexyl C*H$_3$*).

^{13}C NMR (THF d$_8$, 75.5 MHz, RT):

δ [ppm] = 191.77 (1C, *H*C=O), 170.73 (2C, *C*=O), 165.71 (2C, *C*=O), 153.91 (1C, Ph*C*), 152.45 (1C, Ph*C*), 151.67 (2C, α-py*C*), 151.32 (2C, α-py*C*), 147.31 (1C, Ph*C*), 142.58 (1C, Ph*C*), 140.93 (2C, py*C*), 140.74 (1C, Ph*C*), 140.54 (1C, Ph*C*), 137.13 (1C, Ph*C*), 136.94 (1C, Ph*C*), 130.39 (1C, Ph*C*), 130.13 (1C, Ph*C*), 127.28 (1C, Ph*C*), 126.35 (1C, Ph*C*), 124.09 (1C, Ph*C*), 122.47 (1C, Ph*C*), 122.24 (1C, Ph*C*),

121.05 (1C, Ph*C*), 110.36 (2C, β-py*C*), 110.16 (2C, β-py*C*), 56.34 (1C, quat*C*), 50.72 (2C, *CH*$_2$), 40.08 (2C, hexyl*CH*$_2$), 32.33 (2C, hexyl*CH*$_2$), 31.63 (2C, hexyl*CH*$_2$), 30.46 (2C, *C*(CH$_3$)$_3$), 30.00 (6C, *CH*$_3$), 23.30 (2C, hexyl*CH*$_2$), 14.21 (2C, hexyl*CH*$_3$).

MS (FAB, NBA): m/z = 905 [M]$^+$.

IR (diamond, RT): $\tilde{\nu}$ [cm^{-1}] = 2930, 2859, 1685, 1605, 1585, 1505, 1446, 1296, 1242, 800, 748.

EA: calculated for C$_{56}$H$_{68}$N$_6$O$_5$ * 0.5 CH$_2$Cl$_2$ (947.65): C 71.61, H 8.34, N 8.87; found: C 71.99, H 8.44, N 8.51.

[*N,N'*-Bis[6-(3,3-dimethylbutyrylamino)pyridin-2-yl]-5-ethynyl-phenyl-isophthalamide]-*N*-methyl-pyrrolidine fullerene monoadduct (41)

Fullerene monoadduct **41** was synthesized *via* GOP VIII using Hamilton Receptor derivative **21** (100 mg, 0.15 mmol), C$_{60}$ (160 mg, 0.22 mmol) and sarcosine (16 mg, 0.18 mmol). The reaction mixture was refluxed for 22 h. The desired fulleropyrrolidine was eluated with toluene/EtOAc 2:1 (silica gel).
Yield: 115 mg (54 %).

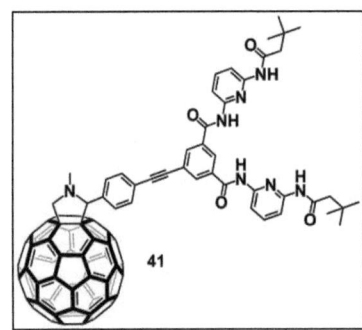

^{1}H NMR (CS$_2$, CDCl$_3$, THF d$_8$, 400 MHz, RT):
δ [ppm] = 9.45 (bs, 2H, N*H*), 8.79 (bs, 2H, N*H*), 8.37 (s, 1H, Ph*H*), 8.18 (d, ^3J = 6.61, 2H, Ph*H*), 7.93 (d, ^3J = 8.54, 4H, Ph*H*), 7.65 (m, 4 H, py*H*), 7.58 (d, ^3J = 8.42 Hz, 2H, py*H*), 5.00 (d, ^2J = 9.23 Hz, 1H, NC*H*$_2$), 4.97 (s, 1H, NC*H*), 4.28 (d, ^2J = 9.4 Hz, 1H, NC*H*$_2$) 2.81 (s, 4H, *CH*$_2$), 2.79 (s, 3H, NC*H*$_3$), 1.04 (s, 18 H, *CH*$_3$).

^{13}C NMR (CS$_2$, CDCl$_3$, THF d$_8$, 100.5 MHz, RT):
δ [ppm] = 170.77 (2C, *C*=O), 164.52 (2C, *C*=O), 151.08 (2C, α-py*C*), 150.62 (2C, α-py*C*), 147.88, 147.84, 147.09, 146.85, 146.78, 146.67, 146.50, 146.32, 146.12, 145.90, 145.83, 145.78, 145.30, 145.12, 144.91, 144.72, 143.57, 143.15, 142.85, 142.74, 142.68, 142.47, 142.13, 140.73, 140.51, 140.45, 138.88 (58C, sp^2*C* C$_{60}$),

Experimental Part

137.60 (2C, PhC), 137.05 (2C, PhC), 134.10 (2C, PhC), 132.53 (2C, PhC), 130.13 (2C, pyC), 129.28 (1C, PhC), 126.84 (1C, PhC), 124.80 (1C, NCH), 123.99 (1C, PhC), 110.47 (2C, β-pyC), 110.24 (2C, β-pyC), 91.69 (1C, ethynylC), 89.02 (1C, ethynylC), 77.52 (1C, NCH$_2$), 70.47 (1C, sp^{3}C C$_{60}$), 69.79 (1C, sp^{3}C C$_{60}$), 50.98 (2C, CH$_2$), 40.14 (1C, NCH$_3$), 31.59 (2C, C(CH$_3$)$_3$), 30.11 (6C, CH$_3$).

MS (FAB, NBA): m/z = 1421 [M]$^+$.

UV/Vis (CH$_2$Cl$_2$, RT): λ [nm] = 432.4, 311.0, 301.7, 255.6, 230.1.

FT-IR (diamond, RT): ṽ [cm^{-1}] = 3006, 2990, 2360, 1698, 1683, 1558, 1541, 1521, 1507, 1473, 1456, 1276, 1261, 764, 750.

[N,N'-Bis[6-(3,3-dimethylbutyrylamino)pyridin-2-yl]-5-ethynyl-phenyl-4-ethynylphenyl-isophthalamide] N-methyl-pyrrolidine fullerene monoadduct (42)
Choosing GOP VIII, fullerene derivative **42** was accessible using Hamilton Receptor compound **26** (40 mg, 0.05 mmol), C$_{60}$ (48 mg, 0.06 mmol) and sarcosine (6 mg, 0.06 mmol). Chromatography was done on silica gel (toluene → toluene/EtOAc 2:1). Yield: 35 mg (44 %).

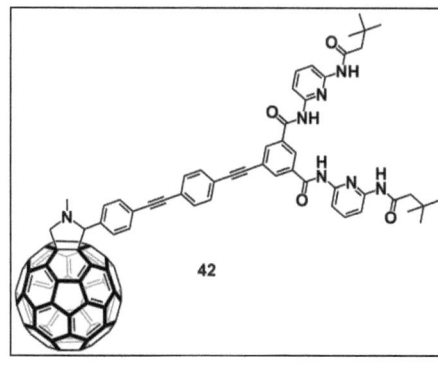

^{1}H NMR (CS$_2$, THF d$_8$, 400 MHz, RT): δ [ppm] = 9.42 (bs, 2H, NH), 8.78 (bs, 2H, NH), 8.37 (s, 3H, PhH), 8.36 (d, ^{3}J = 6.68 Hz, 4H, PhH, pyH), 8.19 (d, ^{3}J = 7.08 Hz, 2H, PhH), 8.17 (d, ^{3}J = 8.72 Hz, 2H, PhH), 7.95 (d, ^{3}J = 7.71 Hz, 2H, pyH), 7.94 (d, ^{3}J = 7.57 Hz, 2H, PhH), 7.92 (t, ^{3}J = 7.32 Hz, 2H, pyH), 5.02 (d, ^{2}J = 9.16 Hz, 1H, NCH$_2$), 5.01 (s, 1H, NCH), 4.31 (d, ^{2}J = 9.39 Hz, 1H, NCH$_2$), 2.19 (m, 7H, NCH$_3$, CH$_2$), 1.05 (s, 18 H, CH$_3$).

^{13}C-NMR (CS$_2$, THF d$_8$, 100.5 MHz, RT):
δ [ppm] = 169.85 (2C, *C*=O), 163.12 (2C, *C*=O), 150.91 (2C, α-py*C*), 150.28 (2C, α-py*C*), 147.60, 147.05, 146.85, 146.62, 146.53, 146.45, 146.25, 146.08, 145.86, 145.68, 145.63, 145.57, 144.93, 144.74, 143.33, 143.01, 142.93, 142.61, 142.42, 142.30, 142.15, 142.01, 141.90, 140.52, 140.47, 140.12, 138.21, (58C, sp^2*C* C$_{60}$), 139.92 (2C, Ph*C*), 138.83 (2C, py*C*), 137.07 (2C, Ph*C*), 136.86 (4C, Ph*C*), 135.86 (2C, Ph*C*), 133.90 (2C, Ph*C*), 132.41 (2C, Ph*C*), 132.18 (2C, Ph*C*), 132.09 (1C, N*C*H), 131.99 (1C, Ph*C*), 129.70 (2C, Ph*C*), 123.49 (1C, Ph*C*), 110.18 (2C, β-py*C*), 109.83 (2C, β-py*C*), 94.68 (1C, ethynyl*C*), 91.30 (1C, ethynyl*C*), 90.24 (1C, ethynyl*C*), 83.41 (1C, ethynyl*C*), 77.60 (1C, N*C*H$_2$), 70.20 (1C, sp^3*C* C$_{60}$), 69.51 (1C, sp^3*C* C$_{60}$), 50.43 (2C, *C*H$_2$), 39.88 (1C, N*C*H$_3$), 31.17 (2C, *C*(CH$_3$)$_3$), 29.86 (6C, *C*H$_3$).

MS (FAB, NBA): m/z = 1519 [M]$^+$.

UV/Vis (CH$_2$Cl$_2$, RT): λ [nm] = 427.2, 308.5, 306.0, 302.5, 237.0.

FT-IR (diamond, RT): $\tilde{\nu}$ [cm^{-1}] = 2957, 2360, 2342, 1698, 1683, 1671, 1545, 1558, 1541, 1521, 1508, 1446, 1298, 1276, 1260, 1242, 799, 763, 750.

N,N-Bis(6-(3,3-dimethylbutanamido)pyridin-2-yl)-5-(3-prop-1-ynyl)isophthalamide-N-methyl-pyrrolidine-fullerene monoadduct (43)

Component **43** was prepared by GOP VIII with Hamilton receptor **29** (50 mg, 0.08 mmol), C$_{60}$ (91 mg, 0.13 mmol) and sarcosine (11 mg, 0.13 mmol). Eluation of the product was achieved with toluene/EtOAc 2:1.
Yield: 45 mg (42 %).

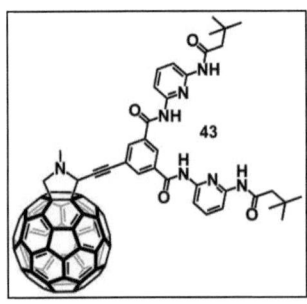

^{1}H NMR (CS$_2$/THF d$_8$, 400 MHz, RT): δ [ppm] = 10.65 (s, 2H, N*H*), 9.46 (s, 2H, N*H*), 8.74 (s, 1H, Ph*H*), 8.38 (s, 2H, Ph*H*), 7.93 (d, ^3J = 8.06 Hz, py*H*), 7.90 (d, ^3J = 8.05 Hz, 2H, py*H*), 7.65 (t, ^3J = 8.06 Hz, py*H*), 5.25 (s, 1H, N*C*H), 4.82 (d, ^2J = 9.40 Hz, 1H, N*C*H$_2$), 4.31 (d, ^2J = 9.40 Hz, 1H, N*C*H$_2$), 3.09 (s, 3H, N*C*H$_3$), 2.31 (s, 4H, *C*H$_2$), 1.02 (s, 18H, *C*H$_3$).

Experimental Part

^{13}C-NMR (CS$_2$/THF d$_8$, 100.5 MHz, RT):

δ [ppm] = 169.71 (2C, C=O), 163.62 (2C, C=O), 150.90 (2C, α-pyC), 150.32 (2C, α-pyC), 147.66, 147.61, 147.07, 146.80, 146.62, 146.57, 146.49, 146.41, 146.36, 146.31, 146.02, 145.93, 145.86, 145.80, 145.70, 145.60, 145.57, 145.55, 144.92, 144.80, 143.33, 142.99, 142.93, 142.55, 142.46, 142.41, 142.28, 142.05, 140.53, 140.48, 140.42 (58C, sp^{2}C C$_{60}$), 140.09 (2C, pyC), 136.89 (2C, PhC), 136.74 (2C, PhC), 134.28 (1C, PhC), 126.90 (1C, NCH), 123.50 (1C, PhC), 110.07 (2C, β-pyC), 109.78 (2C, β-pyC), 89.94 (1C, ethynylC), 87.96 (1C, ethynylC), 75.36 (1C, NCH$_2$), 70.22 (1C, sp^{3}C C$_{60}$), 69.95 (1C, sp^{3}C C$_{60}$), 50.42 (1C, NCH$_3$), 39.34 (2C, CH$_2$), 31.12 (2C, C(CH$_3$)$_3$), 29.84 (6C, CH$_3$).

MS (FAB, NBA): m/z = 1344 [M]$^+$.

UV/Vis (CH$_2$Cl$_2$, rt): λ (nm) = 429.7, 303.9, 255.2, 232.5.

FT-IR (diamond, RT): $\tilde{\nu}$ [cm^{-1}] = 2989, 2360, 2342, 1698, 1684, 1541, 1521, 1507, 1447, 1275, 1261, 765, 750.

[N,N'-Bis[6-(3,3-dimethylbutyrylamino)pyridin-2-yl]-5-ethenyl-phenyll-isophthalamide]-N-methyl-pyrrolidine fullerene monoadduct (44)

Monoadduct **44** could be synthesized *via* GOP VIII using C$_{60}$ (80 mg, 0.11 mmol), Hamilton Receptor derivative **32** (50 mg, 0.07 mmol) and sarcosine (8 mg, 0.09 mmol). The reaction mixture was refluxed for 20 h. Purification of the desired product was done with a silica-gel column (toluene → toluene/EtOAc 2:1).
Yield: 50 mg (47 %).

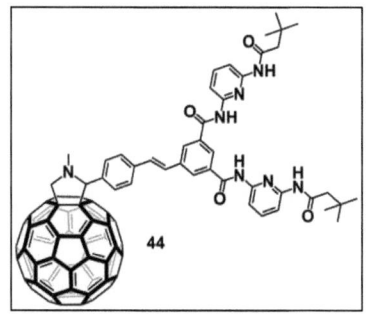

^{1}H NMR (CS$_2$, THF d$_8$, 400 MHz, RT):

δ [ppm] = 9.46 (bs, 2H, NH), 8.88 (bs, 2H, NH), 8.26 (s, 1H, PhH), 8.24 (s, 2H, PhH), 7.96 (d, ^3J = 8.06 Hz, 2H, PhH), 7.94 (d, ^3J = 7.69 Hz, 2H, PhH), 7.65 (t, ^3J = 8.06 Hz, 2H, pyH), 7.63 (d, ^3J = 8.13 Hz, 4H, pyH), 7.34 (d, ^3J = 6.60 Hz, 2H, pyH), 7.23 (d, ^3J = 7.88 Hz, 1H, CH), 7.15 (d, ^3J

= 7.33 Hz, 1H, C*H*), 5.01 (s, 1H, NC*H*), 4.95 (d, ^2J = 12.67 Hz, 1H, NC*H*$_2$), 4.31 (d, ^2J = 9.34 Hz, 1H, NC*H*$_2$), 2.82 (s, 3H, NC*H*$_3$), 2.48 (s, 4H, C*H*$_2$), 1.06 (s, 18 H, C*H*$_3$).

^{13}C NMR (CS$_2$, THF d$_8$, 100.5 MHz, RT):

δ [ppm] = 169.96 (2C, *C*=O), 165.59 (2C, *C*=O), 151.11 (2C, α-py*C*), 150.67 (2C, α-py*C*), 147.68, 147.29, 147.01, 146.75, 146.68, 146.59, 146.33, 146.19, 146.03, 145.90, 145.77, 145.66, 145.59, 145.54, 144.99, 144.84, 144.78, 143.41, 143.09, 142.99, 142.72, 142.49, 142.38, 142.09, 140.59, 140.30, 140.17, 139.88, 138.83 (58C, sp^2*C* C$_{60}$), 143.56 (1C, C*H*), 140.54 (1C, Ph*C*), 137.71 (1C, C*H*), 136.96 (2C, py*C*), 130.95 (2C, Ph*C*), 130.21 (1C, NC*H*), 129.15 (4C, Ph*C*), 128.55 (2 C, Ph*C*), 127.82 (1C, Ph*C*), 127.58 (1C, Ph*C*), 125.55 (1C, Ph*C*), 110.01 (2C, β-py*C*), 109.75 (2C, β-py*C*), 77.87 (1C, NC*H*$_2$), 70.31 (1C, sp^3*C* C$_{60}$), 69.67 (1C, sp^3*C* C$_{60}$), 50.49 (2C, *C*H$_2$), 39.89 (1C, NC*H*$_3$), 31.24 (2C, *C*(CH$_3$)$_3$), 29.89 (6C, *C*H$_3$).

MS (FAB, NBA): m/z = 1423 [M]$^+$.

UV/Vis (CH$_2$Cl$_2$, RT): λ [nm] = 432.4, 310.6, 301.8, 255.6, 230.0.

FT-IR (diamond, RT): $\tilde{\nu}$ [cm^{-1}] = 3008, 2987, 2360, 1692, 1680, 1572, 1539, 1517, 1502, 1479, 1455, 1278, 1262, 763, 751.

***N,N*-Bis(6-(3,3-dimethylbutyrylamino)pyridin-2-yl)-4'-biphenyl-3,5-dicarboxamide)-*N*-methyl-pyrrolidine-fullerene monoadduct (45)**

The synthesis of fulleropyrrolidine **45** was achieved *via* GOP VIII with Hamilton receptor compound **34** (50 mg, 0.08 mmol), C$_{60}$ (116 mg, 0.16 mmol) and sarcosine (15 mg, 0.16 mmol). The reaction mixture was refluxed for 22 h. The product was isolated by eluation with toluene/EtOAc 2:1.

Yield: 40 mg (36 %).

^{1}H NMR (CS$_2$/THF d$_8$, 400 MHz, RT): δ [ppm] = 9.46 (s, 2H, N*H*), 9.14 (s, 2H, N*H*), 8.37 (s, 1H, Ph*H*), 8.34 (s, 2H, Ph*H*), 7.97 (d, ^3J = 7.94 Hz, 2H, Ph*H*), 7.95 (m, ^3J = 7.95 Hz, 2H, Ph*H*), 7.88 (m, 4H, py*H*), 7.61 (m, 2H, py*H*), 5.04 (s, 1H, NC*H*), 5.03 (d,

^2J = 9.52 Hz, 1H, NCH$_2$), 4.32 (d, ^2J = 9.4 Hz, 1H, NCH$_2$), 2.82 (s, 3H, NCH$_3$), 2.16 (s, 4H, CH$_2$), 1.04 (s, 18H, CH$_3$).

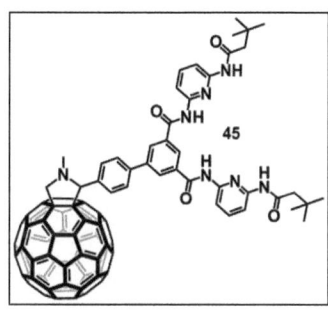

^{13}C NMR (CS$_2$/THF d$_8$, 100.5 MHz, RT):

δ [ppm] = 169.69 (2C, C=O), 164.22 (2C, C=O), 156.73 (2C, α-pyC), 154.39 (2C, α-pyC), 150.85 (2C, pyC), 150.38 (1C, PhC), 147.54(2C, PhC), 147.51, 146.98, 146.78, 146.54, 146.45, 146.36, 146.19, 145.99, 145.84, 145.77, 145.63, 145.57, 145.52, 145.42, 145.39, 145.01, 144.80, 144.69, 144.61, 143.41, 143.29, 142.96, 142.87, 142.82, 142.55, 142.44, 142.40, 142.32, 142.18, 142.18, 142.01, 141.97, 141.81, 141.33, 140.48, 140.45, 140.18, 140.04, 139.73, 139.63 (58C, sp^{2}C C$_{60}$), 137.31 (2C, PhC), 135.92 (1C, PhC), 130.42 (2C, PhC), 129.54 (2C, PhC), 127.66 (1C, PhC), 125.58 (1C, PhC), 123.61 (1C, NCH), 109.91 (2C, β-pyC), 109.64 (2C, β-pyC), 83.43 (1C, NCH$_2$), 70.21 (1C, sp^{3}C C$_{60}$), 69.46 (1C, sp^{3}C C$_{60}$), 50.43 (1C, NCH$_3$), 39.91 (2C, CH$_2$), 31.10 (2C, C(CH$_3$)$_3$), 29.84 (6C, CH$_3$).

MS (FAB, NBA): m/z = 1396 [M]$^+$.

UV/Vis (CH$_2$Cl$_2$, rt): λ (nm) = 430.6, 303.0, 255.2, 227.0.

FT-IR (diamond, RT): $\tilde{v}$ [cm^{-1}] = 2989, 2360, 2342, 1698, 1684, 1541, 1521, 1508, 1447, 1276, 1261, 764, 750.

N,N-Bis(6-(3,3-dimethylbutyrylamino)pyridine-2-yl)-5-(5-thiophen-2-yl)isophthalamide)-N-methyl-pyrrolidine-fullerene monoadduct (46)

Thiophene derivative **46** was prepared according to GOP VIII using Hamilton receptor **36** (40 mg, 0.06 mmol), C$_{60}$ (65 mg, 0.09 mmol) and sarcosine (8 mg, 0.09 mmol). The solution was heated for 20 h under reflux. Eluation of the product was done with toluene/EtOAc 2:1.

Yield: 41 mg (49 %).

Experimental Part

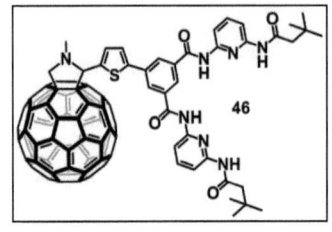

¹H NMR (CS$_2$/THF d$_8$, 400 MHz, RT):

δ [ppm] = 9.46 (s, 2H, N*H*), 8.82 (s, 2H, N*H*), 8.30 (s, 3H, Ph*H*), 7.95 (d, ^3J = 8.06 Hz, 2H, py*H*), 7.91 (d, ^3J = 8.06 Hz, 2H, py*H*), 7.63 (m, 2H, py*H*), 7.50 (d, ^3J = 4.03 Hz, 1H, thio*H*), 7.44 (d, ^3J = 4.03 Hz, 1H, thio*H*), 5.32 (s, 1H, NC*H*), 5.02 (d, ^2J = 9.52 Hz, 1H, NC*H*$_2$), 4.31 (d, ^2J = 9.52 Hz, 1H, NC*H*$_2$), 2.95 (s, 3H, NC*H*$_3$), 2.17 (s, 4H, C*H*$_2$), 1.05 (s, 18H, C*H*$_3$).

¹³C NMR (CS$_2$/THF d$_8$, 100.5 MHz, RT):

δ [ppm] = 169.71 (2C, *C*=O), 164.04 (2C, *C*=O), 150.91 (2C, α-py*C*), 150.42 (2C, α-py*C*), 147.59, 147.57, 147.15, 146.64, 146.60, 146.58, 146.49, 146.37, 146.22, 146.02, 145.90, 145.85, 145.80, 145.75, 145.67, 145.58, 145.54, 145.51, 145.43, 145.02, 144.85, 144.68, 144.62, 143.78, 143.44, 143.30, 143.01, 142.90, 142.53, 142.48, 142.39, 144.35, 142.28, 142.15, 141.98, 140.46 (58C, sp^2*C* C$_{60}$), 140.06 (1C, thio*C*), 139.93 (2C, py*C*), 137.52 (1C, thio*C*), 136.88 (2C, Ph*C*), 132.94 (1C, thio*C*), 129.68 (2C, Ph*C*), 127.86 (2C, Ph*C*, thio*C*), 125.46 (1C, Ph*C*), 124.49 (1C, N*C*H), 109.95 (2C, β-py*C*), 109.68 (2C, β-py*C*), 79.48 (1C, N*C*H$_2$), 70.18 (1C, sp^3*C* C$_{60}$), 69.17 (1C, sp^3*C* C$_{60}$), 50.41 (1C, N*C*H$_3$), 40.29 (2C, *C*H$_2$), 31.11 (2C, *C*(CH$_3$)$_3$), 29.85 (6C, *C*H$_3$).

MS (FAB, NBA): m/z = 1401 [M]$^+$.

UV/Vis (CH$_2$Cl$_2$, rt): λ (nm) = 429.7, 304.7, 255.2, 234.6.

FT-IR (diamond, RT): $\tilde{\nu}$ [cm^{-1}] = 3006, 2989, 2360, 2341, 1698, 1684, 1541, 1521, 1508, 1473, 1447, 1276, 1261, 797, 764, 750.

Fluorene Hamilton receptor fullerene monoadduct (47)

Fluorene fullerene **47** was accessible with GOP VIII using Hamilton receptor **40** (32 mg, 0.04 mmol), C$_{60}$ (40 mg, 0.06 mmol) and sarcosine (5 mg, 0.06 mmol). The reaction mixture was heated at 110 °C for 14 h. The fullerene monoadduct could be eluated with toluene/EtOAc 2:1).

Yield: 25 mg (43 %).

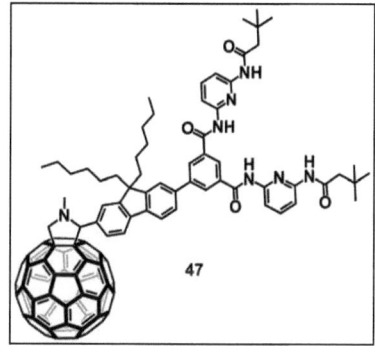

¹H NMR (CDCl₃, 400 MHz, RT):

δ [ppm] = 9.71 (s, 2H, NH), 9.00 (s, 2H, NH), 8.44 (s, 2H, PhH), 8.42 (s, 1H, PhH), 8.05 (m, 8H, PyH, PhH), 7.87 (m, 2H, PyH), 7.74 (m, 2H, PhH), 5.16 (s, 1H, NCH), 5.11 (d, ^2J = 9.42 Hz, 1H, NCH_2), 4.38 (d, ^2J = 9.42 Hz, 1H, NCH_2), 2.87 (s, 3H, NCH_3), 2.25 (s, 4H, CH_2), 1.08 (m, 38H, hexylCH_2, CH_3), 0.77 (m, 6H, hexylCH_3).

¹³C NMR (CDCl₃, 100.5 MHz, RT):

δ [ppm] = 170.66 (2C, C=O), 165.68 (2C, C=O), 157.82 (1C, PhC), 155.45 (1C, PhC), 151.69 (2C, α-py$C$), 151.35 (2C, α-py$C$), 148.04 (1C, PhC), 147.88, 147.53, 147.39, 147.04, 146.84, 146.68, 146.57, 146.39, 146.25, 146.03, 145.87, 145.49, 145.21, 145.13, 143.85, 143.44, 143.31, 143.07, 142.94, 142.87, 142.66, 142.43, 142.05, 140.79, 140.52, 140.14, 140.11, 140.05, 139.92, 139.84, 139.56, 139, 47 (58C, sp²C C_{60}), 139.39 (1C, PhC), 137.69 (2C, py$C$), 137.40 (1C, PhC), 136.92 (1C, PhC), 136.83 (1C, PhC), 136.66 (2C, PhC), 129.92 (1C, PhC), 126.90 (1C, PhC), 126.03 (1C, PhC), 124.95 (1C, NCH), 124.90 (1C, PhC), 124.09 (1C, PhC), 122.15 (1C, PhC), 121.16 (1C, PhC), 121.05 (1C, PhC), 110.29 (2C, β-py$C$), 110.13 (2C, β-py$C$), 84.47 (1C, NCH_2), 78.62 (1C, sp³C C_{60}), 70.57 (1C, sp³C C_{60}), 56.21 (1C, quat$C$), 50.75 (1C, NCH_3), 41.17 (2C, CH_2), 39.73 (2C, hexylCH_2), 32.52 (2C, hexylCH_2), 31.61 (2C, hexylCH_2), 30.65 (2C, C(CH₃)₃), 29.98 (6C, CH_3), 23.41 (2C, hexylCH_2), 14.26 (2C, hexylCH_3).

MS (FAB, NBA): m/z = 1653 [M]⁺.

UV/Vis (THF, rt): λ (nm) = 427.1, 304.7, 255.6, 232.1.

FT-IR (diamond, RT): $\tilde{v}$ [cm⁻¹] = 3006, 2989, 2360, 1648, 1541, 1521, 1508, 1448, 1276, 1261, 764, 750.

3,5-Bis(benzyloxy)benzyl-4-(*tert*-butoxycarbonylamino)butanoate (68)

Compound **68** was synthesized with GOP XI using Boc-GABA **66** (2 g, 9.8 mmol), DMAP (1.2 g, 9.8 mmol), 1-HOBT (1.46 mg, 10.78 mmol), DCC (2.83 g, 13.72 mmol) and (3,5-bis(benzyloxy)phenyl)methanol **67** (3.92 g, 12.25 mmol). The remainder was purified by flash chromatography on silica gel (hexanes/EtOAc 3:1).
Yield: 4.54 g (92 %)

^{1}H NMR (CDCl$_3$, 300 MHz, RT):

δ [ppm] = 7.42 (m, 10H, Ph*H*), 6.60 (s, 3H, Ph*H*), 5.06 (s, 2H, C*H$_2$*Ph), 5.05 (s, 4H, C*H$_2$*Ph), 4.64 (bs, 1H, N*H*), 3.17 (t, ^3J = 6.40 Hz, 2H, NHC*H$_2$*), 2.41 (t, ^3J = 7.44 Hz, 2H, NHCH$_2$CH$_2$C*H$_2$*), 1.84 (m, 2H, NHCH$_2$C*H$_2$*CH$_2$), 1.45 (s, 9H, C*H$_3$*).

^{13}C NMR (CDCl$_3$, 75.5 MHz, RT): δ [ppm] = 173.38 (1C, *C*=O), 160.49 (2C, Ph*C*), 156.32 (1C, *C*=O), 138.57 (1C, Ph), 137.11 (2C, Ph), 129.01 (4C, Ph*C*), 128.44 (2C, Ph*C*), 127.93 (4C, Ph*C*), 107.46 (2C, Ph*C*), 102.15 (1C, Ph*C*), 79.63 (1C, *C*(CH$_3$)$_3$), 70.51 (2C, *C*H$_2$Ph), 65.56 (1C, *C*H$_2$Ph), 40.27 (1C, *C*H$_2$), 31.90 (1C, *C*H$_2$), 28.80 (3C, *C*H$_3$), 21.46 (1C, *C*H$_2$).

MS (MALDI, DHB): m/z = 528 [M+Na]$^+$.

FT-IR (diamond, RT): $\tilde{\nu}$ [cm^{-1}] = 3003, 2362, 2342, 1675, 1598, 1454, 1277, 1261, 1151, 1060, 838, 802, 750, 723.

EA: calculated for C$_{30}$H$_{35}$NO$_6$ (505.60): C 71.27, H 6.98, N 2.77; found: C 71.39, H 7.07, N 2.76.

3,5-Bis(benzyloxy)benzyl-4-(butanylammonium) triflate (69)

Ammonium salt **69** was prepared by using GOP VIb using Frechet dendron **68** (4.35 g, 8.6 mmol).
Yield: 5.69 g (100 %).

¹H NMR (CDCl₃, 300 MHz, RT):

δ [ppm] = 7.74 (bs, 3H, NH_3), 7.40 (m, 10H, PhH), 6.59 (s, 3H, PhH), 5.31 (s, 2H, CH_2Ph), 5.01 (s, 4H, CH_2Ph), 3.10 (m, 2H, NH₃CH₂CH_2CH₂), 2.55 (t, ^3J = 6.50 Hz, 2H, NH₃CH₂CH₂CH_2), 1.98 (t, ^3J = 6.40 Hz, NH₃CH_2CH₂CH₂).

¹³C NMR (CDCl₃, 75.5 MHz, RT): δ [ppm] = 174.33 (1C, C=O), 160.51 (2C, PhC), 137.69 (1C, PhC), 136.99 (2C, PhC), 129.01 (4C, PhC), 128.47 (2C, PhC), 127.93 (4C, PhC), 125.70 (1C, CF₃), 107.60 (2C, PhC), 102.34 (1C, PhC), 70.51 (2C, CH₂Ph), 67.50 (1C, CH₂Ph), 40.30 (1C, CH₂), 32.04 (1C, CH₂), 22.41 (1C, CH₂).

MS (MALDI, DHB): m/z = 429 [M-triflate+Na]⁺, 406 [M-triflate]⁺.

FT-IR (diamond, RT): $\tilde{v}$ [cm⁻¹] = 3007, 2361, 2342, 1673, 1597, 1454, 1276, 1261, 1150, 1060, 837, 800, 750, 724.

(2R,3R)-Dibenzyl-2-(benzoyloxy)-3-(4-(*tert*-butoxycarbonyl)butanoyloxy) succinate (72)

Depsipeptide **72** was synthesized *via* GOP XI with DMAP (1.2 g, 9.8 mmol), 1-HOBT (1.46 g, 10.78 mmol), Boc-GABA **66** (2 g, 9.8 mmol), DCC (2.83 g, 13.72 mmol) and (R,R) 1,2-di(benzyloxy)carbonyl)-2-hydroxyethyl benzoate **71** (5.33 g, 12.25 mg). Purification of the crude product was done with column chromatography on silica gel (hexanes/EtOAc 3:1).
Yield: 4.57 g (75 %).

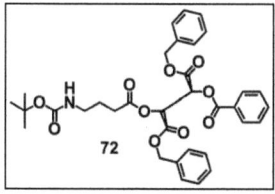

¹H NMR (CDCl₃, 300 MHz, RT):

δ [ppm] = 7.95 (m, 2H, PhH), 7.60 (m, 1H, PhH), 7.42 (t, ^3J = 7.63 Hz, 2H, PhH), 7.34 (m, 4H, PhH), 7.18 (m, 2H, PhH), 7.13 (m, 4H, PhH), 5.94 (d, ^3J = 2.83 Hz, 1H, CH), 5.82 (d, ^3J = 2.83 Hz, 1H, CH), 5.16 (m, 4H, CH_2Ph), 4.60 (bs, 1H, NH), 3.11 (t, ^3J = 6.43 Hz, 2H, NHCH_2), 2.31 (m, 2H, NHCH₂CH_2CH₂), 1.73 (m, 2H, NHCH₂CH_2CH₂), 1.45 (s, 9H, CH_3).

Experimental Part

¹³C NMR (CDCl₃, 75.5 MHz, RT):
δ [ppm] = 172.18 (1C, C=O), 166.04 (1C, C=O), 165.95 (1C, C=O), 165.42 (1C, C=O), 156.27 (1C, C=O), 134.23 (1C, PhC), 134.86 (1C, PhC), 134.05 (2C, PhC), 130.47 (4C, PhC), 129.04 (2C, PhC), 128.95 (2C, PhC), 128.84 (2C, PhC), 128.75 (4C, PhC), 71.57 (1C, C(CH₃)₃), 68.31 (2C, CH), 68.19 (2C, CH₂Ph), 39.91 (1C, CH₂), 31.06 (1C, CH₂), 28.81 (3C, CH₃), 25.35 (1C, CH₂).

MS (MALDI, SIN): m/z = 642 [M+Na]⁺.

FT-IR (diamond, RT): ṽ [cm⁻¹] = 2977, 2358, 1750, 1733, 1713, 1513, 1454, 1366, 1275, 1260, 1194, 1157, 1130, 1095, 1071, 1026, 1002, 963, 750, 715.

EA: calculated for C₃₄H₃₇NO₁₀ (619.66): C 65.90, H 6.02, N 2.26; found: C 65.56, H 6.06, N 2.19.

(2R,3R)-Dibenzyl-2-(benzoyloxy)-3-(4-(butanylammonium)butanoyloxy) succinate triflate (73)

Triflate **73** was accessible by GOP VI b using depsipeptide **72** (4.5 g, 7.25 mmol). Yield: 5.61 g (100 %).

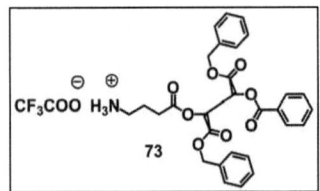

¹H NMR (CDCl₃, 300 MHz, RT):
δ [ppm] = 7.91 (m, 2H, PhH), 7.71 (bs, 3H, NH₃), 7.59 (m, 1H, PhH), 7.42 (m, 2H, PhH), 7.32 (m, 4H, PhH), 7.20 (m, 2H, PhH), 7.11 (m, 4H, PhH), 5.95 (d, ³J = 2.83 Hz, 1H, CH), 5.83 (d, ³J = 2.83 Hz, 1H, CH), 5.20 (m, 4H, CH₂Ph), 3.00 (m, 2H, NH₃CH₂CH₂CH₂), 2.37 (m, 2H, NH₃CH₂CH₂CH₂), 1.90 (t, ³J = 6.40 Hz, NH₃CH₂CH₂CH₂).

¹³C NMR (CDCl₃, 75.5 MHz, RT): δ [ppm] = 172.02 (1C, C=O), 166.13 (1C, C=O), 165.92 (1C, C=O), 165.37 (2C, C=O), 138.28 (1C, PhC), 135.15 (1C, PhC), 134.63 (2C, PhC), 130.39 (4C, PhC), 129.44 (2C, PhC), 129.12 (2C, PhC), 129.06 (2C, PhC), 128.92 (4C, PhC), 125.70 (1C, CF₃), 71.75 (1C, CH), 71.35 (1C, CH), 68.64 (2C, CH₂Ph), 39.76 (1C, CH₂), 30.79 (1C, CH₂), 22.48 (1C, CH₂).

MS (MALDI, SIN): m/z = 543 [M-triflate+Na]⁺, 521 [M-triflate]⁺.

FT-IR (diamond, RT): $\tilde{\nu}$ [cm⁻¹] = 3007, 2989, 2361, 2342, 1766, 1740, 1718, 1673, 1275, 1260, 1195, 1136, 1119, 764, 750, 708.

(3,5-Bis(benzyloxy)benzyl)-4-formyl benzoate (77)

Compound **77** was achieved according to GOP XI with formyl benzoic acid **75** (225 mg, 1.5 mmol), DMAP (202 mg, 1.65 mmol), 1-HOBT (223 mg, 1.65 mmol), DCC (619 mg, 3 mmol) and (3,5-bis(benzyloxy)phenyl)methanol **67** (576 mg, 1.80 mmol). The remaining yellow oil was purified using flash chromatography on silica gel (hexanes/EtOAc 4:1).
Yield: 567 mg (84 %).

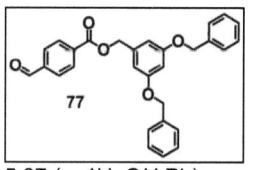

¹H NMR (CDCl₃, 300 MHz, RT):
δ [ppm] = 10.12 (s, 1H, HC=O), 8.22 (d, ³J = 8.29 Hz, 2H, PhH), 7.97 (d, ³J = 8.10 Hz, 2H, PhH), 7.42 (m, 10H, PhH), 6.70 (s, 2H, PhH), 6.63 (s, 1H, PhH), 5.33 (s, 2H, CH₂Ph), 5.07 (s, 4H, CH₂Ph).

¹³C NMR (CDCl₃, 75.5 MHz, RT): δ [ppm] = 192.03 (1C, HC=O), 165.73 (1C, C=O), 160.57 (2C, PhC), 139.63 (1C, PhC), 138.20 (3C, PhC), 137.04 (2C, PhC), 135.38 (2C, PhC), 129.03 (4C, PhC), 128.49 (3C, PhC), 127.94 (4C, PhC), 107.60 (2C, PhC), 102.25 (1C, PhC), 70.56 (2C, CH₂Ph), 67.51 (1C, CH₂Ph).

MS (MALDI, DHB): m/z = 475 [M+Na]⁺.

FT-IR (diamond, RT): $\tilde{\nu}$ [cm⁻¹] = 2360, 1722, 1706, 1596, 1452, 1372, 1341, 1296, 1272, 1204, 1160, 1101, 1042, 846, 758, 747.

EA: calculated for C₂₉H₂₄O₅ (452.50): C 76.98, H 5.35; found: C 76.95, H 5.41.

Experimental Part

(2R,3R)-Dibenzyl-2-(benzoyloxy)-3-(4-formylbenzoyloxy)succinate (78)
Depsipeptide **78** was synthesized by GOP XI with DMAP (887 mg, 7.26 mmol), 1-HOBT (981 mg, 7.26 mmol), 4-formyl benzoic acid **75** (1 g, 6.6 mmol), DCC (2.72 g, 13.2 mmol) and depsipeptide 1st generation alcohol **71** (3.48 g, 8 mmol). The remaining yellow oil was purified by flash chromatography on silica gel (hexanes/EtOAc 4:1).
Yield: 2.4 g (64 %).

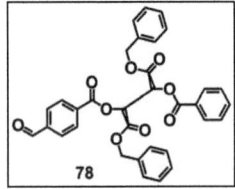

^{1}H NMR (CDCl$_3$, 300 MHz, RT):
δ [ppm] = 10.13 (s, 1H, HC=O), 8.10 (d, ^3J = 8.29 Hz, 2H, PhH), 7.99 (d, ^3J = 7.35 Hz, 2H, PhH), 7.92 (d, ^3J = 8.10 Hz, 2H, PhH), 7.62 (m, 1H, PhH), 7.44 (m, 2H, PhH), 7.21 (m, 4H, PhH), 7.13 (m, 2H, PhH), 7.08 (m, 4H, PhH), 6.06 (s, 4H, CH$_2$Ph), 5.27 (d, ^3J = 2.80 Hz, 1H, CH), 5.13 (d, ^3J = 2.80 Hz, 1H, CH).

^{13}C NMR (CDCl$_3$, 75.5 MHz, RT):
δ [ppm] = 191.94 (1C, HC=O), 165.88 (1C, C=O), 165.66 (1C, C=O), 165.42 (2C, C=O), 139.93 (2C, PhC), 134.96 (2C, PhC), 134.82 (2C, PhC), 134.14 (2C, PhC), 133.73 (2C, PhC), 130.48 (4C, PhC), 129.89 (2C, PhC), 128.90 (2C, PhC), 128.87 (2C, PhC), 128.76 (4C, PhC), 72.25 (1C, CH), 71.68 (1C, CH), 68.33 (2C, CH$_2$Ph).

MS (MALDI, SIN): m/z = 566 [M]$^+$.

FT-IR (diamond, RT): $\tilde{\nu}$ [cm^{-1}] = 2956, 1767, 1741, 1731, 1703, 1274, 1244, 1206, 1195, 1130, 1095, 755, 714.

EA: calculated for C$_{33}$H$_{26}$O$_9$ (566.55): C 69.96, H 4.63; found: C 69.87, H 4.68.

Newkome 1st G-4-formyl benzoate (79)
Newkome dendrimer **79** could be prepared according to GOP XI using DMAP (268 mg, 2.19 mmol), 1-HOBT (296 mg, 2.19 mmol), 4-formyl benzoic acid **75** (298 mg, 1.99 mmol), DCC (821 mg, 3.98 mmol) and di-*tert*-butyl 4-(2-*tert*-butoxycarbonyl)ethyl)-4-aminoheptanedioate **76** (1 g, 2.38 mmol). The crude product was purified using flash chromatography on silica gel (hexanes/EtOAc 3:1).

Yield: 768 mg (70 %).

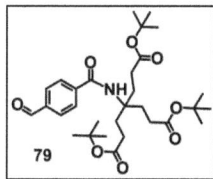

^{1}H NMR (CDCl$_3$, 400 MHz, RT): δ [ppm] = 10.08 (s, 1H, HC=O), 7.96 (d, ^3J = 8.30 Hz, 2H, PhH), 7.93 (d, ^3J = 6.60 Hz, 2H, PhH), 7.42 (s, 1H, NH), 2.33 (t, ^3J = 7.32 Hz, 2H, CH$_2$), 2.15 (t, ^3J = 7.81 Hz, 2H, CH$_2$), 1.43 (s, 27H, CH$_3$).

^{13}C NMR (CDCl$_3$, 100.5 MHz, RT):

δ [ppm] = 191.69 (1C, HC=O), 173.24 (3C, C=O), 165.54 (1C, C=O), 140.31 (1C, PhC), 137.97 (1C, PhC), 129.66 (2C, PhC), 127.75 (2C, PhC), 80.95 (3C, C(CH$_3$)$_3$), 58.11 (1C, NHC(CH$_2$CH$_2$)$_3$), 30.19 (3C, CH$_2$), 29.94 (3C, CH$_2$), 28.04 (9C, CH$_3$).

MS (MALDI, SIN): m/z = 571 [M+Na]$^+$.

FT-IR (diamond, RT): $\tilde{v}$ [cm^{-1}] = 2982, 2935, 2360, 2342, 1720, 1657, 1524, 1367, 1314, 1276, 1260, 1209, 1151, 848, 754.

EA: calculated for C$_{30}$H$_{45}$NO$_8$ (547.68): C 65.79, H 8.28, N 2.56; found: C 65.71, H 8.33, N 2.79.

N-Frechet 1stG-glycine-methylester (80)

Glycine methylester **80** was synthesized with GOP XII using Frechet Dendron **77** (250 mg, 0.55 mmol), NEt$_3$ (772 µL, 5.5 mmol), glycine methylester hydrochloride (729 mg, 5.23 mmol) and NaBH$_3$CN (70 mg, 1.1 mmol).
Yield: 232 mg (80 %).

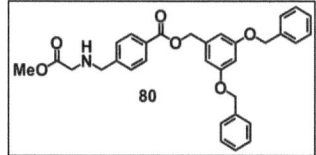

^{1}H NMR (CDCl$_3$, 300 MHz, RT):

δ [ppm] = 8.06 (d, ^3J = 8.35 Hz, 2H, PhH), 8.04 (d, ^3J = 8.15 Hz, 2H, PhH), 7.43 (m, 10H, PhH), 6.71 (s, 2H, PhH), 6.63 (s, 1H, PhH), 5.31 (s, 2H, CH$_2$Ph), 5.06 (s, 4H, CH$_2$Ph), 3.76 (s, 2H, NHCH$_2$Ph), 3.68 (s, 3H, CH$_3$), 3.46 (s, 2H, O=CCH$_2$NH), 1.90 (bs, 1H, NH).

¹³C NMR (CDCl₃, 75.5 MHz, RT): δ [ppm] = 172.63 (1C, *C*=O), 166.49 (1C, *C*=O), 160.41 (2C, Ph*C*), 138.75 (1C, Ph*C*), 138.68 (3C, Ph*C*), 137.02 (2C, Ph*C*), 130.30 (2C, Ph*C*), 128.90 (4C, Ph*C*), 128.33 (3C, Ph*C*), 127.85 (4C, Ph*C*), 107.29 (2C, Ph*C*), 102.00 (1C, Ph*C*), 70.41 (2C, *C*H₂Ph), 66.71 (1C, *C*H₂Ph), 52.22 (1C, *C*H₃), 51.90 (1C, *C*H₂), 50.85 (1*C*, *C*H₂).

MS (MALDI, DHB): m/z = 550 [M+Na]⁺, 526 [M]⁺.

FT-IR (diamond, RT): ṽ [cm⁻¹] = 2361, 1723, 1689, 1615, 1452, 1372, 1342, 1296, 1274, 1200, 1161, 1105, 1046, 846, 752.

EA: calculated for C₃₂H₃₁NO₆ (525.59): C 73.13, H 5.94, N 2.66; found: C 73.10, H 6.19, N 3.00.

N-Depsipeptide 1ˢᵗG-glycine-methylester (81)

Component **81** was accessible *via* GOP XII using depsipeptide **78** (250 mg, 0.44 mmol), NEt₃ (618 µL, 4.4 mmol), glycine methylester hydrochloride (585 mg, 4.19 mmol) and NaBH₃CN (56 mg, 0.88 mmol).
Yield: 233 mg (82 %).

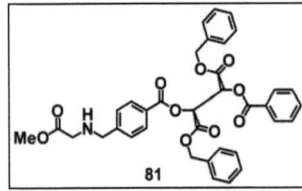

¹H NMR (CDCl₃, 300 MHz, RT):

δ [ppm] = 7.98 (t, ³*J* = 7.91 Hz, 2H, Ph*H*), 7.60 (m, 2H, Ph*H*), 7.42 (m, 2H, Ph*H*), 7.20 (m, 1H, Ph*H*), 7.13 (m, 6H, Ph*H*), 7.21 (m, 6H, Ph*H*), 6.04 (s, 4H, C*H*₂Ph), 5.22 (d, ³*J* = 2.80 Hz, 1H, C*H*), 5.15 (d, ³*J* = 2.80 Hz, 1H, C*H*), 3.90 (s, 2H, NHC*H*₂Ph), 3.76 (s, 3C, C*H*₃), 3.44 (s, 2C, O=CC*H*₂NH), 1.92 (bs, 1H, N*H*).

¹³C NMR (CDCl₃, 75.5 MHz, RT):

δ [ppm] = 173.82 (1C, *C*=O), 166.66 (1C, *C*=O), 166.14 (1C, *C*=O), 166.01 (2C, *C*=O), 145.76 (2C, Ph*C*), 135.63 (2C, Ph*C*), 134.69 (2C, Ph*C*), 131.38 (2C, Ph*C*), 131.15 (2C, Ph*C*), 130.79 (4C, Ph*C*), 129.51 (2C, Ph*C*), 129.34 (2C, Ph*C*), 129.19

(2C, Ph*C*), 128.49 (4C, Ph*C*), 72.51 (1C, *C*H), 72.46 (1C, *C*H), 68.92 (2C, *C*H$_2$Ph), 53.86 (1C, *C*H$_3$), 52.93 (1C, *C*H$_2$), 50.85 (1C, *C*H$_2$).

MS (MALDI, SIN): m/z = 662 [M+Na]$^+$, 640 [M]$^+$.

FT-IR (diamond, RT): $\tilde{v}$ [cm^{-1}] = 2954, 1765, 1731, 1454, 1316, 1245, 1197, 1179, 1130, 1096, 1071, 1056, 1020, 750, 715.

EA: calculated for C$_{36}$H$_{33}$NO$_{10}$ (639.65): C 67.60, H 5.20, N 2.19; found: C 67.11, H 5.34, N 2.17.

N-Newkome 1stG-glycine-methylester (82)

Compound **82** was synthesized using GOP XII with Newkome dendron **79** (250 mg, 0.45 mmol), NEt$_3$ (613 µL, 4.5 mmol), glycine methylester hydrochloride (601 mg, 4.3 mmol) and NaBH$_3$CN (57 mg, 0.90 mmol).
Yield: 271mg (98%).

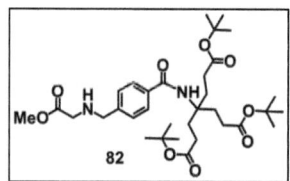

^{1}H NMR (CDCl$_3$, 400 MHz, RT): δ [ppm] = 7.75 (d, ^3J = 8.10 Hz, 2H, Ph*H*), 7.37 (d, ^3J = 8.10 Hz, 2H, Ph*H*), 7.27 (s, 1H, N*H*), 3.84 (s, 2H, NHC*H*$_2$Ph), 3.76 (s, 1H, N*H*), 3.72 (s, 3H, C*H*$_3$), 3.40 (s, 2H, O=CC*H*$_2$NH), 2.29 (t, ^3J = 7.54 Hz, 2H, C*H*$_2$), 2.11 (t, ^3J = 7.54 Hz, 2H, C*H*$_2$), 1.42 (s, 27H, C*H*$_3$).

^{13}C NMR (CDCl$_3$, 100.5 MHz, RT):
δ [ppm] = 173.53 (3C, *C*=O), 173.18 (1C, *C*=O), 166.94 (1C, *C*=O), 143.29 (1C, Ph*C*), 134.32 (1C, Ph*C*), 128.57 (2C, Ph*C*), 127.56 (2C, Ph*C*), 81.19 (3C, *C*(CH$_3$)$_3$), 58.15 (1C, N*H*C(CH$_2$CH$_2$)$_3$), 53.12 (1C, *C*H$_3$), 52.925 (1C, *C*H$_2$), 50.18 (1C, *C*H$_2$), 30.53 (3C, *C*H$_2$), 30.27 (3C, *C*H$_2$), 28.45 (9C, *C*H$_3$).

MS (MALDI, SIN): m/z = 644 [M+Na]$^+$, 621 [M]$^+$.

FT-IR (diamond, RT): $\tilde{v}$ [cm^{-1}] = 2980, 2935, 2361, 1724, 1638, 1540, 1368, 1312, 1276, 1259, 1224, 1153, 1101, 848, 751.

EA: calculated for $C_{33}H_{52}N_2O_9$ (620.77): C 63.81, H 8.44, N 4.51; found: C 63.81, H 8.29, N 5.28.

N-Frechet 1stG-glycine-*tert*-butylester (86)

Tert-butyl ester **86** was synthesized by applying GOP XII with formyl Frechet dendron **77** (250 mg, 0.55 mmol), NEt$_3$ (540 µL, 3.85 mmol), dry MeOH (3 mL), glycine *tert*-butylester hydrochloride (639 mg, 3.45 mmol) and NaBH$_3$CN (70 mg, 1.1 mmol). Yield: 226 mg (72 %).

^{1}H NMR (CDCl$_3$, 300 MHz, RT):

δ [ppm] = 8.06 (d, ^3J = 8.35 Hz, 2H, Ph*H*), 8.04 (d, ^3J = 8.15 Hz, 2H, Ph*H*), 7.43 (m, 10H, Ph*H*), 6.71 (s, 2H, Ph*H*), 6.63 (s, 1H, Ph*H*), 5.31 (s, 2H, C*H*$_2$Ph), 5.06 (s, 4H, C*H*$_2$Ph), 3.85 (s, 2H, NHC*H*$_2$Ph), 3.30 (s, 2C, O=CC*H*$_2$NH), 1.90 (bs, 1H, N*H*), 1.47 (s, 9H, C*H*$_3$).

^{13}C NMR (CDCl$_3$, 75.5 MHz, RT): δ [ppm] = 171.48(1C, *C*=O), 166.18 (1C, *C*=O), 160.02 (2C, Ph*C*), 145.17 (1C, Ph*C*), 138.33 (3C, Ph*C*), 136.64 (2C, Ph*C*), 129.86 (2C, Ph*C*), 128.79 (4C, Ph*C*), 128.53 (3C, Ph*C*), 127.95 (4C, Ph*C*), 106.88 (2C, Ph*C*), 101.63 (1C, Ph*C*), 80.27 (1C, *C*(CH$_3$)$_3$), 70.04 (2C, *C*H$_2$Ph), 66.35 (1C, *C*H$_2$Ph), 52.82 (1C, *C*H$_2$), 50.88 (1C, *C*H$_2$), 28.06 (3C, *C*H$_3$).

MS (MALDI, DHB): m/z = 590 [M+Na]$^+$.

FT-IR (diamond, RT): $\tilde{v}$ [cm^{-1}] = 1720, 1596, 1453, 1368, 1270, 1154, 1103, 1061, 1019, 85, 737, 697.

EA: calculated for $C_{35}H_7NO_6$ (567.26): C 74.05, H 6.57, N 2.47; found: C 73.62, H 6.27, N 2.52.

N-Depsipeptide 1ˢᵗG-glycine-*tert*-butylester (87)

Depsipeptide **87** was prepared according to GOP XII with formyl depsipeptide **78** (250 mg, 0.44 mmol), NEt$_3$ (427 µL, 3.08 mmol), glycine *tert*-butylester hydrochloride (500 mg, 2.76 mmol) and NaBH$_3$CN (56 mg, 0.88 mmol).
Yield: 322 mg (98%).

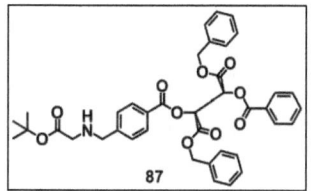

¹H NMR (CDCl$_3$, 300 MHz, RT):

δ [ppm] = 7.98 (t, ^3J = 7.91 Hz, 2H, Ph*H*), 7.60 (m, 2H, Ph*H*), 7.42 (m, 2H, Ph*H*), 7.20 (m, 1H, Ph*H*), 7.13 (m, 6H, Ph*H*), 7.21 (m, 6H, Ph*H*), 6.04 (s, 4H, C*H$_2$*Ph), 5.22 (d, ^3J = 2.80 Hz, 1H, C*H*), 5.15 (d, ^3J = 2.80 Hz, 1H, C*H*), 3.88 (s, 2H, NHC*H$_2$*Ph), 3.32 (s, 2H, O=CC*H$_2$*NH), 1.88 (bs, 1H, N*H*), 1.49 (s, 9H, C*H$_3$*).

¹³C NMR (CDCl$_3$, 75.5 MHz, RT):

δ [ppm] = 171.94 (1C, *C*=O), 166.66 (1C, *C*=O), 166.14 (1C, *C*=O), 166.01 (2C, *C*=O), 145.76 (2C, Ph*C*), 135.63 (2C, Ph*C*), 134.75 (2C, Ph*C*), 131.66 (2C, Ph*C*), 131.02 (2C, Ph*C*), 130.59 (4C, Ph*C*), 129.51 (2C, Ph*C*), 129.36 (2C, Ph*C*), 129.24 (2C, Ph*C*), 128.49 (4C, Ph*C*), 81.85 (1C, *C*(CH$_3$)$_3$), 72.51 (1C, *C*H), 72.46 (1C, *C*H), 68.92 (2C, *C*H$_2$Ph), 53.20 (1C, *C*H$_2$), 51.14 (1C, *C*H$_2$), 28.53 (3C, *C*H$_3$).

MS (MALDI, DCTB): m/z = 705 [M+Na]$^+$.

FT-IR (diamond, RT): $\tilde{\nu}$ [cm^{-1}] = 2954, 1765, 1730, 1454, 1318, 1246, 1192, 1179, 1130, 1095, 1070, 1056, 1024, 750, 714.

EA: calculated for C$_{39}$H$_{39}$NO$_{10}$ * 0.5 NEt$_3$ (731.82): C 66.98, H 6.18, N 2.80; found: C 66.46, H 6.22, N 3.69.

[*N*,*N'*-Bis[6-(3,3-dimethylbutyrylamino)pyridin-2-yl]-5-ethynyl-phenyl-isophthalamide]-pyrrolidine fullerene monoadduct (88)

Fulleropyrrolidine **88** could be synthesized according to GOP VIII using Hamilton receptor derivative **21** (45 mg, 0.07 mmol), C$_{60}$ (58 mg, 0.08 mmol) and glycine (6 mg, 0.08 mmol). The desired product could be isolated by eluation with toluene/EtOAc 1:1 (silica gel).

Yield: 62 mg (63 %).

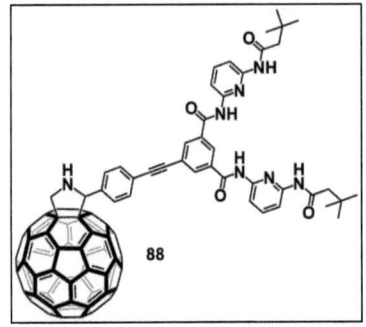

¹H NMR (CS$_2$, THF d$_8$, 400 MHz, RT):

δ [ppm] = 9.44 (bs, 2H, N*H*), 8.80 (bs, 2H, N*H*), 8.37 (s, 1H, Ph*H*), 8.18 (d, ^3J = 6.61 Hz, 2H, Ph*H*), 7.95 (m, 4H, Ph*H*), 7.66 (m, 4 H, py*H*), 7.60 (d, ^3J = 8.30 Hz, 2H, py*H*), 5.83 (s, 1H, NC*H*), 5.11 (d, ^2J = 10.01 Hz, 1H, NC*H*$_2$), 4.89 (d, ^2J = 10.25 Hz, 1H, NC*H*$_2$), 2.20 (s, 4H, C*H*$_2$), 1.07 (s, 18 H, C*H*$_3$).

¹³C NMR (CS$_2$, THF d$_8$, 100.5 MHz, RT):

δ [ppm] = 169.86 (2C, *C*=O), 163.83 (2C, *C*=O), 151.03 (2C, α-py*C*), 150.45 (2C, α-py*C*), 147.48, 147.24, 147.00, 146.66, 146.57, 146.54, 146.41, 146.37, 146.25, 145.80, 145.74, 145.62, 145.59, 145.49, 144.83, 144.70, 143.52, 143.36, 142.96, 142.90, 142.74, 142.70, 142.57, 141.90, 140.51, 140.46, 140.25, 139.93, 139.84, 139.75 (58C, sp^2*C* C$_{60}$), 137.32 (2C, Ph*C*), 136.53 (2C, Ph*C*), 134.05 (2C, Ph*C*), 132.52 (2C, Ph*C*), 132.20 (2C, py*C*), 129.78 (1C, Ph*C*), 128.84 (1C, Ph*C*), 126.62 (1C, Ph*C*), 124.46 (1C, NCH), 122.95 (1C, Ph*C*), 110.11 (2C, β-py*C*), 109.82 (2C, β-py*C*), 91.54 (1C, ethynyl*C*), 89.02 (1C, ethynyl*C*), 78.54 (1C, NC*H*$_2$), 73.44 (1C, sp^3*C* C$_{60}$), 67.46 (1C, sp^3*C* C$_{60}$), 50.45 (2C, *C*H$_2$), 31.19 (2C, *C*(CH$_3$)$_3$), 29.87 (6C, *C*H$_3$).

MS (MALDI, DCTB): m/z = 1408 [M]$^+$.

UV/Vis (CH$_2$Cl$_2$, RT): λ [nm] = 431.0, 306.5, 256.5, 234.5.

FT-IR (diamond, RT): $\tilde{\nu}$ [cm^{-1}] = 3006, 2990, 2360, 1869, 1845, 1793, 1733, 1683, 1585, 1540, 1447, 1297, 1260, 1154, 1020, 898, 826.

3,5-Bis(3,5-bis(benzyloxy)benzyloxybenzoyl chloride (90)

3,5-Bis(3,5-bis(benzyloxy)benzyloxybenzoic acid **89** (50 mg, 0.07 mmol) was suspended in oxalyl chloride (5 mL) and stirred at rt for 2h. The excessive oxalyl chloride was removed under reduced pressure and the occurring acid chloride dried in high vacuum.

Yield: 54 mg (99 %).

¹H NMR (CDCl₃, 400 MHz, RT): δ [ppm] = 7.42 (m, 22H, PhH), 6.83 (t, ^3J = 2.34 Hz, 1H, PhH), 6.70 (d, ^3J = 2.21 Hz, 4H, PhH), 6.60 (t, ^3J = 2.27 Hz, 2H, PhH), 5.05 (s, 8H, OCH₂Ph), 5.03 (s, 4 H, OCH₂Ph).

¹³C NMR (CDCl₃, 100.5 MHz, RT): δ [ppm] = 171.26 (1C, C=O), 160.59 (4C, PhC), 160.12 (2C, PhC), 139.15 (2C, PhC), 137.12 (4C, PhC), 131.38 (1C, PhC), 129.00 (8C, PhC), 128.43 (4C, PhC), 127.96 (8C, PhC), 109.35 (2C, PhC), 106.79 (5C, PhC), 102.13 (2C, PhC), 70.54 (6 C, OCH₂Ph).

MS (MALDI, DHB): m/z = 777 [M]⁺.

FT-IR (diamond, RT): ṽ [cm⁻¹] = 3095, 3072, 3039, 2970, 2902, 2865, 2848, 2642, 1691, 1592, 1448, 1377, 1320, 1259, 1134, 1042, 999, 952, 850.

Frechet 2ⁿᵈ ethynyl Hamilton Receptor fullerene monoadduct (91)

Fullerene derivative **91** was synthesized by GOP III with Fulleropyrrolidine **88** (14 mg, 0.01 mmol), pyridine (1 μL, 0.01 mmol) and benzoyl chloride **90** (8 mg, 0.01 mmol). The reaction mixture was stirred for 2 h. After adding CH₂Cl₂ (50 mL) and water (50 mL) the organic layer was separated and dried over MgSO₄ and the solvent was removed in vacuum. Purification of the crude product was done by column chromatography on silica gel (CH₂Cl₂/EtOAc 4:1).
Yield: 15 mg (70 %).

¹H NMR (THF d₈, 400 MHz, RT):

δ [ppm] = 9.67 (bs, 2H, NH), 9.04 (bs, 2H, NH), 8.46 (s, 1H, PhH), 8.27 (m, 2H, PhH), 8.10 (d, ^3J = 8.10, 2H, PhH), 8.00 (m, 4H, PhH), 7.73 (d, ^3J = 7.54, 4H, pyH), 7.60 (d, ^3J = 8.30 Hz, 2H, pyH), 7.40 (m, 10H, PhH), 7.30 (m, 6H, PhH), 7.11 (s, 2H, PhH), 7.02 (s, 1H, PhH), 6.87 (s, 2H, PhH), 6.75 (s, 4H, PhH), 6.59 (s, 2H, PhH), 5.79 (s, 1H, NCH), 5.75 (d, ^2J = 9.92 Hz, 1H, NCH₂), 5.08 (s, 4H, OCH₂Ph), 5.04 (s, 8 H, OCH₂Ph), 4.77 (d, ^2J = 9.98 Hz, 1H, NCH₂), 2.26 (s, 4H, CH₂), 1.07 (s, 18H, CH₃).

¹³C NMR (THF d₈, 100.5 MHz, RT):

δ [ppm] = 170.72 (3C, C=O), 164.80 (2C, C=O), 161.27 (4C, PhC), 160.90 (2C, PhC), 151.72 (2C, α-pyC), 151.17 (2C, α-pyC), 148.17, 148.08, 147.82, 147.03, 146.91, 146.80, 146.48, 146.40, 146.21, 146.13, 146.02, 145.91, 145.75, 145.36, 145.23, 145.17, 144.96, 143.89, 143.7, 143.37, 143.24, 143.09, 142.92, 142.82, 142.67, 142.40, 142.23, 140.82, 140.71, 140.53, 140.41, 140.34 (58C, sp²C C₆₀), 138.16 (2C, PhC), 138.06 (4C, PhC), 136.90 (2C, PhC), 136.64 (2C, PhC), 135.28 (2C, PhC), 134.25 (2C, PhC), 134.11 (2C, pyC), 133.20 (1C, PhC), 129.04 (1C, PhC), 128.84 (1C, PhC), 128.82 (8C, PhC), 128.39 (4C, PhC), 128.12 (8C, PhC), 124.68 (2C, PhC), 124.32 (1C, NCH), 110.37 (2C, β-pyC), 110.18 (2C, β-pyC), 107.85 (2C, PhC), 106.78 (5C, PhC), 101.90 (2C, PhC), 91.37 (1C, ethynylC), 89.36 (1C, ethynylC), 71.43 (1C, NCH₂), 70.54 (6 C, OCH₂Ph), 68.98 (1C, sp³C C₆₀), 67.72 (1C, sp³C C₆₀), 50.68 (2C, CH₂), 30.52 (2C, C(CH₃)₃), 29.97 (6C, CH₃).

MS (MALDI, DCTB): m/z = 2148 [M]⁺.

UV/Vis (CH₂Cl₂, RT): λ [nm] = 424.5, 302.5, 254.0, 233.5.

FT-IR (diamond, RT): $\tilde{v}$ [cm⁻¹] = 3012, 2991, 2169, 1683, 1585, 1508, 1445, 1296, 1239, 1153, 1042, 801, 526.

N-dodecyl glycine (93)

Dodecyl amine **92** (8.77 g, 0.04 mol) was dissolved in a mixture of EtOH (25 mL) and water (13 mL). After cooling to 0 °C iodo acetic acid (2 g, 0.01 mol) was added slowly portionwise. The reaction mixture was stirred for 20 h reaching room temperature and afterwards poured into 100 mL acetone. The occurring white precipitate was filtered, washed with acetone and dried in vacuum.

Yield: 1.5 g (62 %).

^{1}H NMR (MeOH d$_4$, 300 MHz, RT):

δ [ppm] = 8.38 (bs, 1H, COO*H*), 4.91 (s, 1H, N*H*), 3.46 (s, 2H, NHC*H$_2$*COOH), 2.96 (t, ^3J = 8.38 Hz, 2H, C*H$_2$*NH), 1.68 (m, 2H, C*H$_2$*CH$_2$NH), 1.30 (bs, 18H, C*H$_2$*), 0.90 (t, ^3J = 7.59 Hz, 3H, C*H$_3$*).

^{13}C NMR (MeOH d$_4$, 75.5 MHz, RT):

δ [ppm] = 171.61 (1C, *C*OOH), 51.13 (1C, NH*C*H$_2$COOH), 48.55 (1C, *C*H$_2$NH), 33.47 (1C, *C*H$_2$), 31.15 (1C, *C*H$_2$), 31.05 (2C, *C*H$_2$), 30.92 (1C, *C*H$_2$), 30.88 (1C, *C*H$_2$), 30.66 (1C, *C*H$_2$), 28.02 (1C, *C*H$_2$), 27.78 (1C, *C*H$_2$), 24.14 (1C, *C*H$_2$), 14.84 (1C, *C*H$_3$).

FT-IR (diamond, RT): $\tilde{\nu}$ [cm^{-1}] = 2920, 2848, 2360, 2342, 1619, 1559, 1541, 1507, 1465, 1458, 1397, 1388, 1374, 1276, 1261, 764, 750.

EA: calculated for $C_{14}H_{29}NO_2$ * 0.5 acetone (272.39): C 68.28, H 11.75, N 5.14; found: C 68.83, H 11.80, N 5.65.

N,N'-Bis[6-(3,3-dimethylbutyrylamino)pyridin-2-yl]-5-ethynyl-phenyl-isophthalamide]-*N*-dodecyl-pyrrolidine fullerene monoadduct (94)

Fullerene compound **94** could be prepared according to GOP VIII using 4-formyl-phenyl Hamilton receptor **21** (40 mg, 0.06 mmol), C$_{60}$ (51 mg, 0.07 mmol) and *N*-dodecyl-glycine **93** (17 mg, 0.07 mmol). The desired product was eluated with toluene/EtOAc 3:1.

Yield: 38 mg (40 %).

^{1}H-NMR (THF d$_8$, 400 MHz, RT):

δ [ppm] = 9.63 (s, 2H, N*H*), 8.97 (s, 2H, N*H*), 8.45 (s, 1H, Ph*H*), 8.25 (s, 2H, Ph*H*), 8.01 (d, ^3J = 8.10, 4H, Ph*H*), 7.72 (m, 6 H, py*H*), 5.25 (d, ^2J = 9.96 Hz, 1H, NC*H$_2$*), 5.22 (s, 1H, NC*H*), 5.02 (d, ^2J = 9.98 Hz, 1H, NC*H$_2$*), 4.20 (m, 2H, NC*H$_2$*CH$_2$), 2.25 (s, 4H, C*H$_2$*), 1.31 (m, 20H, dodecylC*H$_2$*), 1.07 (s, 18 H, C*H$_3$*), 0.89 (t, ^3J = 7.62 Hz, 3H, dodecylC*H$_3$*).

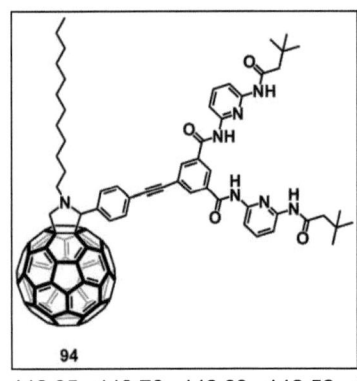

¹³C NMR (THFd₈, 100.5 MHz, RT):

δ [ppm] = 170.65 (2C, C=O), 164.71 (2C, C=O), 158.11, 157.82, 156.57, 156.44, 155.39, 155.30, 154.52, 148.07, 148.03, 147.91, 147.63, 147.42, 147.33, 147.03, 146.97, 146.84, 146.68, 146.57, 146.52, 146.34, 146.22, 146.07, 145.98, 145.89, 145.51, 145.30, 145.18, 145.11, 143.90, 142.13, 143.73, 143.43, 143.31, 143.14, 142.91, 142.85, 142.76, 142.69, 142.52, 142.41, 142.29, 140.88, 140.83, 140.50, 140.14 (58C, sp²C C$_{60}$), 151.67 (2C, α-pyC), 151.13 (2C, α-pyC), 139.60 (2C, PhC), 137.90 (2C, pyC), 137.81 (2C, PhC), 136.66 (2C, PhC), 134.12 (1C, PhC), 132.63 (2C, PhC), 130.57 (1C, PhC), 127.37 (1C, PhC), 124.74 (1C, NCH), 123.59 (1C, PhC), 110.40 (2C, β-pyC), 110.17 (2C, β-pyC), 91.61 (1C, ethynylC), 89.18 (1C, ethynylC), 83.03 (1C, NCH$_2$), 70.13 (1C, sp³C C$_{60}$), 67.85 (1C, sp³C C$_{60}$), 53.92 (1C, NCH$_2$CH$_2$), 53.19 (1C, NCH$_2$CH$_2$), 50.71 (2C, CH$_2$), 32.77 (1C, dodecylCH$_2$), 31.59 (1C, dodecylCH$_2$), 31.53 (2C, C(CH$_3$)$_3$), 30.60 (6C, CH$_3$), 30.54 (1C, dodecylCH$_2$), 29.98 (1C, dodecylCH$_2$), 29.69 (1C, dodecylCH$_2$), 29.24 (1C, dodecylCH$_2$), 28.43 (1C, dodecylCH$_2$), 28.43 (1C, dodecylCH$_2$), 25.73 (1C, dodecylCH$_2$), 25.59 (1C, dodecylCH$_2$), 14.35 (1C, dodecylCH$_3$).

MS (FAB, NBA): m/z = 1574 [M]⁺.

UV/Vis (THF, rt): λ (nm) = 429.3, 306.0, 253.6, 233.0.

FT-IR (diamond, RT): $\tilde{\nu}$ [cm⁻¹] = 3006, 2924, 2853, 2360, 2342, 1698, 1683, 1558, 1541, 1521, 1508, 1448, 1276, 1261, 764, 750.

***N,N'*-Bis[6-(3,3-dimethylbutyrylamino)pyridin-2-yl]-5-ethenyl-phenyl-isophthalamide]-*N*-dodecyl-pyrrolidine fullerene monoadduct (95)**

Fullerene compound **95** could be prepared according to GOP VIII using 4-formyl-phenyl Hamilton receptor **32** (10 mg, 0.02 mmol), C$_{60}$ (16 mg, 0.02 mmol) and *N*-dodecyl-glycine **93** (6 mg, 0.02 mmol). The desired product was eluated with toluene/EtOAc 3:1.

Experimental Part

Yield: 12 mg (40 %).

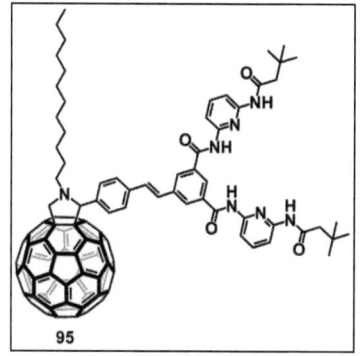

¹H NMR (THF d₈, 400 MHz, RT): δ [ppm] = 9.60 (bs, 2H, NH), 9.01 (bs, 2H, NH), 8.31 (s, 1H, PhH), 8.02 (s, 2H, PhH), 8.05 (d, ^3J = 8.06 Hz, 2H, PhH), 7.99 (d, ^3J = 7.69 Hz, 2H, PhH), 7.75 (m, 2H, py$H$), 7.72 (d, ^3J = 8.13, 2H, py$H$), 7.34 (m, 2H, py$H$), 5.26 (d, ^2J = 9.52 Hz, 1H, NCH_2), 5.20 (s, 1H, NCH), 5.00 (d, ^2J = 9.56 Hz, 1H, NCH_2), 4.21 (m, 2H, NCH$_2$CH_2), 2.25 (s, 4H, CH_2), 1.31 (bs, 20H, dodecylCH_2), 1.08 (s, 18H, CH_3), 0.89 (t, ^3J = 7.62 Hz, 3H, dodecylCH_3).

¹³C NMR (THF d₈, 100.5 MHz, RT): δ [ppm] = 170.62 (2C, C=O), 164.78 (2C, C=O), 158.01, 157.72, 156.53, 156.44, 155.37, 155.30, 154.52, 148.00, 148.03, 147.99, 147.65, 147.42, 147.31, 147.03, 146.95, 146.86, 146.68, 146.52, 146.3´8, 146.22, 146.07, 145.92, 145.89, 145.51, 145.30, 145.18, 145.11, 143.90, 142.13, 143.43, 143.31, 143.14, 142.91, 142.76, 142.69, 142.52, 142.41, 142.29, 140.88, 140.83, 140.50, 140.14 (58C, sp²C C$_{60}$), 151.67 (2C, α-py$C$), 151.13 (2C, α-py$C$), 143.58 (1C, CH), 139.62 (2C, PhC), 137.90 (2C, py$C$), 137.85 (2C, PhC), 137.77 (1C, CH), 136.68 (2C, PhC), 134.15 (2C, PhC), 132.63 (2C, PhC), 130.54 (1C, PhC), 127.37 (1C, PhC), 124.74 (1C, NCH), 123.59 (1C, PhC), 110.41 (2C, β-py$C$), 110.17 (2C, β-py$C$), 83.03 (1C, NCH$_2$), 70.12 (1C, sp³C C$_{60}$), 67.86 (1C, sp³C C$_{60}$), 53.94 (1C, NCH$_2$CH$_2$), 53.19 (1C, NCH$_2$$CH_2$), 50.71 (2C, CH$_2$), 32.77 (1C, dodecyl$CH_2$), 31.59 (1C, dodecyl CH$_2$), 31.50 (2C, C(CH$_3$)$_3$), 30.60 (6C, CH$_3$), 30.52 (1C, dodecyl$CH_2$), 29.99 (1C, dodecyl$CH_2$), 29.69 (1C, dodecyl$CH_2$), 29.24 (1C, dodecyl$CH_2$), 28.39 (1C, dodecyl$CH_2$), 28.34 (1C, dodecyl$CH_2$), 25.70 (1C, dodecyl$CH_2$), 25.60 (1C, dodecyl$CH_2$), 14.30 (1C, dodecyl$CH_3$).

MS (FAB, NBA): m/z = 1576 [M]⁺.

UV/Vis (CH₂Cl₂, RT): λ [nm] = 432.4, 310.4, 302.0, 255.4, 230.2.

FT-IR (diamond, RT): $\tilde{\nu}$ [cm^{-1}] = 3010, 2986, 2365, 1698, 1681, 1578, 1543, 1521, 1500, 1488, 1452, 1283, 1265, 764, 751.

N-(6-(3,3-dimethylbutyrylamino)pyridin-2-yl)3-iodobenzamide (97)

Iodobenzamide **97** was accessible by applying GOP XI with DMAP (247 mg, 2.02 mmol), 1-HOBT (300 mg, 2.22 mmol), 3-iodobenzoic acid **96** (500 mg, 2.02 mmol), DCC (584 mg, 2.83 mmol) and *N*-(6-aminopyridin-2-yl)-3,3-dimethylbutanamide **17** (522 mg, 2.52 mmol). Purification was done by column chromatography on silica gel (CH$_2$Cl$_2$→CH$_2$Cl$_2$/EtOAc 4:1).
Yield: 547 mg (62 %).

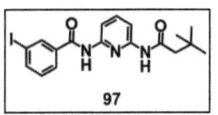

^{1}H NMR (CDCl$_3$, 400 MHz, RT):
δ [ppm] = 8.15 (s, 1H, Ph*H*), 8.12 (s, 1H, N*H*), 7.93 (m, 2H, Ph*H*), 7.80 (m, 2H, py*H*), 7.69 (t, ^3J = 8.01 Hz, 1H, py*H*), 7.51 (s, 1H, N*H*), 7.16 (t, ^3J = 8.12 Hz, 1H, Ph*H*), 2.18 (s, 2H, C*H*$_2$), 1.04 (s, 9H, C*H*$_3$).

^{13}C NMR (CDCl$_3$, 100.5 MHz, RT):
δ [ppm] = 169.81 (1C, *C*=O), 163.34 (1C, *C*=O), 149.96 (1C, α-py*C*), 149.61 (1C, α-py*C*), 141.53 (1C, Ph*C*), 141.37 (1C, Ph*C*), 136.51 (1C, py*C*), 130.89 (1C, Ph*C*), 126.70 (2C, Ph*C*), 110.31 (1C, β-py*C*), 110.00 (1C, β-py*C*), 94.87 (1C, Ph*C*), 52.19 (1C, *C*H$_2$), 31.77 (1C, *C*(CH$_3$)$_3$), 30.20 (3C, *C*H$_3$).

MS (FAB, NBA): m/z = 438 [M]$^+$.

IR (diamond, RT): $\tilde{\nu}$ [cm^{-1}] = 2957, 1672, 1584, 1506, 1446, 1298, 1277, 1259, 1243, 798, 764, 749.

EA: calculated for C$_{18}$H$_{20}$IN$_3$O$_2$ (437.27): C 49.56, H 4.39, N 9.63; found: C 49.35, H 4.10, N 9.81.

Experimental Part

N-(6-(3,3-dimethylbutyrylamino)pyridin-2-yl)-3-((4-formyl-phenyl)ethynyl) benzamide (98)

Pyridin derivative **98** was prepared according to GOP IV using Hamilton receptor analogue **97** (300 mg, 0.68 mmol), Pd(PPh$_3$)$_2$Cl$_2$ (4 mg, 0.01 mmol), CuI (3 mg, 0.02 mmol) and 4-ethynylbenzaldehyde **20** (118 mg, 0.91 mmol) in THF (10 mL) and HNEt$_2$ (5 mL). The residue was cleaned up by column chromatography on silica gel (CH$_2$Cl$_2$/EtOAc 5:1).

Yield: 293 mg (98 %).

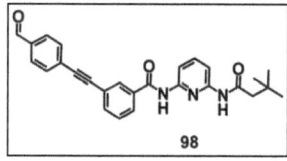

^{1}H NMR (CDCl$_3$, 400 MHz, RT):

δ [ppm] = 10.01 (s, 1H, H*C*=O), 8.25 (s, 1H, Ph*H*), 8.05 (d, ^3J = 8.11 Hz, 2H, Ph*H*), 8.03 (d, ^3J = 7.94 Hz, 1H, Ph*H*), 7.88 (d, ^3J = 8.42 Hz, 2H, Ph*H*), 7.76 (m, 3H, py*H*, Ph*H*), 7.74 (d, ^3J = 8.30 Hz, 1H, py*H*), 7.71 (bs, 2H, N*H*), 7.67 (m, 1H, py*H*), 2.23 (s, 2H, C*H$_2$*), 1.09 (s, 9H, C*H$_3$*).

^{13}C NMR (CDCl$_3$, 100.5 MHz, RT):

δ [ppm] = 191.37 (1C, H*C*=O), 170.24 (1C, *C*=O), 164.43 (1C, *C*=O), 149.58 (1C, α-py*C*), 149.31 (1C, α-py*C*), 140.96 (1C, Ph*C*), 135.70 (1C, Ph*C*), 135.16 (1C, py*C*), 134.64 (1C, Ph*C*), 132.19 (2C, Ph*C*), 130.33 (1C, Ph*C*), 129.62 (2C, Ph*C*) 129.13 (1C, Ph*C*), 128.90 (1C, Ph*C*), 127.46 (1C, Ph*C*), 123.46 (1C, Ph*C*), 109.86 (1C, β-py*C*), 109.61 (1C, β-py*C*), 91.89 (1C, ethynyl*C*), 89.72 (1C, ethynyl*C*), 51.76 (1C, *C*H$_2$), 31.34 (1C, *C*(CH$_3$)$_3$), 29.77 (3C, *C*H$_3$).

MS (FAB, NBA): m/z = 440 [M]$^+$.

IR (diamond, RT): $\tilde{\nu}$ [cm^{-1}] = 1686, 1671, 1601, 1586, 1512, 1504, 1448, 1296, 1240, 1208, 799, 735, 726.

N-(6-(3,3-dimethylbutyrylamino)pyridin-2-yl)-4-iodobenzamide (101)

Benzamide **101** could be prepared using GOP III with 4-iodobenzoyl chloride **99** (1 g, 0.38 mol), *N*-(6-aminopyridin-2-yl)-3,3-dimethylbutanamide **17** (770 mg, 0.38 mol)

and triethylamine (528 μL, 0.38 mol). Column chromatography on silica gel (CH$_2$Cl$_2$→CH$_2$Cl$_2$/EtOAc 4:1) was used to purify the desired product.
Yield: 1.57 g (96 %).

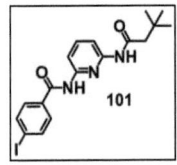

^{1}H NMR (CDCl$_3$, 300 MHz, RT): δ [ppm] = 8.24 (bs, 2H, N*H*), 8.01 (m, 2H, py*H*), 7.86 (d, ^3J= 7.85 Hz, 2H, Ph*H*), 7.75 (m, 1H, py*H*), 7.62 (d, ^3J= 7.85 Hz, 2H, Ph*H*), 2.25 (s, 2H, C*H$_2$*), 1.11 (s, 9H, C*H$_3$*).

^{13}C NMR (CDCl$_3$, 75.5 MHz, RT):
δ [ppm] = 170. 65 (1C, *C*=O), 165.02 (1C, *C*=O), 149.95 (1C, α-py*C*), 149.69 (1C, α-py*C*), 141.33 (2C, Ph*C*), 133.93 (1C, Ph*C*), 129.03 (2C, Ph*C*), 110.26 (1C, β-py*C*), 110.02 (1C, β-py*C*), 99.96 (1C, Ph*C*), 52.14 (1C, *C*H$_2$), 31.77 (1C, *C*(CH$_3$)$_3$), 30.19 (3C, *C*H$_3$).

MS (FAB, NBA): m/z = 437 [M]$^+$.

IR (diamond, RT): $\tilde{\nu}$ [cm^{-1}] = 2959, 1686, 1675, 1582, 1506, 1454, 1308, 1246, 1007, 889, 838, 807, 754.

EA: calculated for C$_{18}$H$_{20}$IN$_3$O$_2$ (437.27): C 49.56, H 4.39, N 9.63; found: C 49.10, H 4.42, N 9.82.

N-(6-(3,3-dimethylbutyrylamino)pyridin-2-yl)-2-iodobenzamide (102)

Compound **102** was synthesized by GOP III with 2-Iodobenzoyl chloride **100** (1g, 0.38 mol), *N*-(6-aminopyridin-2-yl)-3,3-dimethylbutanamide **17** (770 mg, 0.38 mol) and triethylamine (528 μL, 0.38 mol). Column chromatography on silica gel (CH$_2$Cl$_2$→CH$_2$Cl$_2$/EtOAc 4:1) was used to purify the desired product.
Yield: 800 mg (49 %)

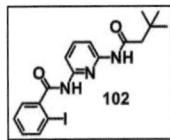

^{1}H NMR (CDCl$_3$, 300 MHz, RT): δ [ppm] = 8.03 (s, 1H, N*H*), 7.99 (d, ^3J= 7.91 Hz, 2H, py*H*), 7.92 (d, ^3J= 7.91 Hz, 1H, Ph*H*), 7.76 (m, 1H, py*H*), 7.60 (s, 1H, N*H*), 7.51 (d, ^3J= 7.89 Hz, 1H,

PhH), 7.42 (t, ^3J= 8.12 Hz, 1H, PhH), 7.16 (t, ^3J= 8.01 Hz, 1H, PhH), 2.24 (s, 2H, CH$_2$), 1.10 (s, 9H, CH$_3$).

^{13}C NMR (CDCl$_3$, 75.5 MHz, RT):
δ [ppm] = 170.70 (1C, C=O), 167.56 (1C, C=O), 150.02 (1C, α-pyC), 149.46 (1C, α-pyC), 141.80 (2C, PhC), 132.18 (1C, PhC), 18.74 (2C, PhC), 110.45 (1C, β-pyC), 110.08 (1C, β-pyC), 92.63(1C, PhC), 52.07 (1C, CH$_2$), 31.75 (1C, C(CH$_3$)$_3$), 30.18 (3C, CH$_3$).

MS (FAB, NBA): m/z = 438 [M]$^+$.

IR (diamond, RT): $\tilde{\nu}$ [cm^{-1}] = 2956, 2361, 2342, 1673, 1583, 1506, 1446, 1299, 1260, 1137, 801, 764, 749.

EA: calculated for C$_{18}$H$_{20}$IN$_3$O$_2$ (437.27): C 49.56, H 4.39, N 9.63; found: C 49.73, H 4.63, N 9.28.

N-(6-(3,3-3,3-dimethylbutyrylamino)pyridin-2-yl)-4-((4-formyl-phenyl)ethynyl) benzamide (103)

Hamilton receptor analogue **103** was accessible by GOP IV with pyridine derivative **101** (250 mg, 0.57 mmol), Pd(PPh$_3$)$_2$Cl$_2$ (4 mg, 0.01 mmol), CuI (3 mg, 0.02 mmol) and 4-ethynylbenzaldehyde **20** (89 mg, 0.91 mmol) in THF/NEt$_3$ (20 mL, 1:2). The crude product was cleaned up by column chromatography on silica gel (CH$_2$Cl$_2$/EtOAc 5:1)
Yield: 240 mg (96 %).

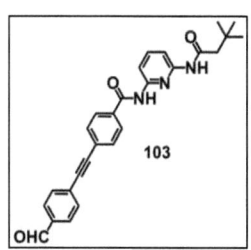

^{1}H NMR (CDCl$_3$, 300 MHz, RT): δ [ppm] = 10.04 (s, 1H, HC=O), 8.26 (bs, 2H, NH), 8.07 (d, ^3J = 7.54 Hz, 2H, PhH), 8.00 (d, ^3J = 8.10 Hz, 2H, PhH), 7.88 (d, ^3J = 8.48 Hz, 4H, PhH), 7.76 (t, ^3J = 8.10 Hz, 1H, pyH), 7.68 (m, 2H, pyH), 2.26 (s, 2H, CH$_2$), 1.12 (s, 9H, CH$_3$).

Experimental Part

^{13}C NMR (CDCl$_3$, 75.5 MHz, RT):
δ [ppm] = 191.76 (1C, HC=O), 170. 68 (1C, C=O), 164.89 (1C, C=O), 149.98 (1C, α-pyC), 149.75 (1C, α-pyC), 141.34 (2C, PhC), 136.18 (1C, PhC), 134.42 (2C, PhC), 132.66 (1C, PhC), 132.54 (2C, PhC), 130.03 (1C, PhC), 129.21 (2C, PhC), 127.64 (1C, PhC), 110.25 (1C, β-pyC), 110.01 (1C, β-pyC), 92.46 (1C, ethynylC), 91.64 (1C, ethynylC), 52.16(1C, CH$_2$), 31.77 (1C, C(CH$_3$)$_3$), 30.19 (3C, CH$_3$).

MS (FAB, NBA): m/z = 439 [M]$^+$.

IR (diamond, RT): $\tilde{\nu}$ [cm^{-1}] = 2959, 1699, 1675, 1603, 1586, 1504, 1489, 1447, 1298, 1241, 1208, 1153, 829, 813, 758.

N-(6-(3,3-Dimethylbutyrylamino)pyridin-2-yl)-2-((4-formyl-phenyl)ethynyl) benzamide (104)

Benzamide **104** could be prepared by applying GOP IV using Hamilton receptor analogue **102** (250 mg, 0.57 mmol), Pd(PPh$_3$)$_2$Cl$_2$ (4 mg, 0.01 mmol), CuI (3 mg, 0.02 mmol) and 4-ethynylbenzaldehyde **20** (89 mg, 0.91 mmol) in THF/NEt$_3$ (20 mL, 5:1). The remainder was purified by column chromatography on silica gel (CH$_2$Cl$_2$/EtOAc 5:1)
Yield: 130 mg (52 %).

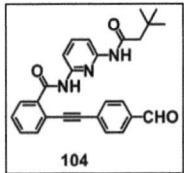

^{1}H NMR (CDCl$_3$, 300 MHz, RT):
δ [ppm] = 10.06 (s, 1H, HC=O), 8.05 (bs, 1H, NH), 8.03 (bs, 1H, NH), 7.99 (d, ^3J = 7.88 Hz, 2H, pyH), 7.88 (d, ^3J = 7.89 Hz, 2H, PhH), 7.74 (m, 1H, pyH), 7.51 (d, ^3J = 7.87 Hz, 2H, PhH), 7.45 (m, 2H, PhH), 7.20 (m, 2H, PhH), 2.25 (s, 2H, CH$_2$), 1.13 (s, 9H, CH$_3$).

^{13}C NMR (CDCl$_3$, 75.5 MHz, RT):
δ [ppm] = 190.75 (1C, HC=O), 170. 75 (1C, C=O), 167.46 (1C, C=O), 150.03 (1C, α-pyC), 149.78 (1C, α-pyC), 141.82 (2C, PhC), 136.25 (1C, PhC), 134.44 (2C, PhC), 132.14 (1C, PhC), 132.02 (2C, PhC), 130.13 (1C, PhC), 128.70 (2C, PhC), 127.72

(1C, PhC), 110.44 (1C, β-pyC), 110.11 (1C, β-pyC), 93.01 (1C, ethynylC), 91.85 (1C, ethynylC), 52.05 (1C, CH_2), 31.76 (1C, $C(CH_3)_3$), 30.20 (3C, CH_3).

MS (FAB, NBA): m/z = 439 [M]$^+$.

IR (diamond, RT): $\tilde{\nu}$ [cm^{-1}] = 1688, 1671, 1603, 1584, 1515, 1502, 1449, 1296, 1241, 1210, 799, 736, 725.

N-(6-(3,3-Dimethylbutyrylamino)pyridin-2-yl)-3-(3,3-diethoxyprop-1-ynyl)benzamide (105)

Benzamide **105** was prepared according to GOP IV with N-(6-(3,3-dimethylbutyrylamino)pyridin-2-yl), 3-iodobenzamide **97** (250 mg, 0.57 mmol), Pd(PPh$_3$)$_2$Cl$_2$ (4 mg, 0.01 mmol), CuI (3 mg, 0.02 mmol) and 3,3-diethoxyprop-1-yne **27** (88 mg, 90 µL, 0.69 mmol) in dry THF (10 mL) and HNEt$_2$ (5 mL). The crude product was purified by column chromatography on silica gel (CH$_2$Cl$_2$→CH$_2$Cl$_2$/EtOAc 4:1).
Yield: 245 mg (98 %).

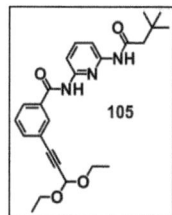

^{1}H NMR (CDCl$_3$, 300 MHz, RT):

δ [ppm] = 8.26 (s, 1H, NH), 8.20 (s, 1H, NH), 7.98 (d, ^3J = 8.01 Hz, 1H, pyH), 7.85 (t, ^3J = 8.25 Hz, 1H, pyH), 7.70 (m, 3H, PhH, pyH), 7.74 (t, ^3J = 7.35 Hz, 1H, PhH), 7.20 (t, ^3J = 7.20 Hz, 1H, PhH), 5.50 (s, 1H, CH), 3.70 (m, 4H, OCH$_2$), 2.24 (s, 2H, CH$_2$), 1.28 (t, ^3J = 7.70 Hz, 6H, OCH$_2$CH$_3$), 1.09 (s, 9H, CH$_3$).

^{13}C NMR (CDCl$_3$, 75.5 MHz, RT): 170.71 (1C, C=O), 164.24 (1C, C=O), 149.73 (1C, α-pyC), 149.62 (1C, α-pyC), 141.10 (1C, PhC), 136.08 (1C, pyC), 135.26 (1C, PhC), 130.38 (1C, PhC), 128.92 (1C, PhC), 127.50 (1C, PhC), 126.24 (1C, PhC), 109.87 (1C, β-pyC), 109.56 (1C, β-pyC), 94.38 (1C, ethynylC), 91.60 (1C, ethynylC), 85.81 (1C, CH), 61.03 (2C, OCH$_2$), 51.60 (1C, CH$_2$), 31.28 (1C, C(CH$_3$)$_3$), 29.73 (3C, CH$_3$), 15.06 (2C, OCH$_2$CH$_3$).

MS (FAB, NBA): m/z = 437 [M]$^+$.

IR (diamond, RT): ṽ [cm^{-1}] = 2958, 1672, 1584, 1509, 1446, 1298, 1241, 799, 734.

N-(6-(3,3-Dimethylbutyrylamino)pyridin-2-yl)3-(3-oxoprop-1-ynyl)benzamide (106)

Pyridine derivative **106** could be synthesized using GOP VIa starting from Hamilton receptor analogue **105** (178 mg, 0.41 mmol).
Yield: 144 mg (97 %).

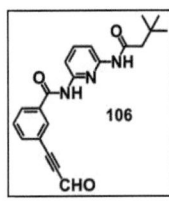

^{1}H NMR (CDCl$_3$, 300 MHz, RT):

δ [ppm] = 9.42 (1s, H*C*=O), 8.28 (s, 1H, N*H*), 8.19 (s, 1H, N*H*), 8.06 (d, ^3J = 8.10 Hz, 1H, py*H*), 7.98 (m, 1H, py*H*), 7.84 (m, 3H, Ph*H*, py*H*), 7.72 (t, ^3J = 7.36 Hz, 1H, Ph*H*), 7.19 (t, ^3J = 7.20 Hz, 1H, Ph*H*), 2.23 (s, 2H, C*H$_2$*), 1.09 (s, 9H, C*H$_3$*).

^{13}C NMR (CDCl$_3$, 75.5 MHz, RT):

δ [ppm] = 176.42 (1C, H*C*=O), 170.28 (1C, *C*=O), 163.79 (1C, *C*=O), 149.54 (1C, α-py*C*), 149.17 (1C, α-py*C*), 141.02 (1C, Ph*C*), 136.28 (1C, py*C*), 136.01 (1C, Ph*C*), 130.37 (1C, Ph*C*), 129.23 (1C, Ph*C*), 126.24 (1C, Ph*C*), 120.21 (1C, Ph*C*), 109.88 (1C, β-py*C*), 109.56 (1C, β-py*C*), 92.88 (1C, ethynyl*C*), 88.69 (1C, ethynyl*C*), 51.56 (1C, *C*H$_2$), 31.27 (1C, *C*(CH$_3$)$_3$), 29.71 (3C, *C*H$_3$).

MS (FAB, NBA): m/z = 363 [M]$^+$.

IR (diamond, RT): ṽ [cm^{-1}] = 2959, 1676, 1603, 1585, 1505, 1490, 1446, 1300, 1241, 802, 758.

N-(6-(3,3-Dimethylbutyrylamino)pyridin-2-yl)-4'-formyl-biphenyl-3-carboxamide (107)

Compound **107** was accessible *via* GOP VII using iodo Hamilton receptor **97** (250 mg, 0.57 mmol), 4-formylphenylboronic acid **33** (103 mg, 0.68 mmol), Pd(PPh$_3$)$_4$ (33 mg, 0.03 mmol) and K$_2$CO$_3$ (189 mg, 1.37 mmol). Column chromatography on silica gel (CH$_2$Cl$_2$/EtOAc 4:1) was used for the purification.
Yield: 130 mg (55 %).

Experimental Part

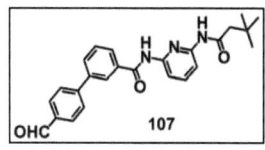

¹H NMR (THF d₈, 300 MHz, RT):

δ [ppm] = 10.79 (s, 1H, HC=O), 9.52 (s, 1H, NH), 9.07 (s, 1H, NH), 8.32 (s, 1H, PhH), 8.03 (m, 6H, PhH, pyH), 8.00 (d, ^3J = 6.61 Hz, 2H, PhH), 7.75 (t, ^3J = 7.72 Hz, 1H, pyH), 6.89 (t, ^3J = 7.79 Hz, 1H, PhH), 2.29 (s, 2H, CH₂), 1.09 (s, 9H, CH₃).

¹³C NMR (THF d₈, 75.5 MHz, RT):

δ [ppm] = 191.12 (1C, HC=O), 170.23 (1C, C=O), 165.31 (1C, C=O), 151.18 (1C, α-pyC), 150.90 (1C, α-pyC), 146.00 (1C, PhC), 140.29 (2C, PhC), 140.01 (1C, pyC), 136.55 (2C, PhC), 130.62 (1C, PhC), 130.22 (2C, PhC), 129.14 (1C, PhC), 127.85 (1C, PhC), 126.78 (2C, PhC), 109.65 (1C, β-pyC), 109.51 (1C, β-pyC), 50.21 (1C, CH₂), 31.14 (1C, C(CH₃)₃), 29.50 (3C, CH₃).

MS (FAB, NBA): m/z = 415 [M]⁺.

IR (diamond, RT): ṽ [cm⁻¹] = 2959, 1681, 1605, 1587, 1511, 1446, 1297, 1243, 1158, 1134, 799, 750.

EA: calculated for C₂₅H₂₅N₃O₃ (415.48): C 72.27, H 6.06, N 10.11; found: C 72.07, H 6.07, N 9.86.

N-(6-(3,3-Dimethylbutyrylamino)pyridine-2-yl)-3-(3-formylthiophen-2-yl)benzamide (108)

Thiophene derivative **108** was prepared according to GOP VII using iodobenzamide **97** (162 mg, 0.37 mmol), 4-formylthiopheneboronic acid **35** (70 mg, 0.48 mmol), Pd(PPh₃)₄ (22 mg, 0.02 mmol) and K₂CO₃ (124 mg, 0.89 mmol). Column chromatography (CH₂Cl₂/EtOAc 4:1) of the residue leads to a white powder. Yield: 65 mg (42 %).

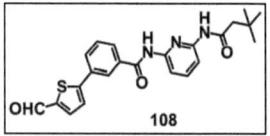

¹H NMR (DMSO d₆, 300 MHz, RT):

δ [ppm] = 9.89 (s, 1H, HC=O), 9.54 (bs, 1H, NH), 9.06 (bs, 1H, NH), 8.30 (s, 1H, PhH), 8.00 (m, 4H, PhH, pyH), 7.87 (d, ^3J = 4.00 Hz, 1H, thioH), 7.70 (m, 1H, pyH), 7.67

(d, ^3J = 4.00 Hz, 1H, thio$H$), 7.58 (m, 1H, py$H$), 2.26 (s, 2H, CH_2), 1.08 (s, 9H, CH_3).

^{13}C NMR (DMSO d$_6$, 75.5 MHz, RT):

δ [ppm] = 183.29 (1C, HC=O), 171.21 (1C, C=O), 165.73 (1C, C=O), 152.93 (1C, α-py$C$), 151.94 (1C, α-py$C$); 151.57 (1C, thio$C$), 144.74 (1C, thio$C$), 140.77 (1C, py$C$), 138.32 (1C, thio$C$), 137.27 (1C, PhC), 134.62 (1C, PhC), 130.42 (1C, PhC), 130.18 (1C, PhC), 129.36 (1C, thio$C$), 128.76 (1C, PhC), 126.37 (1C, PhC), 110.49 (1C, β-py$C$), 110.29 (β-py$C$), 50.97 (1C, CH_2), 31.91 (1C, C(CH$_3$)$_3$), 30.27 (3C, CH_3).

MS (FAB, NBA): m/z = 447 [M-Na]$^+$, 424 [M]$^+$.

IR (diamond, RT): $\tilde{v}$ [cm^{-1}] = 2959, 1684, 1658, 1612, 1583, 1511, 1447, 1398, 1295, 1279, 1260, 1157, 805, 790, 764, 749.

EA: calculated for C$_{23}$H$_{23}$N$_3$O$_3$S * 0.5 DMSO * 0.5 THF (496.63): C 62.98, H 6.24, N 8.39, S 9.02; found: C 63.35, H 6.77, N 8.14, S 8.32.

N-(6-(3,3-Dimethylbutyrylamino)pyridin-2-yl)-3-(4-phenyl-ethynyl)benzamide-*N*-methyl-pyrrolidine-fullerene monoadduct (109)

Fullerene monoadduct could be prepared using GOP VIII with Hamilton receptor analogue **98** (89 mg, 0.2 mmol), C$_{60}$ (189 mg, 0.26 mmol) and sarcosine (23 mg, 0.26 mmol). Column chromatography was performed on silica with toluene/EtOAc 2:1 to isolate the monoadduct.

Yield: 102 mg (43 %).

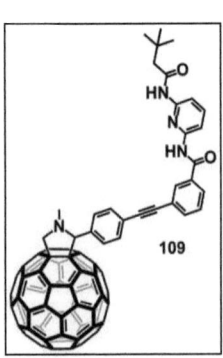

^{1}H NMR (CS$_2$, CDCl$_3$, 400 MHz, RT):

δ [ppm] = 8.16 (s, 2H, NH), 8.03 (d, ^3J = 8.02 Hz, 2H, PhH), 7.97 (d, ^3J = 7.95 Hz, 1H, PhH), 7.96 (s, 1H, PhH), 7.94 (d, ^3J = 8.08 Hz, 3H, PhH), 7.71 (t, ^3J = 8.06 Hz, 1H, PhH), 7.66 (d, ^3J = 7.79 Hz, 2H, py$H$), 7.46 (t, ^3J = 7.79 Hz, 1H, py$H$), 5.00 (d, ^2J = 9.38 Hz, 1H, NCH_2), 4.96 (s, 1H, NCH), 4.29 (d, ^2J = 9.38 Hz, 1H, NCH_2), 2.83 (s, 3H, NCH_3), 2.22 (s, 2H, CH_2), 1.10 (s, 9H, CH_3).

^{13}C NMR (CDCl$_3$, 100.5 MHz, RT):

δ [ppm] = 169.47 (1C, C=O), 163.84 (1C, C=O), 155.83 (1C, α-pyC), 153.59 (1C, α-pyC), 152.62 (1C, PhC), 149.34, 146.17, 147.11, 146.37, 146.18, 146.14, 146.11, 146.02, 145.96, 145.93, 145.77, 145.55, 145.42, 145.39, 145.31, 145.18, 145.15, 145.09, 145.05, 144.99, 144.53, 144.42, 144.22, 144.17, 142.99, 142.85, 142.54, 142.43, 142.06, 141.97, 141.91, 14187, 141.81, 141.71, 141.67, 141.53, 141.39, 140.60, 140.08, 140.04, 139.79, 139.41, 137.64 (58 C, sp^{2}C C$_{60}$), 136.75 (1C, PhC), 136.29 (1C, pyC), 135.77 (1C, PhC), 134.76 (2C, PhC), 131.91 (1C, PhC), 129.92 (2C, PhC) 129.19 (1C, PhC), 128.84 (1C, PhC), 126.76 (1C, PhC), 124.04 (1C, PhC), 122.79 (1C, NCH), 109.55 (1C, β-pyC), 109.37 (1C, β-pyC), 90.72 (1C, ethynylC), 88.95 (1C, ethynylC), 83.06 (1C, NCH$_2$), 69.87 (1C, sp^{3}C C$_{60}$), 68.80 (1C, sp^{3}C C$_{60}$), 51.49 (1C, NCH$_3$), 39.85 (1C, CH$_2$), 30.99 (1C, C(CH$_3$)$_3$), 29.63 (3C, CH$_3$).

MS (FAB, NBA): m/z = 1187 [M]$^+$.

UV/Vis (CH$_2$Cl$_2$, rt): λ (nm) = 431.4, 308.9, 291.7, 255.6, 243.

IR (diamond, RT): $\tilde{\nu}$ [cm^{-1}] = 2360, 2342, 1718, 1655, 1447, 1276, 1261, 764, 750.

N-(6-(3,3-Dimethylbutyrylamino)pyridin-2-yl)3-(3-prop-1-ynyl)benzamido-*N*-methyl-pyrrolidine-fullerene-monoadduct (110)

Monoadduct **110** was synthesized using GOP VIII with Hamilton receptor analogue **106** (40 mg, 0.11 mmol), C$_{60}$ (103 mg, 0.14 mmol) and sarcosine (12 mg, 0.14 mmol). For the separation of the desired product toluene/EtOAc 4:1 on silica gel was used. Yield: 30 mg (25 %)

^{1}H NMR (CDCl$_3$, 400 MHz, RT):

δ [ppm] = 8.15 (s, 2H, N*H*), 8.02 (d, ^3J = 7.32 Hz, 1H, Ph*H*), 7.96 (d, ^3J = 6.47 Hz, 1H, Ph*H*), 7.94 (s, 1H, Ph*H*), 7.85 (d, ^3J = 7.81 Hz, 1H, py*H*), 7.72 (t, ^3J = 8.06 Hz, 1H, py*H*), 7.65 (d, ^3J = 7.69 Hz, 1H, py*H*), 7.46 (m, 1H, Ph*H*), 5.12 (s, 1H, NC*H*), 4.82 (d, ^2J = 9.28 Hz, 1H, NC*H$_2$*), 4.26 (d, ^2J = 9.52 Hz, 1H, NC*H$_2$*), 3.11 (s, 3H, NC*H$_3$*), 2.22 (s, 2H, C*H$_2$*), 1.09 (s, 9H, C*H$_3$*).

Experimental Part

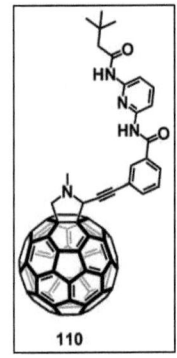

110

¹³C NMR (CDCl₃, 100.5 MHz, RT):

δ [ppm] = 170.18 (1C, C=O), 164.36 (1C, C=O), 153.20, 153.73, 153.12, 152.86, 147.48, 147.39, 146.82, 146.47, 146.43, 146.32, 146.16, 146.08, 145.82, 145.62, 145.56, 145.47, 145.37, 145.28, 144.67, 144.61, 144.52, 144.44, 143.07, 142.74, 142.68, 142.33, 142.22, 142.10, 142.06, 141.86 (58 C, sp²C C_{60}), 149.56 (1C, α-py$C$), 149.31 (1C, α-py$C$), 141.81 (1C, PhC), 136.54 (1C, py$C$), 135.35 (1C, PhC), 134.60 (1C, PhC), 130.40 (1C, PhC), 19.17 (1C, PhC), 127.64 (1C, PhC), 123.17 (1C, NCH), 109.85(1C, β-py$C$), 109.60 (1C, β-py$C$), 90.07 (1C, ethynyl$C$), 87.25 (1C, ethynyl$C$), 75.07 (NCH₂), 70.15 (sp³C C_{60}), 68.30 (sp³C C_{60}), 51.73 (1C, NCH₃), 39.89 (1C, CH₂), 31.34 (1C, C(CH₃)₃), 29.81 (3C, CH₃).

MS (FAB, NBA): m/z = 1110 [M]⁺.

UV/Vis (CH₂Cl₂):): λ (nm) = 429.7, 306.4, 257.7, 236.7.

IR (diamond, RT): $\tilde{\nu}$ [cm⁻¹] = 3006, 2360, 2342, 1652, 1447, 1276, 1261, 764, 750.

N-(6-(3,3-Dimethylbutyrylamino)pyridin-2-yl)-4'-biphenyl-3-carboxamide)-N-dodecyl-pyrrolidine-fullerene monoadduct (111)

Fulleropyrrolidine **111** was prepared using GOP VIII with Hamilton receptor analogue **107** (63 mg, 0.15 mmol), C_{60} (131 mg, 0.18 mmol) and N-dodecyl glycine **93** (40 mg, 0.17 mmol). The product could be isolated by column chromatography on silica gel (toluene/EtOAc 4:1).
Yield: 86 mg (44 %).

¹H NMR (THF d₈, 400 MHz, RT):

δ [ppm] = 9.39 (s, 1H, NH), 8.99 (s, 1H, NH), 8.34 (s, 1H, PhH), 8.07 (m, 6H, PhH, py$H$), 8.02 (d, ³J = 6.65 Hz, 2H, PhH), 7.77 (t, ³J = 7.70 Hz, 1H, py$H$), 6.93 (t, ³J = 7.79 Hz, 1H, PhH), 5.26 (d, ²J = 9.92 Hz, 1H, NCH_2), 5.24 (s, 1H, NCH), 5.03 (d, ²J = 9.98 Hz, 1H, NCH_2), 4.21 (m, 2H, NCH_2CH₂), 2.24 (s, 2H, CH_2), 1.30 (bs, 20H, dodecylCH_2), 1.07 (s, 9H, CH_3), 0.88 (t, ³J = 7.65 Hz, 3H, dodecylCH_3).

Experimental Part

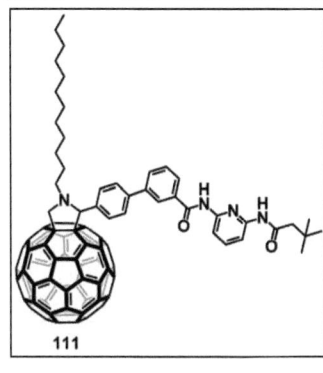

^{13}C NMR (THF d_8, 100.5 MHz, RT):

δ [ppm] = 170.64 (1C, C=O), 165.84 (1C, C=O), 158.13, 157.95, 156.44, 155.56, 154.90, 154.83, 151.61, 151.40, 148.06, 148.02, 147.87, 147.76, 147.69, 147.47, 147.23, 147.01, 146.95, 146.90, 146.85, 146.65, 146.52, 146.36, 146.21, 146.17, 146.01, 145.95, 145.92, 145.89, 145.50, 145.28, 145.16, 145.12, 143.89, 143.72, 143.41, 143.30, 143.15, 143.10, 142.94, 142.91, 142.85, 142.75, 142.66, 142.50, 142.40, 142.23, 141.36, 141.06, 140.86, 140.80, 140.54, 140.41 (58C, sp^{2}C C$_{60}$), 151.61 (1C, α-pyC), 150.40 (1C, α-pyC), 140.30 (1C, PhC), 140.10 (1C, PhC), 138.07 (1C, pyC), 137.90 (1C, PhC), 137.06 (1C, PhC), 136.49 (1C, PhC), 136.32 (1C, PhC), 130.75 (1C, PhC), 129.96 (2C, PhC), 127.87 (1C, PhC), 127.51 (1C, NCH), 126.67 (2C, PhC), 110.03 (1C, β-pyC), 109.97 (1C, β-pyC), 83.10 (1C, NCH$_2$), 70.19 (1C, sp^{3}C C$_{60}$), 68.02 (1C, sp^{3}C C$_{60}$), 53.17 (1C, NCH$_2$CH$_2$), 50.65 (1C, CH$_2$), 32.78 (1C, dodecylCH$_2$), 31.60 (1C, dodecylCH$_2$), 31.51 (1C, C(CH$_3$)$_3$), 30.61 (3C, CH$_3$), 30.58 (1C, dodecylCH$_2$), 29.97 (1C, dodecylCH$_2$), 29.68 (1C, dodecylCH$_2$), 29.25 (1C, dodecylCH$_2$), 28.47 (1C, dodecylCH$_2$), 28.37 (1C, dodecylCH$_2$), 25.90 (1C, dodecylCH$_2$), 25.72 (1C, dodecylCH$_2$), 14.36 (1C, dodecylCH$_3$).

MS (FAB, NBA): m/z = 1317 [M]$^+$.

UV/Vis (THF, rt): λ (nm) = 430.6, 305.5, 258.1.

IR (diamond, RT): $\tilde{v}$ [cm^{-1}] = 2919, 2848, 2360, 2342, 1618, 1559, 1542, 1521, 1508, 1465, 1458, 1397, 1387, 1373, 1276, 1261, 764, 750.

N-(6-(3,3-Dimethylbutyrylamino)pyridine-2-yl)-3-(thiophen-2-yl)benzamide-N-methyl-pyrrolidine-fullerene-monoadduct (112)

Hamilton receptor analogue **108** (30 mg, 0.07 mmol), C$_{60}$ (63 mg, 0.09 mmol) and sarcosine (8 mg, 0.09 mmol) were used as it is mentioned for GOP VIII. Isolation of the monoadduct could be achieved by eluating with clumn chromatography on silica gel (toluene/EtOAc 3:1).

Yield: 35 mg (43 %)

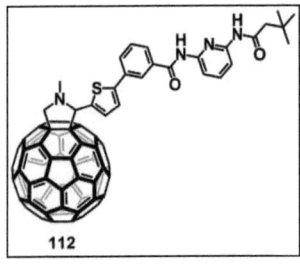

112

¹H-NMR (CS₂/THF d₈, 400 MHz, RT):

δ [ppm] = 9.13 (s, 1H, NH), 8.76 (s, 1H, NH), 8.12 (s, 1H, PhH), 7.99 (d, ^3J = 4.00 Hz, 1H, thio$H$), 7.94 (d, ^3J = 4.00 Hz, 1H, thio$H$), 7.77 (m, 3H, PhH), 7.64 (m, 1H, py$H$), 7.43 (m, 2H, py$H$), 5.32 (s, 1H, NCH), 5.01 (d, ^2J = 9.50 Hz, 1H, NCH_2), 4.30 (d, ^2J = 9.6 Hz, 1H, NCH_2), 2.95 (s, 3H, NCH_3), 2.18 (s, 2H, CH_2), 1.06 (s, 9H, CH_3).

¹³C NMR (CS₂/THF d₈, 100.5 MHz, RT):

δ [ppm] = 169.61 (1C, C=O), 164.48 (1C, C=O), 156.50 (1C, α-py$C$), 154.35 (1C, α-py$C$), 153.60 (1C, thio$C$), 150.84 (1C, thio$C$), 150.50, 147.59, 147.21, 146.97, 146.68, 146.63, 146.59, 146.50, 146.40, 146.30, 146.24, 146.06, 145.90, 145.86, 145.81, 145.68, 145.63, 145.58, 145.54, 145.45, 145.04, 144.90, 144.76, 144.69, 144.66, 143.46, 143.31, 143.02, 142.90, 142.55, 142.49, 142.41, 142.37, 142.32, 142.20, 141.99, 141.91, 140.46, 140.23, 140.04, 139.98, 137.51, 136.93, 136.29, 136.01, 134.87 (58C, sp²C C₆₀), 140.72 (1C, py$C$), 136.68 (1C, thio$C$), 136.23 (1C, thio$C$), 135.54 (1C, PhC), 130.25 (1C, PhC), 129.94 (1C, PhC), 129.51 (1C, PhC), 127.53 (1C, PhC), 125.58 (1C, PhC), 124.35 (1C, PhC), 124.12 (1C, NCH), 110.43 (1C, β-py$C$), 110.28 (1C, β-py$C$), 80.15 (1C, NCH_2), 70.76 (1C, sp³C C₆₀), 69.79 (sp³C C₆₀), 51.00 (1C, NCH_3), 40.80 (1C, CH₂), 31.64 (1C, C(CH₃)₃), 30.39 (3C, CH₃).

MS (FAB, NBA): m/z = 1168 [M]⁺.

UV/Vis (THF, rt): λ (nm) = 431.0, 303.5, 255.0.

IR (diamond, RT): ṽ [cm⁻¹] = 3006, 2989, 2360, 1652, 1523, 1447, 1276, 1261, 764, 750.

Cyanuric acid OPA (113)

OPA **113** was accessible according to GOP IV using 4-iodophenyl isocyanuric acid **63** (150 mg, 0.45 mmol), Pd(PPh$_3$)$_2$Cl$_2$ (3 mg, 0.01 mmol), CuI (3 mg, 0.02 mmol) and 4-ethynylbenzaldehyde **20** (70 mg, 0.54 mmol) in THF (10 mL) before NEt$_3$ (4 mL) was added. The crude product was cleaned by column chromatography on silica gel (CH$_2$Cl$_2$→CH$_2$Cl$_2$/THF 1:1)

Yield: 110 mg (73 %).

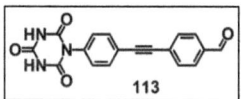

^{1}H-NMR (THF d$_8$, 400 MHz, RT): δ [ppm] = 11.59 (bs, 2H, N*H*), 10.93 (s, 1H, H*C*=O), 7.98 (d, ^3J = 6.68 Hz, 2H, Ph*H*), 7.80 (d, ^3J = 8.67 Hz, 2H, Ph*H*), 7.45 (d, ^3J = 8.48 Hz, 2H, Ph*H*), 7.17 (d, ^3J = 7.02 Hz, 2H, Ph*H*).

^{13}C NMR (THF d$_8$, 100.5 MHz, RT):

δ [ppm = 192.78 (1C, H*C*=O), 147.99 (2C, N*C*=O), 147.95 (1C, N*C*=O), 136.06 (1C, Ph*C*), 134.03 (2C, Ph*C*), 133.43 (2C, Ph*C*), 130.51 (2C, Ph*C*), 130.38 (1C, Ph*C*), 130.14 (2C, Ph*C*), 126.26 (1C, Ph*C*), 120.16 (1C, Ph*C*), 90.56 (1C, ethynyl*C*), 87.54 (1C, ethynyl*C*).

IR (diamond, RT): ṽ [cm^{-1}] = 3406, 2361, 1765, 1714, 1294, 1177, 1152, 1025, 1004, 821, 763, 733.

Cyanuric acid OPA *N*-methyl-pyrrolidine-fullerene monoadduct (114)

Cyanuric acid bearing fullerene derivative **114** was synthesized according to GOP VIII with Cyanuric acid derivative **113** (50 mg, 0.15 mmol), C$_{60}$ (152 mg (0.22 mmol) and sarcosine (18 mg, 0.22 mmol). Separation of the monoadduct was done with column chromatography on silica gel (toluene/EtOAc 1:1).

Yield: 46 mg (29 %).

^{1}H NMR (CS$_2$, THF d$_8$, 400 MHz, RT):

δ [ppm] = 10.51 (s, 2H, N*H*), 7.77 (d, ^3J = 6.68 Hz, 2H, Ph*H*), 7.54 (d, ^3J = 8.55 Hz, 2H, Ph*H*), 7.46 (d, ^3J = 8.79 Hz, 2H, Ph*H*), 7.12 (d, ^3J = 8.55 Hz, 2H, Ph*H*), 4.98 (d, ^2J = 9.52 Hz, 1H, NC*H*$_2$), 4.97 (s, 1H, NC*H*), 4.77 (d, ^2J = 9.52 Hz, 1H, NC*H*$_2$), 2.94 (s, 3H, C*H*$_3$).

Experimental Part

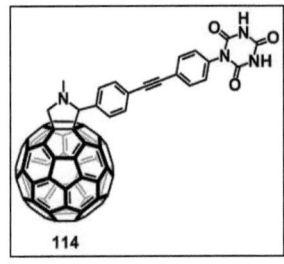

^{13}C NMR (CS$_2$, THF d$_8$, 100.5 MHz, RT):

δ [ppm] = 148.58 (2C, N*C*=O), 147.56 (1C, N*C*=O), 147.22, 147.00, 146.76, 146.31, 146.25, 146.14, 146.09, 145.99, 145.82, 145.66, 145.51, 145.38, 145.33, 145.23, 144.84, 144.77, 144.73, 144.70, 144.45, 143.26, 143.13, 142.75, 142.69, 142.28, 142.17, 141.97, 141.75, 141.66, 141.34, 140.35, 140.28, 140.07 (58C, sp^2*C* C$_{60}$), 136.64 (1C, Ph*C*), 136.47 (1C, Ph*C*), 136.02 (1C, Ph*C*), 135.66 (1C, Ph*C*), 133.72 (1C, Ph*C*), 131.80 (2C, Ph*C*), 129.37 (1C, Ph*C*), 129.08 (2C, Ph*C*), 123.67 (1C, N*C*H$_2$), 123.50 (2C, Ph*C*), 90.57 (1C, ethynyl*C*), 90.12 (1C, ethynyl*C*), 78.16 (1C, N*C*H), 70.40 (1C, sp^3*C* C$_{60}$), 69.10 (1C, sp^3*C* C$_{60}$), 62.01 (1C, *C*H$_3$).

MS (FAB, NBA): m/z = 1073 [M]$^+$.

UV/Vis (CH$_2$Cl$_2$, rt): λ (nm) = 430.5, 318.5, 314.8, 239.2, 224.5.

IR (diamond, RT): $\tilde{\nu}$ [cm^{-1}] = 2949, 2781, 2361, 2342, 1716, 1645, 1457, 1435, 1276, 1261, 764, 750.

6.2.2 Supramolecular Porphyrin Building Blocks and Their Precursors

5-(4-Ethynyl-phenyl-cyanuric-phenyl)-10,15,20-tris(3,5-dimethoxy-phenyl)-porphyrinato zinc II (7)

Cyanuric acid bearing porphyrin **7** was accessible *via* GOP IV with 4-Iodophenyl isoyanuric acid **63** (11 mg, 0.03 mmol), Pd$_2$dba$_3$ (1.0 mg, 0.01 mmol), AsPh$_3$ (1.0 mg, 0.01 mmol), CuI (1.0 mg, 0.01 mmol) and porphyrin **62** (31 mg, 0.04 mmol) in dry THF (8 mL) and NEt$_3$ (3 mL). The reaction mixture was stirred under inert conditions and with exclusion of light for 3 d. The residue was purified by column chromatography (CH$_2$Cl$_2$→CH$_2$Cl$_2$/EtOAc/THF 2:1:1).
Yield: 28 mg (81 %).

¹H NMR (THF d₈, 400 MHz, RT):

δ [ppm] = 10.69 (bs, 2H, N*H*), 8.98 (d, ^3J = 4.6 Hz, 2H, pyr*H*), 8.95 (s, 4H, pyr*H*), 8.87 (d, ^3J = 4.6 Hz, 2H, pyr*H*), 8.25 (d, ^3J = 8.1 Hz, 2H, Ph*H*), 7.96 (d, ^3J = 8.1 Hz, 2H, Ph*H*), 7.78 (d, ^3J = 8.7 Hz, 2H, Ph*H*), 7.38 (s, 6H, Ph*H*), 7.07 (d, ^3J = 8.6 Hz, 2H, Ph*H*), 6.91 (s, 3H, Ph*H*), 3.94 (s, 18H, OC*H₃*).

¹³C NMR (THF d₈, 100.5 MHz, RT):

δ [ppm] = 159.8 (3C, Ph*C*), 150.8 (2C, N*C*=O), 150.7 (1C, N*C*=O), 150.5 (2C, α-pyr*C*), 150.1 (4C, α-pyr*C*), 149.9 (1C, Ph*C*), 146.0 (6C, Ph*C*), 144.9 (2C, Ph*C*), 138.6 (1C, Ph*C*), 135.6 (2C, Ph*C*), 132.2 (4C, β-pyr*C*), 131.8 (4C, β-pyr*C*), 131.7 (2C, Ph*C*) 130.4 (2C, α-pyr*C*), 130.1 (2C, Ph*C*), 124.3 (3C, meso*C*), 123.0 (1C, Ph*C*), 121.4 (1C, Ph*C*), 120.3 (1C, meso*C*), 114.6 (6C, Ph*C*), 100.2 (3C, Ph*C*), 94.8 (1C, ethynyl*C*), 90.2 (1C, ethynyl*C*), 55.6 (6C, O*C*H₃).

MS (FAB, NBA): m/z = 1084 [M]⁺.

UV/Vis (CH₂Cl₂): λ_{max} (log ε) [nm] = 588.0 (3.10), 548.0 (3.82), 430.0 (5.16).

FT-IR (diamond, RT): $\tilde{v}$ [cm⁻¹] = 3000, 2970, 2949, 2360, 2342, 1795, 1738, 1721, 1590, 1548, 1515, 1488, 1489, 1421, 1366, 1351, 1229, 1216, 1204, 1153, 1107, 1062, 1026, 999, 953, 937, 825, 797, 757, 740, 723.

5-(*p*-Bromo-phenyl)-10,15,20-tri(*p-tert*-butyl-phenyl)-porphyrin (51)

Porphyrin **51** was synthesized according to GOP IXa with pyrrole **50** (2.18 g, 2.25 mL, 0.03 mol), 4-*tert*-butylbenzaldehyde **48** (3.94 g, 4.06 mL, 0.02 mol), 4-bromo-benzaldehyde **23** (1.5 g, 0.01 mol), tetraphenyl phosphonium chloride (50 mg, 0.13 mmol), boron trifluoride etherate (2.01 mL, 0.02 mol) and DDQ (5.516 g, 0.02 mol). The residue was purified with column chromatography (CH₂Cl₂/hexanes 2:1). Yield: 909 mg (13 %).

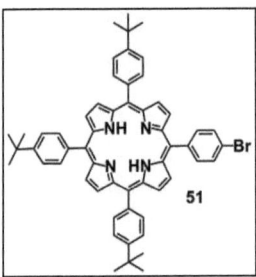

¹H NMR (CDCl₃, 400 MHz, RT): δ [ppm] = 8.89 (d, ³J = 4.77 Hz, 2H, pyr$H$), 8.87 (m, 4H, pyr$H$), 8.81 (d, ³J = 7.95, 2H, pyr$H$), 8.15 (d, ³J = 7.95, 6H, PhH), 8.09 (d, ³J = 8.35, 2H, PhH), 7.88 (d, ³J = 7.95, 2H, PhH), 7.76 (d, ³J = 7.95, 6H, PhH), 1.60 (s, 27 H, CH_3), -2.77 (s, 2H, NH).

¹³C NMR (CDCl₃, 100.5 MHz, RT):

δ [ppm] = 150.54 (3C, PhC), 150.42 (1C, PhC), 141.31 (3C, PhC), 139.23 (1C, PhC), 139.11 (6C, PhC), 135.89 (2C, PhC), 131.14 (16C, pyr$C$), 129.87 (2C, PhC), 123.58 (6 C, PhC), 120.40 (3C, meso$C$), 120.11 (1C, meso$C$), 34.89 (3C, C(CH₃)₃), 31.69 (9C, CH₃).

MS (FAB, NBA): m/z = 863 [M]⁺.

UV/Vis (CH₂Cl₂): λ$_{max}$ (log ε) [nm] = 645 (4.00), 592.0 (4.07), 550.0 (4.19), 514.5 (4.37), 417.5 (5.50).

IR (diamond, RT): ṽ [cm⁻¹] = 2961, 2361, 2341, 1474, 1275, 1261, 1073, 1014, 1002, 981, 966, 798, 750.

5-(p-Iodo-phenyl)-10,15,20-tri(p-tert-butyl-phenyl)-porphyrin (52)

Iodo porphrin **52** was synthesized by GOP IXb using pyrrole **50** (2 g, 0.03 mol), 4-iodo-benzaldehyde **49** (1.74 g, 0.01 mol), 4-tert-butylbenzaldehyde **48** (1.25 g, 0.01 mol), TFA (1.73 mL, 2.56 g, 0.01 mol), TEA (3.76 mL, 2.73 g, 0.01 mol) and DDQ (5.1 g, 0.01 mol). The crude product was purified with column chromatography (CH₂Cl₂/hexanes 2:1).

Yield: 1.0 g (11 %).

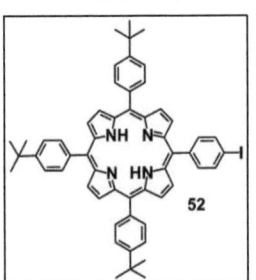

¹H NMR (CDCl₃, 400 MHz, RT): δ [ppm] = 8.90 (m, 8H, pyr$H$), 8.83 (d, ³J = 4.90 Hz, 2H, PhH), 8.19 (d, ³J = 7.95, 6H, PhH), 8.12 (d, ³J = 8.29, 2H, PhH), 7.79 (d, ³J = 8.10, 6H, PhH), 1.63 (s, 27 H, CH_3), -2.72 (s, 2H, NH).

Experimental Part

^{13}C NMR (CDCl$_3$, 100.5 MHz, RT):

δ [ppm] = 150.83 (3C, PhC), 150.42 (1C, PhC), 141.31 (3C, PhC), 139.65 (1C, PhC), 139.11 (6C, PhC), 134.93 (3C, PhC), 131.20 (16C, pyrC), 123.99 (6 C, PhC), 120.53 (3C, mesoC), 120.23 (1C, mesoC), 91.52 (1C PhC), 35.31(3C, C(CH$_3$)$_3$), 32.11 (9C, CH$_3$).

MS (MALDI, DCTB): m/z = 910 [M]$^+$.

UV/Vis (CH$_2$Cl$_2$): λ$_{max}$ (log ε) [nm] = 646.5 (4.14), 593.0 (4.23), 551.5 (4.41), 517.0 (4.57), 419.0 (5.76).

IR (diamond, RT): $\tilde{\nu}$ [cm^{-1}] = 2955, 2860, 2166, 1999, 1976, 1472, 1107, 965, 847, 733.

5-(p-Bromo-phenyl)-10,15,20-tri(p-tert-butyl-phenyl)-porphyrinato zinc (II) (53)

Metallation of free base porphyrin **51** (800 mg, 0.93 mmol) was done under GOP X with zinc acetate (612 mg, 2.29 mmol). Pure CH$_2$Cl$_2$ was chosen as eluent and silica gel as stationary phase for chromatography.

Yield: 844 mg (98 %).

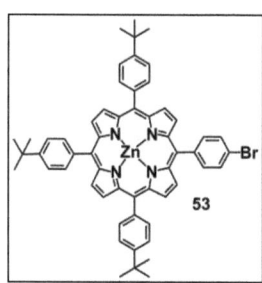

^{1}H NMR (CDCl$_3$, 400 MHz, RT): δ [ppm] = 8.96 (d, ^3J = 4.77 Hz, 2H, pyrH), 8.94 (s, 4H, pyrH), 8.88 (d, ^3J = 8.05, 2H, pyrH), 8.11 (d, ^3J = 8.07, 6H, PhH), 8.06 (d, ^3J = 8.15, 2H, PhH), 7.85 (d, ^3J = 7.95, 2H, PhH), 7.72 (d, ^3J = 7.90, 6H PhH), 1.59 (s, 27 H, CH$_3$).

^{13}C NMR (CDCl$_3$, 100.5 MHz, RT):

δ [ppm] = 150.15 (3C, PhC), 150.05 (1C, PhC), 149.82 (4C, α-pyrC), 149.46 (4C, α-pyrC) 140.32 (3C, PhC), 140.16 (2C, PhC), 135.88 (6C, PhC), 134.41 (2C, PhC), 131.92 (2C, β-pyrC), 131.55 (2C, β-pyrC), 130.95 (4C, β-pyrC), 129.49 (1C, PhC), 123.23 (10 C, PhC, mesoC), 34. 80 (3C, C(CH$_3$)$_3$), 31.66 (9C, CH$_3$).

MS (FAB, NBA): m/z = 926 [M]$^+$.

UV/Vis (CH$_2$Cl$_2$): λ$_{max}$ (log ε) [nm] = 592.0 (4.01), 550.5 (4.40), 423.0 (5.65).

IR (diamond, RT): $\tilde{\nu}$ [cm^{-1}] = 2961, 2361, 232, 1487, 1364, 1339, 1275, 1266, 1206, 1110, 1070, 999, 801, 764, 749, 723.

5-(*p*-Iodo-phenyl)-10,15,20-tri(*p*-*tert*-butyl-phenyl)-porphyrinato zinc(II) (54)

Free base porphyrin **52** (800 mg, 0.88 mmol) could be metallated using GOP X with zinc acetate (581 mg, 2.64 mmol). Purification was done with column chromatography on silica gel (CH$_2$Cl$_2$).
Yield: 835 mg (98%).

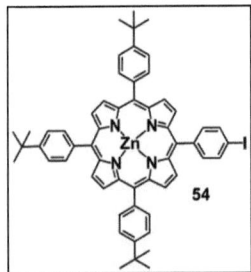

^{1}H NMR (CDCl$_3$, 400 MHz, RT): δ [ppm] = 8.94 (s, 4H, pyr*H*), 8.92 (d, ^3J = 4.78 Hz, 2H, pyr*H*), 8.88 (d, ^3J = 8.09, 2H, pyr*H*), 8.13 (d, ^3J = 8.05, 6H, Ph*H*), 8.06 (d, ^3J = 8.16, 2H, Ph*H*), 7.74 (d, ^3J = 7.95, 2H, Ph*H*), 7.72 (d, ^3J = 7.90, 6H Ph*H*), 1.62 (s, 27 H, C*H*$_3$).

^{13}C NMR (CDCl$_3$, 100.5 MHz, RT):
δ [ppm] = 150.60 (3C, Ph*C*), 150.34 (1C, Ph*C*), 149.99 (4C, α-pyr*C*), 149.87 (4C, α-pyr*C*) 139.74 (3C, Ph*C*), 139.58 (2C, Ph*C*), 136.00 (6C, Ph*C*), 135.57 (2C, Ph*C*), 134.23 (2C, β-pyr*C*), 132.04 (2C, β-pyr*C*), 131.90 (4C, β-pyr*C*), 131.02 (1C, Ph*C*), 121.01 (10 C, Ph*C*, meso*C*), 34.63 (3C, *C*(CH$_3$)$_3$), 31.62 (9C, *C*H$_3$).

MS (MALDI, DCTB): m/z = 975 [M]$^+$.

UV/Vis (CH$_2$Cl$_2$): λ$_{max}$ (log ε) [nm] = 592.0 (4.09), 548.5 (4.47), 421.0 (5.63).

IR (diamond, RT): $\tilde{\nu}$ [cm^{-1}] = 2956, 2031, 2010, 1461, 1337, 1267, 1205, 1109, 1066, 997, 798, 720.

5-(*p*-Formyl-phenyl)-10,15,20-tri(*p*-*tert*-butyl-phenyl)-porphyrin (55)

Bromo porphyrin **51** (400 mg, 0.46 mmol) was dissolved in dry THF (50 mL) and cooled down to -78 °C. Then *n*-BuLi (1 mL, 1.6 mmol, 1.6 M in hexanes) was added dropwise. The reaction mixture was then stirred for 2 h at -78 °C before DMF (71 µL, 0.92 mmol) was added dropwise. After another stirring for 30 min the solution was allowed to reach room temperature and 1 M HCl in H$_2$O (20 mL) was added. The mixture was neutralized with NaOH, extracted with CH$_2$Cl$_2$, washed with water, dried over MgSO$_4$ and then the solvent was removed. The residue was purified with column chromatography on silica gel (CH$_2$Cl$_2$).

Yield: 297 mg (80 %).

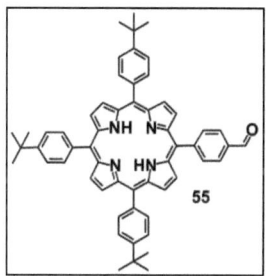

^{1}H NMR (CDCl$_3$, 400 MHz, RT): δ [ppm] = 10.32 (s, 1H, HC=O), 8.90 (d, ^3J = 4.77 Hz, 6H, pyrH), 8.76 (m, 2H, pyrH), 8.37 (d, ^3J = 7.32, 2H, PhH), 8.23 (d, ^3J = 6.96, 2H, PhH), 8.15 (d, ^3J = 7.95, 6H, PhH), 7.76 (d, ^3J = 7.95, 6H, PhH), 1.61 (s, 27 H, CH$_3$), -2.71 (s, 2H, NH).

^{13}C NMR (CDCl$_3$, 100.5 MHz, RT):

δ [ppm] = 192.13 (1C, HC=O), 150.54 (3C, PhC), 150.42 (1C, PhC), 141.31 (3C, PhC), 139.23 (2C, PhC), 139.11 (6C, PhC), 135.89 (2C, PhC), 131.14 (16C, pyrC), 129.87 (1C, PhC), 123.58 (6 C, PhC), 120.40 (3C, mesoC), 120.11 (1C, mesoC), 34.89 (3C, C(CH$_3$)$_3$), 31.69 (9C, CH$_3$).

MS (FAB, NBA): m/z = 811 [M]$^+$.

UV/Vis (CH$_2$Cl$_2$): λ$_{max}$ (log ε) [nm] = 645.5 (4.02), 592.5 (4.11), 553.0 (4.27), 515.5 (4.42), 421.0 (5.53).

EA: calculated for C$_{57}$H$_{54}$ON$_4$ * CH$_2$Cl$_2$ (896.00): C 77.75, H 6.30, N 6.25; found: C 77.23, H 5.95, N 6.54.

IR (diamond, RT): $\tilde{\nu}$ [cm^{-1}] = 2921, 2360, 1700, 1602, 1582, 1485, 1394, 1350, 1302, 1275, 1261, 1233, 1185, 1073, 1010, 987, 966, 819, 800, 764, 750, 724, 708.

5-(4-Ethynyl-trimethylsilyl-phenyl)-10,15,20-tri(3,5-dimethoxy-phenyl)-porphyrin (60)

Porphyrin **60** was synthesized according to GOP IXb with pyrrole **50** (1.39 mL, 0.02 mol), 3,5-dimethoxy-benzaldehyde **59** (2.46 g, 0.02 mol), 4-trimethylsilyl-ethynyl-benzaldehyde **58** (1 g, 0.01 mol), TFA (578 µL, 0.02 mol), NEt$_3$ (1.25 mL, 0.05 mol) and DDQ (3.405 g, 0.02 mol). The residue was purified with column chromatography on silica (hexanes/EtOAc 4:1).
Yield: 200 mg (5 %).

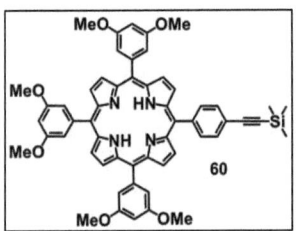

^{1}H NMR (CDCl$_3$, 400 MHz, RT):

δ [ppm] = 8.98 (d, ^3J = 4.6 Hz, 2H, pyr$H$), 8.97 (s, 4H, pyr$H$), 8.83 (d, ^3J = 4.6 Hz, 2H, pyr$H$), 8.20 (d, ^3J = 8.3 Hz, 2H, PhH), 7.91 (d, ^3J = 8.3, 2H, PhH), 7.41 (s, 6H, PhH), 6.92 (s, 3H, PhH), 3.98 (s, 18H, OCH_3), 0.40 (s, 9 H, CH_3), -2.80 (s, 2H, NH).

^{13}C NMR (CDCl$_3$, 100.5 MHz, RT):

δ [ppm] = 158.90 (3C, PhC), 144.01 (6C, PhC), 142.53 (2C, PhC), 134.44 (1C, PhC), 131.12 (2C, PhC) 130.46 (16C, pyr$C$), 122.67 (3C, PhC), 119.90 (1C, meso$C$), 119.21 (1C, meso$C$), 113.93 (4C, PhC), 105.00 (1C, ethynyl$C$), 100.28 (3C, PhC), 95.64 (1C, ethynyl$C$), 55.61 (6C, OCH$_3$), 0.19 (3C, CH$_3$).

MS (FAB, NBA): m/z = 891 [M]$^+$.

UV/Vis (CH$_2$Cl$_2$): λ$_{max}$ (log ε) [nm] = 645.0 (3.89), 592.0 (4.18), 555.5 (4.29), 516.0 (4.52), 422.0 (5.77).

FT-IR (diamond, RT): $\tilde{\nu}$ [cm^{-1}] = 3008, 2970, 2927, 2854, 2360, 2341, 2155, 1738, 1590, 1454, 1420, 1365, 1296, 1229, 1216, 1205, 1154, 1063, 1020, 973, 919, 860, 801, 758, 736, 711.

5-(4-Ethynyl-trimethylsilyl-phenyl)-10,15,20-tri(3,5-dimethoxy-phenyl)-porphyrinato zinc(II) (31)

Zinc porphyrin **61** was prepared with GOP X using free base porphyrin **60** (176 mg, 0.198 mmol) and zinc(II)acetate (130 mg, 0.59 mmol). Purification was achieved by column chromatography on silica gel (CH$_2$Cl$_2$/EtOAC 5:1).
Yield: 164 mg (87 %).

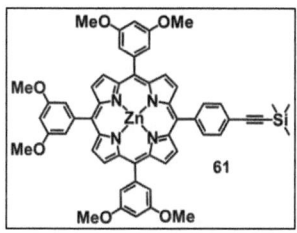

^{1}H NMR (CDCl$_3$, 400 MHz, RT):

δ [ppm] = 9.02 (d, ^3J = 4.6 Hz, 2H, pyr*H*), 9.01 (m, 4H, pyr*H*), 8.90 (d, ^3J = 4.6 Hz, 2H, pyr*H*), 8.16 (d, ^3J = 8.3 Hz, 2H, Ph*H*), 7.87 (d, ^3J = 8.3, 2H, Ph*H*), 7.30 (s, 6H, Ph*H*), 6.74 (s, 3H, Ph*H*), 3.83 (s, 18H, OC*H$_3$*), 0.38 (s, 9 H, C*H$_3$*).

^{13}C NMR (CDCl$_3$, 100.5 MHz, RT):

δ [ppm] = 158.63 (3C, Ph*C*), 150.09 (4C, α-pyr*C*), 150.04 (4C, α-pyr*C*), 144.54 (6C, Ph*C*), 143.18 (2C, Ph*C*), 134.26 (1C, Ph*C*), 132.15 (2C, β-pyr*C*), 132.05 (2C, β-pyr*C*), 131.63 (2C, Ph*C*) 130.21 (4C, β-pyr*C*), 122.35 (3C, Ph*C*), 120.91 (1C, meso*C*), 120.27 (1C, meso*C*), 113.8 (4C, Ph*C*), 105.10 (1C, ethynyl*C*), 100.00 (3C, Ph*C*), 95.48 (1C, ethynyl *C*), 55.62 (6C, O*C*H$_3$), 0.13 (3C, *C*H$_3$).

MS (FAB, NBA): m/z = 953 [M]$^+$.

UV/Vis (CH$_2$Cl$_2$): λ$_{max}$ (log ε) [nm] = 591.0 (3.64), 548.0 (4.38), 422.0 (5.69).

FT-IR (diamond, RT): $\tilde{\nu}$ [cm^{-1}] = 3006, 2970, 2945, 2360, 2342, 2155, 1738, 1589, 1508, 1494, 1454, 1421, 1365, 1351, 1228, 1217, 1205, 1153, 1063, 1028, 1001, 938, 860, 810, 797, 762, 740, 720.

5-(4-Ethynyl-phenyl)-10,15,20-tri(3,5-dimethoxy-phenyl)-porphyrinato zinc(II) (62)

Acetylenic porphyrin **62** could be synthesized by GOP V using porphyrin **61** (160 mg, 0.17 mmol) and TBAF (200 μL, 0.19 mmol of a 1M solution in THF). After evaporation of the solvent column chromatography on silica gel (CH$_2$Cl$_2$/EtOAc 4:1) was used to purify the desired product.

Yield: 135 mg (91 %).

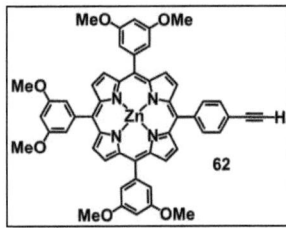

¹H NMR (CDCl$_3$, 400 MHz, RT):

δ [ppm] = 9.02 (d, ^3J = 4.6 Hz, 2H, pyr*H*), 9.01 (m, 4H, pyr*H*), 8.90 (d, ^3J = 4.6 Hz, 2H, pyr*H*), 8.17 (d, ^3J = 8.3 Hz, 2H, Ph*H*), 7.87 (d, ^3J = 8.3 Hz, 2H, Ph*H*), 7.33 (s, 6H, Ph*H*), 6.79 (s, 3H, Ph*H*), 3.86 (s, 19H, OC*H$_3$*, ethynyl*H*).

¹³C NMR (CDCl$_3$, 100.5 MHz, RT):

δ [ppm] = 158.62 (3C, Ph*C*), 150.10 (4C, α-pyr*C*), 150.04 (4C, α-pyr*C*), 144.53 (6C, Ph*C*), 143.52 (2C, Ph*C*), 134.38 (1C, Ph*C*), 132.29 (2C, β-pyr*C*), 132.11 (2C, β-pyr*C*), 131.62 (2C, Ph*C*) 130.45 (4C, β-pyr*C*), 121.37 (3C, Ph*C*), 120.90 (3C, meso*C*), 120.27 (1C, meso*C*), 113.82 (4C, Ph*C*), 100.04 (3C, Ph*C*), 83.36 (1C, ethynyl*C*), 78.12 (1C, ethynyl*C*), 55.63 (6C, O*C*H$_3$).

MS (FAB, NBA): m/z = 881 [M]$^+$.

UV/Vis (CH$_2$Cl$_2$): λ$_{max}$ (log ε) [nm] = 593.0 (3.81), 547.0 (4.49), 422.0 (5.78).

FT-IR (diamond, RT): $\tilde{v}$ [cm^{-1}] = 2999, 2970, 2940, 2836, 2580, 2360, 2342, 1738, 1589, 1525, 1494, 1453, 1420, 1365, 1350, 1317, 1230, 1216, 1203, 1152, 1063, 1001, 954, 938, 860, 811, 797, 762, 738, 720.

Bis-[5-(4-ethynyl-phenyl)-10,15,20-tri(3,5-dimethoxy-phenyl)-porphyrinato zinc(II)] (64)

Bisporphyrin **64** is a side product of the porphyrin **7** synthesis.

Yield: 14 mg (20 %)

Experimental Part

¹H NMR (CDCl₃, 400 MHz, RT): δ [ppm] = 9.02 (d, ^3J = 4.64 Hz, 4H, pyrH), 9.01 (s, 8H, pyrH), 8.90 (d, ^3J = 4.64 Hz, 4H, pyrH), 8.17 (d, ^3J = 8.30 Hz, 6H, PhH), 7.87 (d, ^3J = 8.30 Hz, 2H, PhH), 7.33 (s, 12H, PhH), 6.79 (s, 6H, PhH), 3.86 (s, 36H, OCH₃).

¹³C NMR (CDCl₃, 100.5 MHz, RT):

δ [ppm] = 158.64 (6C, PhC), 150.08 (8C, α-pyrC), 150.04 (8C, α-pyrC), 144.52 (12C, PhC), 143.47 (4C, PhC), 134.28 (2C, PhC), 132.16 (4C, β-pyrC), 132.06 (4C, β-pyrC), 131.61 (4C, PhC) 130.37 (8C, β-pyrC), 121.34 (6C, mesoC), 120.93 (4C, PhC), 120.28 (2C, mesoC), 113.76 (12C, PhC), 100.00 (6C, PhC), 95.22 (2C, ethynylC), 78.13 (2C, ethynylC), 55.57 (12C, OCH₃).

MS (FAB, NBA): m/z = 1762 [M]⁺.

UV/Vis (CH₂Cl₂): λ$_{max}$ (log ε) [nm] = 592.0 (4.05), 548.5 (4.64), 424.0 (5.80).

IR (diamond, RT): $\tilde{v}$ [cm⁻¹] = 2924, 2360, 2342, 1718, 1590, 1558, 1541, 1522, 1508, 1457, 1419, 1348, 1276, 1261, 1203, 1153, 1063, 1000, 797, 764, 751, 723.

Bis-[5-(phenyl)-10,15,20-tri(3,5-dimethoxy-phenyl)-thiopheno-porphyrinato zinc(II)] (65)

Bis-[5-(4-ethynyl-phenyl)-10,15,20-(3,5-dimethoxy-phenyl)-porphyrinato zinc(II)] **64** (10 mg, 0.01 mmol) was dissolved in dry DMF (10 mL) before Na₂S*9H₂O (7 mg, 0.03 mmol) was added. The reaction mixture was heated at 100 °C for 2h and stirred at rt over night. The solution was poored into CH₂Cl₂ (50 mL) and washed with water (100 mL) three times. After drying with MgSO₄ and evaporation of the solvent the crude product was purified by column chromatography on silica gel (CH₂Cl₂/EtOAc 5:1).

Yield: 8 mg (79%).

¹H NMR (CDCl₃, 400 MHz, RT):

δ [ppm] = 9.10 (d, ³J = 3.77 Hz, 4H, pyrH), 9.06 (s, 12H, pyrH), 8.30 (d, ³J = 7.91 Hz, 4H, PhH), 8.10 (d, ³J = 7.91 Hz, 4H, PhH), 7.69 (s, 2H, thioH), 7.43 (s, 8H, PhH), 7.40 (s, 4H, PhH), 6.89 (s, 4H, PhH), 6.86 (s, 2H, PhH), 3.96 (s, 24H, OCH₃), 3.93 (s, 12H, OCH₃).

MS (MALDI, DHB): m/z = 1792 [M]⁺.

UV/Vis (CH₂Cl₂): λ_max (log ε) [nm] = 592.5 (4.15), 548.5 (4.73), 423.0 (5.81).

IR (diamond, RT): ṽ [cm⁻¹] = 1987, 1583, 1416, 1340, 1190, 1025, 998, 934, 793.

EA: calculated for C₁₀₄H₈₀N₈O₁₂Zn₂S * CH₂Cl₂ (1881.61): C 67.02; H 4.39, N 5.96, S 1.70; found: C 67.19, H 4.48, N 5.75, S 2.07.

5-(4-Ethynyl-phenyl-Hamilton receptor)-10,15,20-tris(3,5-dimethoxy-phenyl)-porphyrinato zinc(II) (115)

Hamilton Receptor bearing porphyrin **115** was prepared with GOP IV using Iodo Hamilton receptor **19** (22 mg, 0.03 mmol), Pd₂dba₃ (1 mg, 0.01 mmol), AsPh₃ (1 mg, 0.01 mmol) and 5-(4-ethynyl-phenyl)-10,15,20-(3,5-dimethoxy-phenyl)-porphyrinato zinc(II) **62** (30 mg, 0.035 mmol) in dry THF (8 mL) and NEt₃ (3 mL). The residue was purified by column chromatography on silica (CH₂Cl₂→CH₂Cl₂/MeOH 99:1).
Yield: 40 mg (88 %).

¹H NMR (THF d₈, 400 MHz, RT):

δ [ppm] = 9.81 (bs, 2H, NH), 9.13 (bs, 2H, NH), 9.00 (d, ³J = 4.64 Hz, 2H, pyrH), 8.96 (s, 4H, pyrH), 8.88 (d, ³J = 4.64 Hz, 2H, pyrH), 8.57 (s, 1H, PhH),

8.47 (d, ^3J = 8.68 Hz, 2H, PhH), 8.31 (d, ^3J = 7.94 Hz, 2H, PhH), 8.10 (m, 2H, py$H$), 8.05 (m, 2H, py$H$), 7.78 (t, ^3J = 8.06 Hz, 2H, py$H$), 7.40 (s, 3H, PhH), 7.39 (s, 8H, o-PhH), 6.93 (s, 2H, PhH), 3.95 (s, 18H, OCH_3), 2.29 (s, 4H, CH_2), 1.11 (s, 18H, CH_3).

^{13}C NMR (THF d$_8$, 100.5 MHz, RT):

δ [ppm] = 170.78 (2C, C=O), 164.91 (2C, C=O), 159.85 (6C, PhC), 151.78 (2C, α-pyr$C$), 150.87 (4C, α-pyr$C$), 150.85 (2C, α-py$C$), 150.65 (4C, α-py$C$), 146.00 (6C, PhC), 145.43 (2C, PhC), 140.58 (2C, py$C$), 136.81 (2C, PhC), 132.39 (2C, β-pyr$C$), 132.20 (2C, β-pyr$C$), 131.59 (2C, PhC), 130.47 (4C, β-pyr$C$), 129.45 (2C, PhC), 127.57 (2C, PhC), 125.00 (3C, meso$C$), 122.49 (1C, meso$C$), 121.48 (1C, PhC), 121.45 (1C, PhC), 114.61 (6C, PhC), 110.42 (2C, β-py$C$), 110.23 (2C, β-py$C$), 100.17 (3C, PhC), 92.04 (1C, ethynyl$C$), 89.44 (1C, ethynyl$C$), 55.63 (6C, OCH$_3$), 50.72 (2C, CH$_2$), 31.65 (2C, C(CH$_3$)$_3$), 30.01 (6C, CH$_3$).

MS (FAB, NBA): m/z = 1424 [M]$^+$.

UV/Vis (CH$_2$Cl$_2$): λ_{max} (log ε) [nm] = 585.5 (3.89), 548.5 (4.40), 422.5 (5.63), 399.0 (4.73), 300.5 (4.89).

IR (diamond, RT): $\tilde{\nu}$ [cm^{-1}] = 2954, 2360, 2333, 1685, 1589, 1499, 1448, 1421, 1348, 1298, 1277, 1260, 1241, 1204, 1154, 1064, 1027, 1000, 938, 798, 763, 750.

5,10,15,20-Tetra(p-Bromo-phenyl)-porphyrin (116)

The synthesis of porphyrin **116** was done according to GOP IXa with pyrrole **50** (2.25 mL, 0.03 mol), 4-bromo-benzaldehyde **23** (6 g, 0.03 mol), tetraphenyl phosphonium chloride (50 mg, 0.13 mmol), boron trifluoride etherate (2.01 mL, 0.02 mol) and 2,3-dichloro-5,6-dicyano-1,4-benzoquinone (5.516 g, 0.02 mol). The residue was purified with column chromatography on silica gel with dichloromethane as eluent.

Yield: 4.47 g (16 %).

^{1}H NMR (THF d$_8$, 400 MHz, RT): δ [ppm] = 8.87 (s, 8H, pyr$H$), 8.12 (d, ^3J = 8.30, 8H, PhH), 7.97 (d, ^3J = 8.42, 8H, PhH), -2.79 (s, 2H, NH).

¹³C NMR (THF d₈, 100.5 MHz, RT):
δ [ppm] = 141.98 (8C, α-pyrC), 136.72 (16C, PhC), 130.79 (4C, PhC), 123.25 (4 C, PhC), 119.77 (8C, β-pyrC), 79.36 (4C, mesoC).

MS (FAB, NBA): m/z = 931 [M]⁺.

UV/Vis (CH₂Cl₂): λ_{max} (log ε) [nm] = 645.0 (3.84), 592.0 (3.94), 550.0 (4.08), 515.0 (4.30), 419.0 (5.48).

IR (diamond, RT): $\tilde{\nu}$ [cm⁻¹] = 1486, 1474, 1391, 1070, 1011, 992, 984, 967, 852, 842, 797, 756, 728.

EA: calculated for C₄₄H₂₆N₄Br₄ * (CH₂Cl₂) (1015.25): C 53.24, H 2.78, N 5.52; found: C 54.86, H 3.29, N 5.35.

5,10,15,20-Tetra(p-Bromo-phenyl)-porphyrinato zinc (II) (117)

Metallation of free base porphyrin **116** (520 mg, 0.56 mmol) was done under GOP X with zinc acetate (368 mg, 1.68 mmol). Purification was achieved with column chromatography on silica gel (CH₂Cl₂).
Yield: 544 mg (98 %).

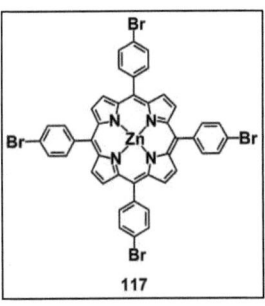

¹H NMR (THF d₈, 400 MHz, RT): δ [ppm] = 8.81 (s, 8H, pyrH), 8.04 (d, ³J = 8.25 Hz, 8H, PhH), 7.67 (d, ³J = 8.33 Hz, 8H, PhH).

¹³C NMR (THF d₈, 100.5 MHz, RT):
δ [ppm] = 150.79 (8C, α-pyrC), 136.62 (8C, PhC), 135.14 (8C, PhC), 135.11 (4C, PhC), 130.14 (4C, PhC), 123.85 (8C, β-pyrC), 108.13 (4C, mesoC).

MS (FAB, NBA): m/z = 994 [M]⁺.

UV/Vis (CH$_2$Cl$_2$): λ_{max} (log ε) [nm] = 591.5 (3.91), 548.5 (4.47), 420.5 (5.68).

IR (diamond, RT): $\tilde{v}$ [cm^{-1}] = 2359, 1521, 1479, 1390, 1336, 1276, 1261, 1204, 1177, 1098, 1070, 1042, 1007, 994, 949, 883, 850, 793, 764, 751, 720.

EA: calculated for C$_{44}$H$_{24}$N$_4$Br$_4$Zn * (CH$_2$Cl$_2$) (1078.65): C 50.11; H 2.43; N 5.19; found: C 50.86, H 3.19, N 4.79.

5,10,15,20-Tetra(4-ethynyl-trimethylsilyl-phenyl)-porphyrinato zinc (II) (118)

GOP IV was applied to prepare porphyrin **118** with 5,10,15,20-tetra(*p*-bromo-phenyl)-porphyrinato zinc(II) **117** (400 mg, 0.4 mmol), Pd(PPh$_3$)Cl$_2$ (12 mg, 0.016 mmol), CuI (9 mg, 0.048 mmol), PPh$_3$ (8 mg, 0.032 mmol) and trimethylsilyl-acetylene (251 µL, 1.78 mmol) in THF/HNEt$_2$ (25 mL, 1:2). The reaction mixture was heated to 80 °C for 2 d. For purification column chromatography on silica gel (CH$_2$Cl$_2$) was used.
Yield: 363 mg (85 %)

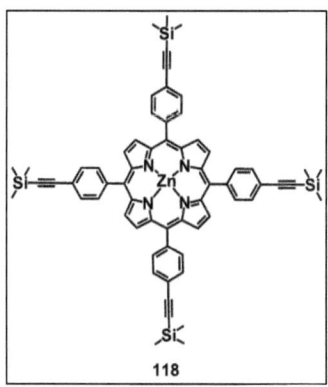

^{1}H NMR (CDCl$_3$, 400 MHz, RT): δ [ppm] = 8.86 (s, 8 H, pyr*H*), 8.10 (d, ^3J= 6.02 Hz, 8H, Ph*H*), 7.87 (d, ^3J = 7.32 Hz, 8H, Ph*H*), 0.38 (s, 36 H, C*H*$_3$).

^{13}C NMR (CDCl$_3$, 100.5 MHz, RT):
δ [ppm] = 149.92 (4C, Ph*C*), 142.84 (8C, Ph*C*), 135.71 (8C, Ph*C*), 134.27 (8C, α-pyr*C*), 131.95 (4C, Ph*C*), 130.26 (8C, ß-pyr*C*), 122.54 (4C, meso*C*), 105.06 (4C, ethynyl*C*) 95.52 (4C, ethynyl*C*), 0.09 (12C, *C*H$_3$).

MS (FAB, NBA): m/z = 1061 [M]$^+$.

UV/Vis (CH$_2$Cl$_2$): λ_{max} (log ε) [nm] = 590.0 (4.25), 550.0 (3.46), 422.0 (4.74), 312.0 (3.65), 260.0 (3.13).

IR (diamond, RT): $\tilde{v}$ [cm^{-1}] = 2157, 1486, 1276, 1249, 1072, 1000, 862, 842, 810, 796, 760, 720.

5,10,15,20-Tetra(4-ethynyl-phenyl)-porphyrinato zinc (II) (119)

Deprotection of porphyrin **118** (152 mg, 0.14 mmol) was achieved by GOP V using TBAF in THF (127 µL of 1M, 0.63 mmol). The crude product was cleaned by column chromatography on silica gel (CH$_2$Cl$_2$).
Yield: 120 mg (100 %).

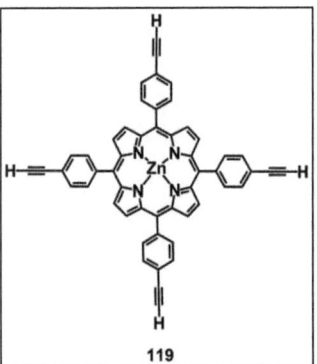

^{1}H NMR (CDCl$_3$, 400 MHz, RT): δ [ppm] = 8.87 8.86 (s, 8 H, pyr*H*), 8.16 (d, ^3J = 8.06 Hz, 8H, Ph*H*), 7.87 (d, ^3J = 8.06 Hz, 8H, Ph*H*), 3.29 (s, 4H, ethynyl *H*).

^{13}C NMR (CDCl$_3$, 100.5 MHz, RT):
δ [ppm] = 149.90 (4C, Ph*C*), 143.58 (8C, Ph*C*), 135.77 (8C, Ph*C*), 134.37 (8C, α-pyr*C*), 131.89 (4C, Ph*C*), 130.34 (8C, β-pyr*C*), 129.70 (4C, meso*C*), 83.75 (4C, ethynyl*C*) 78.15 (4C, ethynyl*C*).

MS (FAB, NBA): m/z = 773 [M]$^+$.

UV/Vis (CH$_2$Cl$_2$): λ$_{max}$ (log ε) [nm] = 590.0 (4.33), 550.0 (4.76), 422.0 (5.07), 306.0 (3.85), 248.0 (4.25).

IR (diamond, RT): $\tilde{v}$ [cm^{-1}] = 2360, 1276, 1261, 1097, 1069, 998, 796, 764, 750, 720, 708.

5,10,15,20-Tetra(4-Hamilton receptor-ethynyl-phenyl)-porphyrinato zinc (II) (120)

GOP IV was used for the preparation of tetra Hamilton receptor porphyrin **120**: iodo Hamilton receptor **19** (81 mg, 0.12 mmol), Pd(PPh$_3$)$_2$Cl$_2$ (2 mg, 0.01 mmol), CuI (2 mg, 0.01 mmol) and tetraacetylene porphyrin **119** (20 mg, 0.03 mmol) in a mixture

of THF (10 mL) and NEt$_3$ (5 mL) at 80 °C for 3 d. Column chromatography on silica gel (CH$_2$Cl$_2$/MeOH 3%) of the residue yielded the desired product
Yield: 56 mg (74 %)

^{1}H-NMR (THF d$_8$, 400 MHz, RT): δ [ppm] = 9.79 (s, 8H, N*H*), 9.09 (s, 8H, N*H*), 8.96 (s, 8H, pyr*H*), 8.55 (s, 4H, Ph*H*), 8.45 (s, 8H, Ph*H*), 8.32 (d, ^3J = 8.10 Hz, 8H, Ph*H*), 8.06 (m, 24H, Ph*H*, py*H*), 7.77 (t, ^3J = 8.01, 8H, py*H*), 2.28 (s, 16H, C*H*$_2$), 1.09 (s, 72 H, C*H*$_3$).

^{13}C-NMR (THF d$_8$, 100.5 MHz, RT):
δ [ppm] = 171.05 (8C, *C*=O), 165.19 (8C, *C*=O), 152.02 (8C, α-pyr*C*), 151.52 (8C, α-py*C*), 150.98 (8C, α-py*C*), 143.34 (8C, Ph*C*), 140.87 (8C, py*C*), 137.09 (8C, Ph*C*), 135.87 (8C, Ph*C*), 134.63 (8C, Ph*C*), 132.58 (8C, Ph*C*), 130.85 (4C, Ph*C*), 129.40 (8C, β-pyr*C*), 125.20 (4C, meso *C*), 122.99 (4C, Ph*C*), 110.72 (8C, β-py *C*), 110.50 (8C, β-py*C*), 92.16 (4C, ethynyl*C*), 89.85 (4C, ethynyl*C*), 50.98 (8C, *C*H$_2$), 31.93 (8C, *C*(CH$_3$)$_3$), 30.27 (32C, *C*H$_3$).

120

MS (FAB, NBA): m/z = 2941 [M]$^+$.

UV/Vis (THF): λ_{max} (log ε) [nm] = 597.5 (4.15), 557.0 (4.41), 430.5 (5.68), 299.0 (5.17).

IR (diamond, RT): $\tilde{\nu}$ [cm^{-1}] = 2960, 1674, 1584, 1506, 1396, 1295, 1239, 1154, 1129, 996, 796, 718, 594, 540.

6.2.3 Building Blocks for Supramolecular Wires and Their Precursors

***N,N'*-Bis[6-(3,3-dimethylbutyrylamino)pyridin-2-yl]-5-trimethylsilylethynyl-isophthalamide (121)**

The synthesis was done according to GOP IV using Iodo-Hamilton Receptor **19** (500 mg, 0.75 mmol), Pd(PPh$_3$)Cl$_2$ (6 mg, 0.01 mmol), CuI (4 mg, 0.02 mmol) and trimethylsilylacetylene (88 mg, 0.90 mmol) in dry THF (12 mL) and NEt$_3$ (7 mL). The reaction mixture was stirred for another 24 hours at rt. Purification of the raw product was achieved by column chromatography (CH$_2$Cl$_2$/EtOAc 9:1).
Yield: 463 mg (97 %).

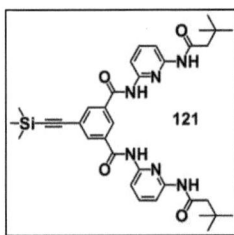

^{1}H NMR (THF d$_8$, 400 MHz, RT):
δ [ppm] = 9.68 (bs, 2H, N*H*), 9.09 (bs, 2H, N*H*), 8.47 (s, 1H, Ph*H*), 8.19 (s, 2H, Ph*H*), 8.03 (m, 4 H, py*H*), 7.72 (d, ^3J = 8.0 Hz, 2H, py*H*), 2.63 (s, 4H, C*H*$_2$), 1.08 (s, 18 H, C*H*$_2$), 0.28 (s, 9H, Si(C*H*$_3$)$_3$).

^{13}C NMR (THF d$_8$, 100.5 MHz, RT):
δ [ppm] = 170.7 (2C, *C*=O), 164.7 (2C, *C*=O), 151.7 (2C, α-py*C*), 151.1 (2C, α-py*C*), 140.5 (2C, Ph*C*), 136.5 (2C, Ph*C*), 134.4 (1C, Ph*C*), 127.9 (1C, Ph*C*), 124.6 (1C, py*C*), 110.4 (2C, β-py*C*), 110.2 (2C, β-py*C*), 104.3 (1C, ethynyl*C*), 96.5 (1C, ethynyl*C*), 50.7 (2C, *C*H$_2$), 31.6 (2C, *C*(CH$_3$)$_3$), 30.0 (6C, *C*H$_3$), -0.3 (3C, Si(*C*H$_3$)$_3$).

MS (FAB, NBA): m/z = 642 [M]$^+$.

Experimental Part

FT-IR (diamond, RT): $\tilde{v}$ [cm^{-1}] = 3294, 2956, 2869, 2355, 1673, 1585, 1510, 1444, 1397, 1366, 1297, 1244, 1156, 802.

N,N'-Bis[6-(3,3-dimethylbutyrylamino)pyridin-2-yl]-5-ethynyl-isophthalamide (122)

Cleavage of the protecting group was achieved by GOP V using TBAF in THF (0.23 mL, 0.76 mmol, 1 M solution) and TMS protected acetylene Hamilton receptor **121** (443 mg, 0.69 mmol) in dry THF (10 mL) at room temperature. Eluation with CH$_2$Cl$_2$/EtOAc 4:1 yielded the desired product.
Yield: 378 mg (96 %).

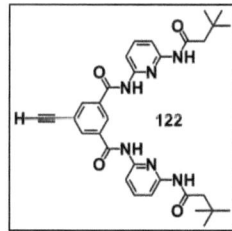

^{1}H NMR (THF d$_8$, 400 MHz, RT):

δ [ppm] = 9.70 (bs, 2H, N*H*), 9.08 (bs, 2H, N*H*), 8.48 (s, 1H, Ph*H*), 8.22 (s, 2H, Ph*H*), 8.03 (m, 4 H, py*H*), 7.73 (d, ^3J = 8.0 Hz, 2H, py*H*), 3.83 (s,1H, ethynyl*H*), 2.70 (s, 4H), C*H$_2$*, 1.08 (s, 18 H, C*H$_3$*).

^{13}C NMR (THF d$_8$, 100.5 MHz, RT):

δ [ppm] = 170.8 (2C, *C*=O), 164.7 (2C, *C*=O), 151.7 (2C, α-py*C*), 151.2 (2C, α-py*C*), 140.5 (2C, Ph*C*), 136.6 (2C, Ph*C*), 134.6 (2C, Ph*C*), 127.9 (2C, py*C*), 124.0 (1C, Ph*C*), 110.4 (2C, β-py*C*), 110.2 (2C, β-py*C*), 82.7 (1C, ethynyl*C*), 80.8 (1C, ethynyl*C*), 50.7 (2C, *C*H$_2$), 31.6 (2C, *C*(CH$_3$)$_3$), 30.0 (6C, *C*H$_3$).

MS (FAB, NBA): m/z = 570 [M]$^+$.

FT-IR (diamond, RT): $\tilde{v}$ [cm^{-1}] = 3295, 2954, 2865, 2360, 1673, 1555, 1512, 1446, 1396, 1297, 1240, 1156, 1132, 800.

1,4-Bis-N,N'-bis[6-(3,3-dimethylbutyrylamino) pyridin-2-yl]-5-ethynyl-isophthalamide (123)

Double Hamilton receptor **123** was accessible acording to GOP IV using iodo Hamilton Receptor **19** (107 mg, 0.16 mmol), Pd(PPh$_3$)$_4$ (4 mg, 0.003 mmol), CuI (1 mg, 0.005 mmol) and ethynyl Hamilton receptor **122** (100 mg, 0.18 mmol) in dry THF (15 mL) and HN*i*Pr$_2$ (4 mL). The mixture was stirred for 40 h at room

temperature and the remainder was purified by column chromatography on silica gel (CH$_2$Cl$_2$/EtOAc/MeOH 4:1:0.02).
Yield: 122 mg (69 %).

^{1}H NMR (DSMO d$_6$, 400 MHz, RT):

δ [ppm] = 10.69 (bs, 4H, N*H*), 10.04 (bs, 4H, N*H*), 8.53 (s, 2H, Ph*H*), 8.34 (s, 4H, Ph*H*), 7.86 (m, 8 H, py*H*), 7.72 (m, 4H, py*H*), 2.31 (s, 8H, C*H*$_2$), 1.02 (s, 36H, C*H*$_3$).

^{13}C NMR (DMSO d$_6$, 100.5 MHz, RT):

δ [ppm] = 171.29 (4C, *C*=O), 164.86 (4C, *C*=O), 150.91 (4C, α-py*C*), 150.37 (4C, α-py*C*), 140.38 (4C, Ph*C*), 135.28 (4C, Ph*C*), 134.16 (2C, Ph*C*), 132.32 (2C, Ph*C*), 122.87 (4C, py*C*), 110.95 (4C, β-py*C*), 110.46 (4C, β-py*C*), 90.50 (2C, ethynyl*C*), 49.42 (4C, CH$_2$), 31.25 (4C, *C*(CH$_3$)$_3$), 29.92 (12C, CH$_3$).

MS (FAB, NBA): m/z = 1111 [M]$^+$.

FT-IR (diamond, RT): $\tilde{\nu}$ [cm^{-1}] = 3730, 3272, 2962, 1701, 1686, 1675, 1655, 1639, 1543, 1440, 1295, 1238, 797.

Bis-*N,N'*-bis[6-(3,3-dimethylbutyrylamino) pyridin-2-yl]-5-ethynyl-isophthalamide (124)

Using GOP IV OPA wire **124** was accessible with the following reagents: Pd(PPh$_3$)$_2$Cl$_2$ (2 mg, 0.01 mmol), CuI (2 mg, 0.01 mmol) and Hamilton receptor **122** (172 mg, 0.3 mmol) in dry THF (12 mL) and NEt$_3$ (5 mL). The mixture was stirred at room temperature for 72 h and the crude product was purified using column chromatography (CH$_2$Cl$_2$/MeOH 95:5).
Yield: 163 mg (95 %).

^{1}H NMR (THF d$_8$, 400 MHz, RT): δ [ppm] = 9.82 (bs, 4H, N*H*), 9.16 (bs, 4H, N*H*), 8.58 (s, 2H, Ph*H*), 8.35 (s, 4H, Ph*H*), 8.04 (m, 8 H, py*H*), 7.74 (d, ^3J = 8.0 Hz, 4H, py*H*), 2.27 (s, 8H, C*H*$_2$), 1.09 (s, 36H, C*H*$_3$).

Experimental Part

^{13}C NMR (THF d_8, 100.5 MHz, RT):

δ [ppm] = 170.8 (4C, C=O), 164.8 (4C, C=O), 151.8 (4C, α-pyC), 151.1 (4C, α-pyC), 140.6 (4C, PhC), 136.8 (4C, PhC), 135.3 (2C, PhC), 129.1 (4C, pyC), 122.8 (2C, PhC), 110.5 (4C, β-pyC), 110.2 (4C, β-pyC), 81.4 (2C, ethynylC), 75.2 (2C, ethynylC), 50.7 (4C, CH_2), 31.6 (4C, $C(CH_3)_3$), 30.0 (12C, CH_2).

MS (FAB, NBA): m/z = 1135 [M]$^+$.

FT-IR (diamond, RT): $\tilde{v}$ [cm^{-1}] = 3726, 3307, 2970, 2360, 2342, 1738, 1680, 1585, 1513, 1446, 1365, 1298, 1233, 1156, 1131, 1085, 799.

1,4-[Bis-*N,N'*-bis[6-(3,3-dimethylbutyrylamino) pyridin-2-yl]-5-ethynyl-isophthalamide] benzene (126)

OPA **126** was synthesized with GOP IV using 1,4-diiodobenzene **125** (27 mg, 0.08 mmol), PPh$_3$ (1 mg, 0.01 mmol), Pd(PPh$_3$)Cl$_2$ (1 mg, 0.01 mmol), CuI (1 mg, 0.01 mmol) and Hamilton receptor **122** (100 mg, 0.176 mmol) in dry THF (10 mL) and HN*i*Pr$_2$ (4 mL). The mixture was stirred for 60 h at room temperature and the remainder was purified by column chromatography on silica gel (CH$_2$Cl$_2$/EtOAc/MeOH 4:1:0.02).
Yield: 152 mg (72 %).

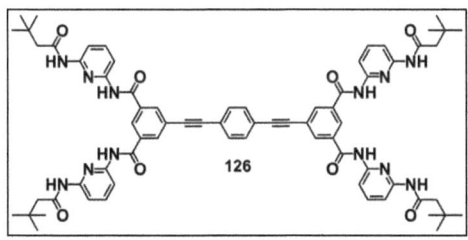

^{1}H NMR (THF d_8, 400 MHz, RT):

δ [ppm] = 9.79 (bs, 4H, N*H*), 9.12 (bs, 4H, N*H*), 8.55 (s, 2H, Ph*H*), 8.33 (s, 4H, Ph*H*), 8.04 (m, 8 H, py*H*), 7.73 (m, 4H, py*H*), 7.63 (s, 4H, Ph*H*), 2.28 (s, 8H, C*H*$_2$), 1.09 (s, 36H, C*H*$_3$).

^{13}C NMR (THF d$_8$, 100.5 MHz, RT):

δ [ppm] = 170.8 (4C, C=O), 164.8 (4C, C=O), 151.7 (4C, α-pyC), 151.1 (4C, α-pyC), 140.6 (4C, PhC), 136.8 (4C, PhC), 135.3 (4C, PhC), 133.8 (2C, PhC), 132.6 (2C, PhC), 129.5 (4C, pyC), 122.8 (2C, PhC), 110.5 (4C, β-pyC), 110.2 (4C, β-pyC), 81.4 (2C, ethynylC), 75.3 (2C, ethynylC), 50.7 (4C, CH$_2$), 31.6 (4C, C(CH$_3$)$_3$), 30.0 (12C, CH$_3$).

MS (FAB, NBA): m/z = 1210 [M]$^+$.

FT-IR (diamond, RT): $\tilde{v}$ [cm^{-1}] = 3732, 3286, 2970, 2360, 2342, 1738, 1679, 1585, 1510, 1446, 1365, 1297, 1231, 1218, 1156, 1131, 799.

1,4-[Bis-N,N'-Bis[6-(3,3-dimethylbutyrylamino) pyridin-2-yl]-5-ethynyl-isophthalamide]-2,5-bisoctyloxy benzene (128)

GOP IV was applied for the preparation of OPA **128** using 1,4-Diiodo-2,5-bisoctyloxybenzene **127** (32 mg, 0.052 mmol), Pd(PPh$_3$)$_2$Cl$_2$ (3 mg, 0.004 mmol), CuI (2 mg, 0.01 mmol) and Hamilton receptor **122** (65 mg, 0.115 mmol) in dry THF (15 mL) and HNEt$_2$ (5 mL). The reaction mixture was then stirred for 3.5 h at 70 °C and then for 48 h at room temperature. The crude product was purified by column chromatography on silica gel (CH$_2$Cl$_2$/EtOAc/MeOH 5:1:0.01).
Yield: 102 mg (75 %).

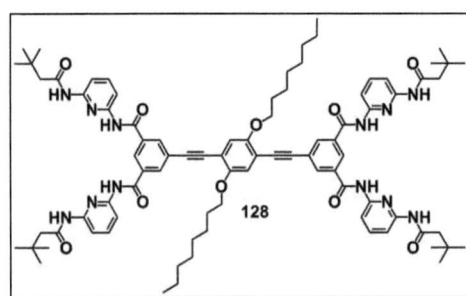

^{1}H NMR (THF d$_8$, 400 MHz, RT):

δ [ppm] = 9.77 (bs, 4H, NH), 9.12 (bs, 4H, NH), 8.48 (s, 2H, PhH), 8.28 (s, 4H, PhH), 8.06 (d, ^3J = 6.70 Hz, 4H, pyH), 8.03 (t, ^3J = 6.80 Hz, 4H, pyH), 7.74 (d, ^3J = 8.12 Hz, 4H, pyH), 7.19 (s, 2H, PhH), 4.12 (t, ^3J = 6.4 Hz, 4H, OCH$_2$), 2.27 (s, 4H, CH$_2$), 1.73 (m, 12H, CH$_2$CH$_2$), 1.30 (m, 12H, CH$_2$CH$_2$), 1.09 (s, 36 H, CH$_3$), 0.82 (t, ^3J = 6.73 Hz, 6H, CH$_2$CH$_3$).

¹³C NMR (THF d₈, 100.5 MHz, RT):

δ [ppm] = 170.8 (4C, C=O), 164.9 (4C, C=O), 154. 8 (2C, PhC), 151.7 (4C, α-pyC), 151.1 (4C, α-pyC), 140.5 (4C, PhC), 136.6 (4C, PhC), 134.2 (2C, PhC), 129.5 (4C, pyC), 125.2 (2C, PhC), 117.6 (2C, PhC), 114.7 (2C, PhC), 110.4 (4C, β-pyC), 110.2 (4C, β-pyC), 93.9 (2C, ethynylC), 88.6 (2C, ethynylC), 70.2 (2C, OCH₂) 50.7 (4C, CH₂), 31.6 (4C, C(CH₃)₃), 32.7 (4C, CH₂CH₂), 31.6 (12C, CH₃), 30.0 (2C, CH₂CH₂), 29.8 (2C, CH₂CH₂), 25.5 (2C, CH₂CH₂), 23.4 (2C, CH₂CH₂), 14.4 (2C, CH₂CH₃).

MS (FAB, NBA): m/z = 1468 [M]⁺.

FT-IR (diamond, RT): ṽ [cm⁻¹] = 3303, 2954, 2868, 2359, 1738, 1681, 1586, 1681, 1589, 1503, 1446, 1366, 1323, 1297, 1239, 1156, 1083, 800.

6.2.4 Porphyrin Precursors for Wittig-Horner Olefinations

5-(4-Methoxy-methyl-phenyl)-10,15,20-tri(p-tert-butyl-phenyl)-porphyrin (130)

Porphyrin **130** could be synthesized using GOP IX with pyrrole **50** (2.25 mL, 0.03 mol), 4-*tert*-butylbenzaldehyde **48** (3.94 g, 0.02 mol), 4-methoxy-methyl-benzaldehyde **129** (1.215 g, 0.01 mol), tetraphenyl phosphonium chloride (50 mg, 0.13 mmol), boron trifluoride etherate (2.01 mL, 0.02 mol) and 2,3-dichloro-5,6-dicyano-1,4-benzoquinone (5.516 g, 0.02 mol). The residue was purified with column chromatography on silica gel (CH₂Cl₂).
Yield: 870 mg (15 %)

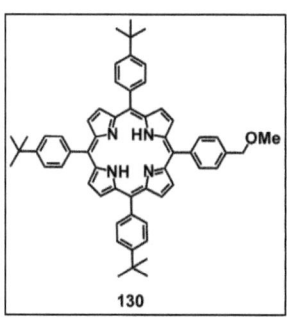

¹H NMR (CDCl₃, 400 MHz, RT):

δ [ppm] = 8.90 (m, 6H, pyrH), 8.86 (d, ³J = 4.48 Hz, 2H, pyrH), 8.24 (d, ³J = 7.81, 2H, PhH), 8.15 (d, ³J = 7.95 Hz, 6H, PhH), 7.79 (d, ³J = 8.06 Hz, 6H, PhH), 7.75 (d, ³J = 7.81 Hz, 2H, PhH), 4.82 (s, 2H, CH₂), 3.67 (s, 3H, OCH₃), 1.63 (s, 27 H, CH₃), -2.72 (s, 2H, NH).

¹³C NMR (CDCl₃, 100.5 MHz, RT):

δ [ppm] = 150.56 (3C, PhC), 141.82 (1C, PhC), 139.29 (3C, PhC), 139.25 (1C, PhC), 137.60 (6C, PhC), 134.74 (2C, PhC), 131.18 (16C, pyrC), 126.11 (2C, PhC), 123.65 (6 C, PhC), 120.28 (3C, mesoC), 119.52 (1C, mesoC), 74.88 (1C, CH_2), 58.63 (1C, OCH_3), 34.84 (3C, $C(CH_3)_3$), 31.63 (9C, CH_3).

MS (FAB, NBA): m/z = 826 [M]⁺.

UV/Vis (CH₂Cl₂): λ_{max} (log ε) [nm] = 647.0 (3.77), 591.5 (3.62), 551.5 (3.96), 516.5 (4.17), 419.0 (5.51).

IR (diamond, RT): $\tilde{v}$ [cm⁻¹] = 2964, 2925, 2870, 2821, 1916, 1731, 1609, 1560, 1503, 1474, 1398, 1363, 1352, 1267, 1221, 1195, 1154, 1109, 1050, 1024, 982, 967, 923, 881, 849.

EA: calculated for $C_{58}H_{58}N_4O$ * 0.5 CH_2Cl_2 (869.57): C 80.80, H 6.84, N 6.44; found: C 81.31, H 7.29, N 6.00.

5-(4-Bromo-methyl-phenyl)-10,15,20-tri(*p-tert*-butyl-phenyl)-porphyrin (131)

After dissolving *p*-methoxy-methyl porphyrin **130** (150 mg, 0.18 mmol) in CH₂Cl₂ (50 mL) HBr (10 mL, 33 wt% in glacial acetic acid) was added. The solution was stirred for 4 h before water and saturated NaHCO₃ solution was added until the green color flips to violet. The organic layer was separated and washed three times with water. After drying with MgSO₄ and removal of the solvent the desired product could be isolated.
Yield: 153 mg (97 %).

¹H NMR (CDCl₃, 400 MHz, RT): δ [ppm] = 8.90 (m, 6H, pyr*H*), 8.84 (d, ^3J = 4.46 Hz, 2H, pyr*H*), 8.22 (d, ^3J = 8.30 Hz, 2H, Ph*H*), 8.17 (d, ^3J = 7.95, 6H, Ph*H*), 7.80 (d, ^3J = 8.06, 6H, Ph*H*), 7.79 (d, ^3J = 7.81, 2H, Ph*H*), 4.86 (s, 2H, CH_2), 1.62 (s, 27 H, CH_3), -2.74 (s, 2H, N*H*).

Experimental Part

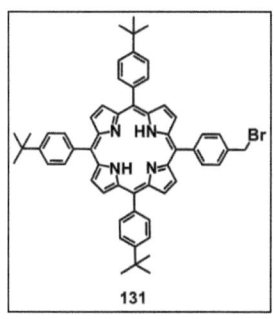
131

¹³C NMR (CDCl₃, 100.5 MHz, RT):
δ [ppm] = 150.59 (3C, PhC), 141.54 (1C, PhC), 139.29 (3C, PhC), 139.28 (1C, PhC), 137.60 (6C, PhC), 134.74 (2C, PhC), 131.18 (16C, pyrC), 124.82 (2C, PhC), 123.65 (6C, PhC), 120.28 (3C, mesoC), 119.52 (1C, meso), 74.88 (1C, CH_2), 34.74 (3C, $C(CH_3)_3$), 31.62 (9C, CH_3).

MS (FAB, NBA): m/z = 874 [M]⁺.

UV/Vis (CH₂Cl₂): λ_{max} (log ε) [nm] = 670.5 (4.02), 591.5 (3.82), 551.5 (4.07), 516.5 (4.21), 420.5 (5.43).

IR (diamond, RT): $\tilde{\nu}$ [cm⁻¹] = 2960, 2361, 1608, 1526, 1485, 1463, 1438, 1394, 1363, 1339, 1267, 1206, 1178, 1113, 1071, 999, 853, 801, 749, 722.

EA: calculated for $C_{57}H_{55}N_4Br$ * CH_2Cl_2 (960.91): C 72.50, H 5.98, N 5.83; found: C 72.02, H 6.25, N 6.07.

5-(4-Bromo-methyl-phenyl)-10,15,20-tri(*p-tert*-butyl-phenyl)-porphyrinato zinc (II) (132)

Metallation of porphyrin **131** (95 mg, 0.11 mol) with zinc(II)acetate (72 mg, 0.33 mmol) was accomplished by GOP X.
Yield: 99 mg (96 %).

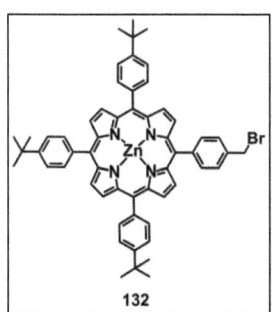
132

¹H NMR (CDCl₃, 400 MHz, RT):
δ [ppm] = 8.99 (m, 6H, pyrH), 8.93 (d, ³J = 4.71 Hz, 2H, pyrH), 8.23 (d, ³J = 7.72 Hz, 2H, PhH), 8.17 (d, ³J = 7.54, 6H, PhH), 7.80 (d, ³J = 8.06, 6H, PhH), 7.79 (d, ³J = 7.94, 2H, PhH), 4.88 (s, 2H, CH_2), 1.63 (s, 27 H, CH_3).

¹³C NMR (CDCl₃, 100.5 MHz, RT):
δ [ppm] = 150.29 (3C, PhC), 142.94 (1C, PhC), 139.55

(3C, PhC), 139.51 (1C, PhC), 137.52 (6C, PhC), 135.36 (2C, PhC), 134.89 (8C, α-pyrC), 134.87 (2C, β-pyrC), 127.75 (6C, β-pyrC), 123.99 (2C, PhC), 123.93 (6 C, PhC), 120.69 (3C, mesoC), 119.52 (1C, mesoC), 77.52 (1C, CH_2), 35.16 (3C, C(CH_3)$_3$), 31.62 (9C, CH_3).

MS (FAB, NBA): m/z = 938 [M]$^+$.

UV/Vis (CH$_2$Cl$_2$): λ_{max} (log ε) [nm] = 588.5 (3.75), 549.0 (4.22), 421.0 (5.64).

IR (diamond, RT): $\tilde{v}$ [cm^{-1}] = 2960, 2361, 2328, 1697, 1609, 1508, 1474, 1398, 1363, 1349, 1313, 1267, 1222, 1194, 1160, 1109, 1070, 1051, 1024, 983, 968, 850, 800, 764, 748.

EA: calculated for C$_{57}$H$_{53}$N$_4$BrZn * (0.5 CHCl$_3$) (995.23): C 69.13, H 5.40, N 5.61; found: C 69.40, H 5.52, N 5.26.

5-(4-Diethyl-phosponato-methyl-phenyl)-10,15,20-tri(p-tert-butyl-phenyl)-porphyrin (133)

Bromomethyl porphyrin **131** (90 mg, 0.096 mmol) was suspended in triethyl phosphate (10 mL) and carefully heated up to 160 °C for 1 h. After cooling down to room temperature the excessive triethyl phosphate was distilled. The occurring oil was treated with n-pentane to give a violet solid. Unfortunately porphyrin **133** is rather bad soluble even in THF or DMSO. Therefore only broad signals in the ^{1}H-NMR spectrum could be detected.

Yield: 524 mg (99 %)

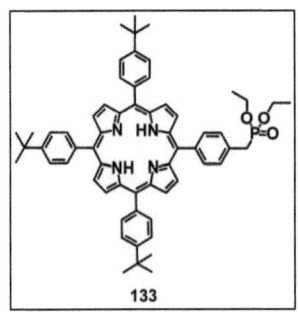

^{1}H NMR (DMSO d$_6$, 400 MHz, RT):
δ [ppm] = 8.88 (s, 6H, pyrH), 8.69 (d, ^3J = 4.46 Hz, 2H, pyrH), 8.03 (d, ^3J = 8.30 Hz, 2H, PhH), 7.70 (d, ^3J = 7.95, 6H, PhH), 7.50 (d, ^3J = 8.06, 6H, PhH), 7.25 (d, ^3J = 7.81, 2H, PhH), 5.87 (s, 2H, CH$_2$), 2.39 (m, 4H, OCH$_2$), 1.62 (bs, 33 H, CH$_3$), -2.77 (s, 2H, NH).

MS (FAB, NBA): m/z = 996 [M]$^+$.

Experimental Part

IR (diamond, RT): $\tilde{\nu}$ [cm^{-1}] = 2960, 2361, 1597, 1444, 1338, 1285, 1191, 993, 795.

5-(4-Trisphenylphosphino-methyl-phenyl)-10,15,20-tri(*p-tert*-butyl-phenyl)-porphyrinato bromide (134)

Bromo methyl porphyrin **131** (100 mg, 0.11 mmol) and triphenylphosphine (90 mg, 0.34 mmol) were dissolved in THF (100 mL) and refluxed for 19 h. After cooling down to room temperature the solvent was reduced and toluene was added. After 2d at -18 °C a violet solid precipitates.

Yield: 102 mg (82 %)

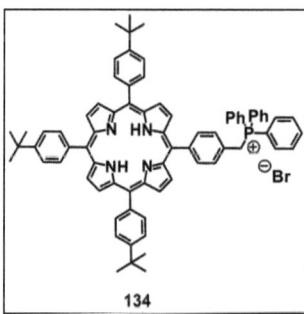

^{1}H NMR (CD$_2$Cl$_2$, 400 MHz, RT):

δ [ppm] = 8.90 (m, 4H, pyr*H*), 8.86 (d, ^3J = 4.47 Hz, 2H, pyr*H*), 8.73 (d, ^3J = 4.80 Hz, 2H, pyr*H*), 8.14 (d, ^3J = 8.30 Hz, 2H, Ph*H*), 8.01 (d, ^3J = 8.08 Hz, 6H, Ph*H*), 7.97 (m, 6H, Ph*H*), 7.88 (d, ^3J = 7.81, 2H, Ph*H*), 7.77 (m, 6H, Ph*H*), 7.66 (m, 6H, Ph*H*), 7.48 (m, 3H, Ph*H*), 5.65 (s, 1H, C*H*$_2$), 1.60 (s, 9H, C*H*$_3$), 1.56 (s, 18H, C*H*$_3$), -2.83 (s, 2H, N*H*).

^{13}C NMR (CD$_2$Cl$_2$, 100.5 MHz, RT):

δ [ppm] = 150.65 (3C, Ph*C*), 138.95 (1C, Ph*C*), 135.25 (3C, Ph*C*), 134.95 (1C, Ph*C*), 134.25 (6C, Ph*C*), 131.97 (2C, Ph*C*), 130.30 (3C, Ph*C*), 130.17 (3C, meso*C*), 128.55 (4C, α-pyr*C*), 128.43 (6C, Ph*C*), 127.10 (4C, α-pyr*C*), 123.72 (2C, β-pyr*C*), 123.65 (2C, β-pyr*C*), 120.69 (4C, β-pyr*C*), 120.40 (1C, Ph*C*), 118.38 (6C, Ph*C*), 118.30 (6C, Ph*C*), 117.40 (3C, Ph*C*), 107.87 (1C, meso*C*), 67.65 (1C, *C*H$_2$), 34.73 (2C, *C*(CH$_3$)$_3$), 33.32 (1C, *C*(CH$_3$)$_3$), 31.37 (6C, CH$_3$), 29.16 (3C, CH$_3$).

MS (FAB, NBA): m/z = 1056 [M-Br]$^+$.

UV/Vis (CH$_2$Cl$_2$): λ_{max} (log ε) [nm] = 647.5 (4.04), 592.0 (4.05), 553.0 (4.21), 515.5 (4.39), 420.0 (5.70).

IR (diamond, RT): $\tilde{v}$ [cm^{-1}] = 2961, 2206, 2193, 2182, 2162, 2023, 1998, 1981, 1192, 1106, 1021, 979, 964, 846, 789, 728, 692.

6.2.5 OPV Rods and Their Precursors

4,4'-(1E,1'E)-2,2'-(2,5-Bis(octyloxy)-1,4-phenylene)bis(ethene-2,1-diyl)-4-bromobenzene (136)

Bis Bromo OPV was synthesized using GOP XIII with 4-bromo benzaldehyde **23** (613 mg, 3.31 mmol), 1,4-methylene-bisphosphonate-2,5-bis(octyloxy)benzene **135** (1 g, 1.58 mmol), KO^{t}Bu (408 mg, 3.63 mmol) and 18-crown-6 (13 mg, 0.05 mmol). The crude product was purified by column chromatography on silica gel (toluene/hexanes 1:3).
Yield: 920 mg (81 %).

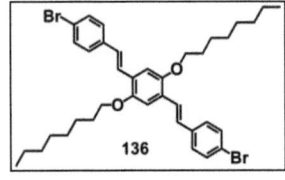

^{1}H NMR (CDCl$_3$, 400 MHz, RT):

δ [ppm] = 7.59 (s, 2H, Ph*H*), 7.52 (d, ^3J = 8.18 Hz, 4H, Ph*H*), 7.46 (d, ^3J = 8.30 Hz, 4H, Ph*H*), 7.24 (d, ^3J = 13.49 Hz, 2H, C*H*), 7.12 (d, ^3J = 13.71 Hz, 2H, C*H*), 4.06 (t, ^3J = 6.47 Hz, 4H, OC*H$_2$*), 1.88 (m, 8H, C*H$_2$*), 1.53 (m, 8H, C*H$_2$*), 1.35 (m, 8H, C*H$_2$*), 0.86 (t, ^3J = 6.96 Hz, 6H, C*H$_3$*).

^{13}C NMR (CDCl$_3$, 100.5 MHz, RT):

δ [ppm] = 151.45 (2C, Ph*C*), 139.52 (2C, Ph*C*), 138.88 (2C, Ph*C*), 129.98 (4C, Ph*C*), 127.47 (2C, Ph*C*), 126.99 (4C, C*H*), 126.83 (2C, Ph*C*), 124.52 (2C, Ph*C*), 110.52 (2C, Ph*C*), 69.54 (2C, OC*H$_2$*), 31.82 (2C, C*H$_2$*), 29.41 (2C, C*H$_2$*), 29.50 (2C, C*H$_2$*), 29.33 (2C, C*H$_2$*), 26.15 (2C, C*H$_2$*), 22.65 (2C, C*H$_2$*), 14.03 (2C, C*H$_3$*).

MS (FAB, NBA): m/z = 694 [M]$^+$.

UV/Vis (CH$_2$Cl$_2$): λ_{max} [nm] = 397.0, 330.5, 232.5.

FT-IR (diamond, RT): $\tilde{v}$ [cm^{-1}] = 2943, 2923, 2901, 2856, 2364, 1484, 1469, 1422, 1257, 1204, 1069, 1025, 1006, 962, 851, 807, 765, 750.

EA: calculated for $C_{38}H_{48}Br_2O_2 \cdot 0.3\ CH_2Cl_2$ (724.91): C 63.51, H 6.77; found: C 63.44, H 7.13.

Bis-dibromo-ethenyl OPV rod (138)

Tetrabromomethane (615 mg, 1.85 mmol) was dissolved in dry CH_2Cl_2 (50 mL) and cooled to 0 °C. Then triphenylphosphine (1.08 g, 4.12 mmol) was added. Afterwards 4,4'- (1E,1'E)- 2,2'- (2,5-bis(octyloxy) -1,4-phenylene) bis (ethene-2,1-diyl) dibenzaldehyde **137** (250 mg, 0.42 mmol) dissolved in dry CH_2Cl_2 (50 mL) was dropped over a period of 30 min to the rapidly stirring solution. After stirring for another 20 min water (100 mL) was added, the organic layer was separated and the aqueous layer was extracted three times with CH_2Cl_2. The combined organic layers were dried over $MgSO_4$ and the solvent was removed in vacuum. Finally the crude product was purified by column chromatography on silica gel (hexanes/CH_2Cl_2 1:1). Yield: 338 mg (89 %).

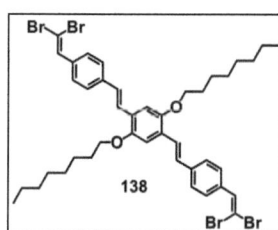

^{1}H NMR (CDCl$_3$, 300 MHz, RT):

δ [ppm] = 7.60 (s, 2H, Ph*H*), 7.56 (d, ^3J = 7.16 Hz, 4H, Ph*H*), 7.51 (d, ^3J = 7.25 Hz, 4H, Ph*H*), 7.49 (s, 2H, C*H*), 7.13 (t, ^3J = 8.29 Hz, 4H, C*H*), 4.07 (t, ^3J = 6.40 Hz, 4H, OC*H$_2$*), 1.89 (m, 8H, C*H$_2$*), 1.55 (m, 8H, C*H$_2$*), 1.33 (m, 8H, C*H$_2$*), 0.91 (t, ^3J = 6.98 Hz, 6H, C*H$_3$*).

^{13}C NMR (CDCl$_3$, 75.5 MHz, RT):

δ [ppm] = 151.58 (2C, Ph*C*), 138.67 (2C, Ph*C*), 136.96 (2C, Ph*C*), 129.16 (4C, Ph*C*), 128.50 (4C, Ph*C*), 127.30 (2C, Ph*C*), 126.83 (4C, CH), 124.89 (2C, CH), 111.06 (2C, Ph*C*), 89.43 (2C, *C*Br$_2$), 69.95 (2C, OC*H$_2$*), 31.25 (2C, C*H$_2$*), 29.91 (2C, C*H$_2$*), 29.84 (2C, C*H$_2$*), 29.74 (2C, C*H$_2$*), 26.72 (2C, C*H$_2$*), 23.11 (2C, C*H$_2$*), 14.55 (2C, C*H$_3$*).

MS (FAB, NBA): m/z = 906 [M]$^+$.

UV/Vis (CH$_2$Cl$_2$): λ_{max} [nm] = 413.0, 339.5, 264.0, 229.0.

FT-IR (diamond, RT): $\tilde{v}$ [cm^{-1}] = 2917, 2854, 2361, 1509, 1491, 1470, 1428, 1388, 1349, 1340, 1276, 1262, 1205, 1067, 1020, 958, 868, 856, 839, 502, 762, 751, 722, 707.

EA: calculated for C$_{42}$H$_{50}$Br$_4$O$_2$ (906.46): C 55.65, H 5.56; found: C 55.21, H 4.98.

Bis-methylthioether OPV rod (140)

Thioether OPV rod **140** was prepared according to GOP XIII with 4-methylthio benzaldehyde **139** (1 g, 6.57 mmol), 1,4-methylene-bisphosphonate-2,5-bis (octyloxy) benzene **135** (2g, 3.13 mmol), KO^{t}Bu (808 mg, 7.2 mmol) and 18-crown-6 (25 mg, 0.09 mmol). The crude product was purified by column chromatography on silica gel (toluene/hexanes 1:1).

Yield: 982 mg (50 %).

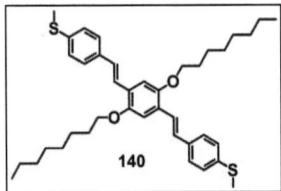

^{1}H NMR (CDCl$_3$, 400 MHz, RT):

δ [ppm] = 7.68 (m, 2H, Ph*H*), 7.45 (d, ^3J = 7.35 Hz, 4H, Ph*H*), 7.32 (d, ^3J = 7.40 Hz, 4H, Ph*H*), 7.27 (d, ^3J = 13.49 Hz, 4H, C*H*), 4.26 (s, 4H, OC*H*$_2$), 2.72 (s, 6H, SC*H*$_3$), 2.08 (m, 8H, C*H*$_2$), 1.76 (m, 8H, C*H*$_2$), 1.53 (m, 8H, C*H*$_2$), 1.11 (s, 6H, C*H*$_3$).

^{13}C NMR (CDCl$_3$, 100.5 MHz, RT):

δ [ppm] = 151.46 (2C, Ph*C*), 137.91 (2C, Ph*C*), 135.40 (2C, Ph*C*), 128.47 (4C, Ph*C*), 127.29 (2C, Ph*C*), 127.11 (4C, C*H*), 126.21 (2C, Ph*C*), 123.29 (2C, Ph*C*), 110.93 (2C, Ph*C*), 69.97 (2C, OC*H*$_2$), 32.25 (2C, C*H*$_2$), 29.93 (2C, C*H*$_2$), 29.84 (2C, C*H*$_2$), 29.75 (2C, C*H*$_2$), 26.72 (2C, C*H*$_2$), 23.11 (2C, C*H*$_2$), 16.26 (2C, SC*H*$_3$), 14.54 (2C, C*H*$_3$).

MS (MALDI, DHB): m/z = 631 [M]$^+$.

UV/Vis (CH$_2$Cl$_2$): λ$_{max}$ [nm] = 405.0, 339.5, 234.5.

FT-IR (diamond, RT): $\tilde{v}$ [cm^{-1}] = 2919, 2852, 2361, 2342, 1497, 1488, 1465, 1426, 1276, 1257, 1197, 1094, 1061, 966, 858, 807, 764, 750.

EA: calculated for C$_{40}$H$_{54}$S$_2$O$_2$ (630.99): C 76.14, H 8.63, S 10.16; found: C 75.39, H 8.63, S 9.35.

Bisthiol OPV rod (141)

OPV rod **140** (200 mg, 0.32 mmol) and sodium thiomethoxide (221 mg, 3.16 mmol) were dissolved in dimethylacetamide (25 mL) under inert conditions. The reaction mixture was then heated at 140 °C for 16 h. After cooling down to 0 °C water (100 mL) was added. The occurring yellow precipitate was filtered, washed with water and dried in high vacuum.
Yield: 144 mg (75 %)

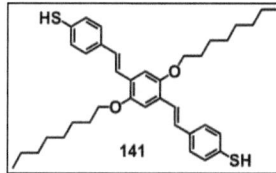

^{1}H NMR (CDCl$_3$, 400 MHz, RT):
δ [ppm] = 7.63 (s, 2H, Ph*H*), 7.59 (d, ^3J = 7.68 Hz, 4H, Ph*H*), 7.45 (d, ^3J = 7.69 Hz, 4H, Ph*H*), 7.27 (d, ^3J = 12.82 Hz, 2H, C*H*), 7.23 (d, ^3J = 12.82 Hz, 2H, C*H*), 4.21 (t, ^3J = 6.47 Hz, 4H, OC*H*$_2$), 3.66 (s, 2H, S*H*), 2.04 (m, 8H, C*H*$_2$), 1.54 (m, 8H, C*H*$_2$), 1.49 (m, 8H, C*H*$_2$), 1.08 (t, ^3J = 7.94 Hz, 6H, C*H*$_3$).

^{13}C NMR (CDCl$_3$, 100.5 MHz, RT):
δ [ppm] = 151.04 (2C, Ph*C*), 137.31 (2C, Ph*C*), 135.83 (2C, Ph*C*), 129.61 (4C, Ph*C*), 129.56 (2C, Ph*C*), 127.91 (4C, C*H*), 127.08 (2C, Ph*C*), 123.24 (2C; Ph*C*), 110.54 (2C, Ph*C*), 69.51 (2C, OC*H*$_2$), 31.80 (2C, C*H*$_2$), 29.47 (2C, C*H*$_2$), 29.40 (2C, C*H*$_2$), 29.30 (2C, C*H*$_2$), 26.27 (2C, C*H*$_2$), 22.67 (2C, C*H*$_2$), 14.10 (2C, C*H*$_3$).

MS (MALDI, DHB): m/z = 602 [M]$^+$.

UV/Vis (CH$_2$Cl$_2$): λ_{max} [nm] = 403.5, 338.5, 235.5.

FT-IR (diamond, RT): $\tilde{v}$ [cm^{-1}] = 2927, 2856, 2360, 2342, 1491, 1420, 1276, 1261, 1203, 1032, 1013, 969, 764, 750, 724.

EA: calculated for $C_{38}H_{50}S_2O_2$ * CH_2Cl_2 (687.86): C 68.10, H 7.62, S 9.32; found: C 68.30, H 7.58, S 9.39.

4-Methylthiobenzyl-triphenylphosphonium bromide (143)

Triphenylphosphine (1.2 g, 4.6 mmol) was dissolved in a mixture of Et_2O (25 mL) and CH_2Cl_2 (10 mL) at room temperature. Then 4-methylthiobenzylbromide **142** (1 g, 4.6 mmol) was added at once and the reaction mixture was stirred for 2 h. The occurring white precipitate was collected, washed with Et_2O three times and dried in vacuum.
Yield: 1.6 g (73 %).

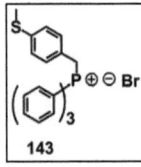

^{1}H NMR (CDCl$_3$, 300 MHz, RT): δ [ppm] = 7.73 (m, 10H, Ph*H*), 7.59 (m, 5H, Ph*H*), 7.00 (d, ^3J = 8.29 Hz, 2H, Ph*H*), 6.90 (d, ^3J = 8.16 Hz, 2H, Ph*H*), 5.27 (d, $^2J_{(H, P)}$ = 18.0 Hz, 2H, PhC*H*$_2$P), 2.34 (s, 3H, SC*H*$_3$).

^{13}C-NMR (CDCl$_3$, 75.5 MHz, RT):

δ [ppm] = 139.62 (1C, Ph*C*), 135.40 (3C, Ph*C*), 134.84 (6C, Ph*C*), 134.71 (2C, Ph*C*), 132.30 (6C, Ph*C*), 130.62 (2C, Ph*C*), 126.51 (1C, Ph*C*), 117.50 (3C, Ph*C*), 30.93 (1C, $^2J_{(C, P)}$ = 47.64 Hz, Ph*C*H$_2$P), 15.69 (1C, S*C*H$_3$).

MS (MALDI, DHB): m/z = 388 [M-Br]$^+$.

IR (diamond, RT): ṽ [cm^{-1}] = 3006, 2879, 2360, 2342, 1495, 1438, 1275, 1261, 1113, 836, 823, 764, 750, 720.

Extended OPV-thioether (144)

OPV-thioether **144** was prepared using GOP XIII with potassium *tert*-butoxide (109 mg, 0.97 mmol), 18-crown-6 (4 mg, 0.01 mmol), 4-methylthiobenzyl-triphenyl-phosphonium bromide **142** (443 mg, 0.95 mmol) and bisformyl OPV **137** (250 mg, 0.42 mmol). Column chromatography (toluene/cyclohexane 2:1) was used for separation of the desired product.
Yield: 175 mg (50 %).

Experimental Part

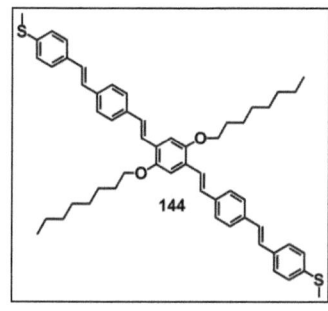

144

¹H NMR (CDCl₃, 400 MHz, RT):

δ [ppm] = 7.74 (m, 2H, PhH), 7.71 (d, ^3J = 7.35 Hz, 4H, PhH), 7.67 (d, ^3J = 7.40 Hz, 4H, PhH), 7.63 (m, 4H, PhH), 7.45 (d, ^3J = 13.49 Hz, 4H, CH), 7.38 (d, ^3J = 13.49 Hz, 4H, CH), 7.33 (m, 2H, PhH), 7.30 (s, 2H, PhH), 4.26 (t, ^3J = 6.41 Hz, 4H, OCH_2), 2.71 (s, 6H, SCH_3), 2.08 (m, 8H, CH_2), 1.75 (m, 8H, CH_2), 1.51 (m, 8H, CH_2), 1.09 (m, 6H, CH_3).

¹³C NMR (CDCl₃, 100.5 MHz, RT):

δ [ppm] = 150.17 (2C, PhC), 136.81 (2C, PhC), 136.30 (2C, PhC), 135.43 (2C, PhC), 133.31 (2C, PhC), 128.01 (4C, PhC), 127.33 (2C, PhC), 127.20 (2C, CH), 127.03 (6C, CH), 126.71 (4C, PhC), 125.85 (2C, PhC), 125.72 (2C, PhC), 125.66 (4C, PhC), 124.27 (2C, PhC), 122.33 (2C, PhC), 69.15 (2C, OCH_2), 30.91 (2C, CH_2), 28.68 (2C, CH_2), 28.43 (2C, CH_2), 28.33 (2C, CH_2), 25.30 (2C, CH_2), 21.69 (2C, CH_2), 14.79 (2C, SCH_3), 13.12 (2C, CH_3).

MS (MALDI, DHB): m/z = 835 [M]⁺.

UV/Vis (CH₂Cl₂): λ_{max} [nm] = 430.5, 362.0, 311.0, 232.0.

IR (diamond, RT): $\tilde{\nu}$ [cm⁻¹] = 2920, 2854, 2361, 2345, 1494, 1489, 1465, 1431, 1276, 1258, 1197, 1098, 1061, 965, 858, 812, 764, 752.

EA: calculated for $C_{56}H_{66}S_2O_2$ * 0.5 CHCl₃ (894.94): C 75.74, H 7.54, 7.15; found: C 75.52, H 7.62, N 7.94.

6.2.6 Porphyrin-Fullerene-Triads and Their Precursors

Perylene-Porphyrin-C₆₀ triad (9)

C₆₀ (11 mg, 0.02 mmol), the perylene-porphyrin malonate **96** (20 mg, 0.01 mmol), CBr₄ (5 mg, 0.01 mmol) and DBU (4 μL, 0.03 mmol) were used according to

GOP XIV for the synthesis of triad **97**. The reaction mixture was stirred for 24 h. Eluation of the triad was achieved by toluene/EtOAc 8:1.
Yield: 12 mg (44 %).

^{1}H NMR (CDCl$_3$, 400 MHz, RT):

δ [ppm] = 8.87 (m, 8H, pyr*H*), 8.51 (m, 8H, Ph*H*), 8.14 (d, ^3J = 8.30 Hz, 1H, Ph*H*), 8.09 (m, 6H, Ph*H*), 8.02 (d, ^3J = 6.84 Hz, 1H, Ph*H*), 7.97 (m , 6H, Ph*H*), 7.81 (m, 1H, Ph*H*), 7.75 (t, ^3J = 7.20 Hz , 1H, Ph*H*), 5.16 (m, 1H, C*H*), 4.57 (t, ^3J = 7.13 Hz , 2H, NC*H$_2$*), 4.49 (m, 4H, OC*H$_2$*), 4.43 (t, ^3J = 7.62 Hz, 2H, OC*H$_2$*), 4.37 (t, ^3J = 7.22 Hz, 2H, PhOC*H$_2$*), 4.25 (m, 2H, OC*H$_2$*), 2.48 (t, ^3J = 6.96 Hz, 4H, C*H$_2$*), 2.23 (m, 4H, C*H$_2$*), 2.09 (m, 4H, C*H$_2$*), 1.62 (s, 9H, C*H$_3$*), 1.63 (s, 18 H, C*H$_3$*), 1.24 (m, 16H, C*H$_2$*), 0.84 (m, 6H, C*H$_3$*).

^{13}C NMR (CDCl$_3$, 100.5 MHz, RT):

δ [ppm] = 167.75 (1C, C=O), 165.42 (1C, C=O), 164.52 (1C, C=O), 162.44 (4C=O), 150.26, 150.15, 149.83, 144.56, 144.48, 144.12, 143.92, 143.56, 143.22, 143.53, 143.31, 143.22, 143.10, 142.87, 142.70, 142.63, 142.59, 142.33, 141.98, 141.82, 141.85, 141.75, 141.44, 141.31, 141.25, 141.09, 140.79, 140.64, 140.53, 140.45, 140.10, 140.03, 139.77, 139.61, 139.60, 139.49, 139.12, 138.25, 137.53, (66C, sp^{2}C C$_{60}$,Ph*C*), 134.30 (4C, Ph*C*), 134.06 (2C, Ph*C*), 132.43 (6C, Ph*C*), 131.89 (2C, Ph*C*), 131.53 (1C, Ph*C*), 131.46 (8C, α-pyr*C*), 130.90 (2C, β-pyr*C*), 130.87 (2C, Ph*C*), 128.79 (2C, Ph*C*), 123.53 (6C, pyr*C*), 123.41 (1C, Ph*C*), 122.85 (2C, Ph*C*), 122.04 (2C, Ph*C*), 121.33 (2C, Ph*C*), 121.19 (2C, Ph*C*), 120.98 (6C, Ph*C*), 120.57 (3C, meso*C*), 120.44 (1C, meso*C*), 113.95 (2C, Ph*C*), 68.14 (2C, sp^{2}C C$_{60}$), 65.98 (2C,

OCH$_2$), 63.15 (1C, OCH$_2$), 61.99 (1C, PhOCH$_2$), 54.62 (1C, NCH$_2$), 52.55 (1C, CH), 51.02 (1C,C), 38.70 (2C, CH$_2$), 34.87 (3C, C(CH$_3$)$_3$), 32.33 (1C, CH$_2$), 31.82 (2C, CH$_2$), 31.71 (1C, CH$_2$), 30.34 (9C, CH$_3$), 29.69 (2C, CH$_2$), 29.52 (2C, CH$_2$), 27.00 (2C, CH$_2$), 22.61 (2C, CH$_2$), 14.09 (2C, CH$_3$).

MS (MALDI, DCTB): m/z = 2423 [M]$^+$.

UV/Vis (CH$_2$Cl$_2$): λ_{max} (log ε) [nm] = 530.0 (4.21), 491.0 (4.10), 425.5 (4.90), 328.0 (3.94), 260.0 (4.47).

IR (diamond, RT): $\tilde{v}$ [cm^{-1}] = 2954, 2928, 1744, 1697, 1594, 1578, 1435, 1403, 1340, 1247, 1204, 1172, 1109, 1068, 998, 851, 796, 745, 719, 526.

5-(3-Ethoxyoxy-phenyl)-10,15,20-tri(*p-tert*-butyl-phenyl)-porphyrin (146)
Pyrrole **50** (2.013 g, 0.03 mol), 3-ethoxyoxy-benzaldehyde **144** (1.246 g, 0.01 mol), 4-*tert*-butyl-benzaldehyde **48** (3.65 g, 0.02 mol), TFA (1.73 mL, 2.56 g, 0.02 mol), TEA (3.76 mL, 2.73 g, 0.03 mol) and DDQ (5.10 g, 0.03 mol) were used according to GOP IXb for the synthesis of porphyrin **146**. Purification was done with column chromatography on silica gel (CH$_2$Cl$_2$).
Yield: 840 mg (13 %).

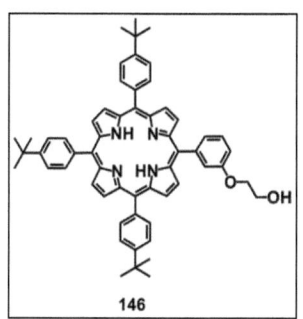

^{1}H NMR (CDCl$_3$, 400 MHz, RT):
δ [ppm] = 8.93 (m, 8 H, pyr*H*), 8.20 (d, ^3J = 8.48 Hz, 6H, Ph*H*), 7.91 (d, ^3J = 7.54 Hz, 1H, Ph*H*), 7.86 (s, 1H, Ph*H*), 7.80 (d, ^3J = 8.48 Hz, 6H, Ph*H*), 7.68 (t, ^3J = 8.29 Hz, 1H, Ph*H*), 7.35 (d, ^3J = 7.06 Hz, 1H, Ph*H*), 4.28 (t, ^3J = 7.26 Hz, 2H, PhOC*H$_2$*), 4.04 (t, ^3J = 7.58 Hz, 2H, HOC*H$_2$*), 2.15 (bs, 1H, O*H*), 1.65 (s, 27 H, C*H$_3$*), -2.68 (s, 2H, N*H*).

^{13}C NMR (CDCl$_3$, 100.5 MHz, RT):
δ [ppm] = 157.39 (1C, Ph*C*), 150.92 (2C, Ph*C*), 144.19 (1C, Ph*C*), 139.63 (3C, Ph*C*), 139.58 (1C, Ph*C*), 134.92 (6C, Ph*C*), 131.52 (1C, Ph*C*), 128.56 (16C, pyr*C*), 127.98 (1C, Ph*C*), 124.05 (6 C, Ph*C*), 120.71 (3C, meso*C*), 119.65 (1C, meso*C*), 114.47

(2C, Ph*C*), 69.78 (1C, HO*C*H$_2$), 61.96 (1C, PhO*C*H$_2$), 35.33 (3C, *C*(CH$_3$)$_3$), 32.13 (9C, *C*H$_3$).

MS (MALDI, OM): m/z = 843 [M]$^+$.

UV/Vis (CH$_2$Cl$_2$): λ$_{max}$ (log ε) [nm] = 647.0 (3.95), 590.5 (3.99), 551.5 (4.15), 516.5 (4.38), 419.0 (5.62).

IR (diamond, RT): ṽ [cm^{-1}] = 2965, 2360, 2342, 1275, 1261, 969, 799, 764, 750.

EA: calculated for C$_{58}$H$_{58}$N$_4$O$_2$ (842.46): C 82.63, H 6.93, N 6.65; found: C 82.91, H 5.95, N 6.68.

5-(3-Ethoxyoxy-phenyl)-10,15,20-tri(p-*tert*-butyl-phenyl)-porphyrinato zinc (II) (147)

Porphyrin **146** (780 mg, 0.93 mmol) was metallated with zinc acetate (624 mg, 2.84 mmol) according to GOP X. Separation of the product was achieved by column chromatography on silica gel (CH$_2$Cl$_2$/EtOAc 5:1).
Yield: 823 mg (98 %).

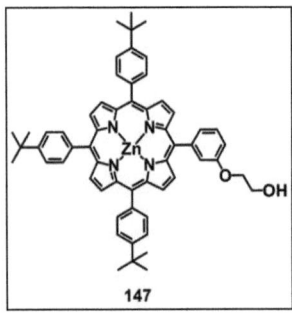

^{1}H NMR (CDCl$_3$, 400 MHz, RT):

δ [ppm] = 9.00 (m, 6 H, pyr*H*), 8.93 (d, ^{3}*J* = 4.46 Hz , 2H, pyr*H*), 8.16 (d, ^{3}*J* = 7.58 Hz, 6H, Ph*H*), 7.84 (d, ^{3}*J* = 7.54 Hz, 1H, Ph*H*), 7.77 (d, ^{3}*J* = 8.48 Hz, 6H, Ph*H*), 7.62 (s, 1H, Ph*H*), 7.52 (t, ^{3}*J* = 8.29 Hz, 1H, Ph*H*), 7.57 (d, ^{3}*J* = 7.06 Hz, 1H, Ph*H*), 7.13 (d, ^{3}*J* = 7.15 Hz, 1H, Ph*H*), 3.65 (t, ^{3}*J* = 7.20 Hz, 2H, PhOC*H$_2$*), 3.22 (m, 2H, HOC*H$_2$*), 2.03 (bs, 1H, O*H*), 1.63 (s, 27 H, C*H$_3$*).

^{13}C NMR (CDCl$_3$, 100.5 MHz, RT):

δ [ppm] = 155.53 (1C, Ph*C*), 149.40 (2C, Ph*C*), 143.41(1C, Ph*C*), 138.90 (3C, Ph*C*), 138.86 (1C, Ph*C*), 133.38 (6C, Ph*C*), 131.16 (1C, Ph*C*), 131.03 (8C, α-pyr*C*), 130.65 (2C, β-pyr*C*), 127.03 (6C, β-pyr*C*), 126.31 (1C, Ph*C*), 122.48 (6 C, Ph*C*), 119.69 (3C, mesoC), 119.15 (1C, mesoC), 112.88 (2C, PhC), 67.50 (1C, HOCH$_2$), 59.78 (1C, PhOCH$_2$), 33.92 (3C, C(CH$_3$)$_3$), 30.76 (9C, CH$_3$).

MS (MALDI, DCTB): m/z = 905 [M]$^+$.

UV/Vis (CH$_2$Cl$_2$): λ$_{max}$ (log ε) [nm] = 587.0 (3.87), 548.0 (4.36), 420.5 (5.63).

IR (diamond, RT): $\tilde{\nu}$ [cm^{-1}] = 2965, 2360, 2342, 1508, 1458, 1276, 1261, 1001, 812, 799, 764, 750, 721.

EA: calculated for C$_{58}$H$_{56}$N$_4$O$_2$Zn (904.37): C 76.85, H 6.23, N 6.18; found: C 76.20, H 6.40, N 5.91.

Perylene bisimide (150)

Perylene derivative **150** was prepared using GOP XI starting with 2-((3-(tert-butoxycarbonyl)propoxy)carbonyl)acetic acid **148** (12 mg, 0.05 mmol), DMAP (6 mg, 0.05 mmol), 1-HOBT (7 mg, 0.05 mmol), DCC (20 mg, 0.1 mmol) and perylene alcohol **149** (34 mg, 0.06 mmol) The desired perylene could be isolated by column chromatography (silica gel, CH$_2$Cl$_2$/EtOAc 4:1) of the crude product.
Yield: 25 mg (57 %)

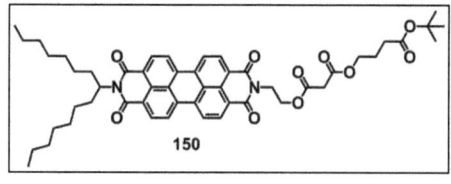

^{1}H NMR (CDCl$_3$, 400 MHz, RT):

δ [ppm] = 8.57 (m, 8H, PhH), 5.19 (m, 1H, CH), 4.55 (m, 2H, NCH$_2$), 4.19 (m, 4H, OCH$_2$), 3.40 (s, 2H, COCH$_2$CO), 2.28 (m, 4H, CH$_2$), 1.92 (m, 4H, CH$_2$), 1.45 (m, 4H, CH$_2$), 1.42 (s, 9H, C(CH$_3$)$_3$), 1.26 (m, 16H, CH$_2$), 0.83 (t, ^3J = 8.35 Hz, 6H, CH$_3$).

^{13}C NMR (CDCl$_3$, 100.5 MHz, RT):

δ [ppm] = 172.42 (1C, C=O), 166.86 (1C, C=O), 166.69 (1C, C=O), 163.75 (4C, C=O), 135.17 (4C, PhC), 134.49 (2C, PhC), 131.84 (2C, PhC), 129.81 (2C, PhC), 129.76 (2C, PhC), 126.51 (2C, PhC), 123.51 (2C, PhC), 123.29 (2C, PhC), 123.14 (2C, PhC), 80.90 (1C, C(CH$_3$)$_3$), 64.97 (1C, OCH$_2$), 63.09 (1C, OCH$_2$), 55.22 (1C,

NCH$_2$), 52.93 (1C, CH), 41.78 (1C, COCH$_2$CO), 41.70 (2C, CH$_2$), 32.75 (1C, CH$_2$), 32.15 (2C, CH$_2$), 30.09 (1C, CH$_2$), 29.63 (3C, C(CH$_3$)$_3$), 28.47 (2C, CH$_2$), 27.39 (2C, CH$_2$), 24.35 (2C, CH$_2$), 23.01 (2C, CH$_2$), 14.47 (2C, CH$_3$).

MS (MALDI, DCTB): m/z = 873 [M]$^+$.

UV/Vis (CH$_2$Cl$_2$): λ$_{max}$ (log ε) [nm] = 525.0 (4.73), 488.5 (4.52), 457.0 (4.11).

IR (diamond, RT): $\tilde{v}$ [cm^{-1}] = 2958, 2927, 2855, 2360, 2340, 1731, 1697, 1655, 1344, 1275, 1260, 1150, 810, 764, 749.

Perylene bisimide (151)
Ester cleavage of perylene **150** (20 mg, 0.02 mmol) was achieved by GOP VI b.
Yield: 17 mg (100 %).

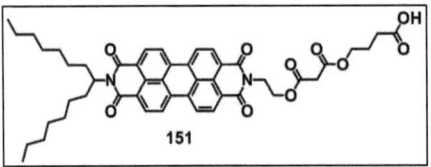

^{1}H NMR (CDCl$_3$, 400 MHz, RT):
δ [ppm] = 8.33 (m, 8H, PhH), 5.18 (m, 1H, CH), 4.58 (m, 2H, NCH$_2$), 4.24 (m, 4H, OCH$_2$), 3.44 (s, 2H, COCH$_2$CO), 2.49 (m, 4H, CH$_2$), 2.26 (m, 4H, CH$_2$), 2.02 (m, 4H, CH$_2$), 1.26 (m, 16H, CH$_2$), 0.85 (t, ^3J = 8.35 Hz, 6H, CH$_3$).

^{13}C NMR (CDCl$_3$, 100.5 MHz, RT):
δ [ppm] = 177.30 (1C, C=O), 166.96 (1C, C=O), 166.46 (1C, C=O), 163.21 (4C, C=O), 134.36 (4C, PhC), 133.63 (2C, PhC), 131.18 (2C, PhC), 129.11 (2C, PhC), 128.96 (2C, PhC), 125.84 (2C, PhC), 122.92 (2C, PhC), 122.67 (2C, PhC), 122.34 (2C, PhC), 64.47 (1C, OCH$_2$), 64.32 (1C, OCH$_2$), 54.87 (1C, NCH$_2$), 52.57 (1C, CH), 41.44 (1C, COCH$_2$CO), 41.23 (2C, CH$_2$), 32.33 (1C, CH$_2$), 31.81 (2C, CH$_2$), 30.42 (1C, CH$_2$), 29.53 (2C, CH$_2$), 27.04 (2C, CH$_2$), 23.65 (2C, CH$_2$), 22.61 (2C, CH$_2$), 14.07 (2C, CH$_3$).

MS (MALDI, DCTB): m/z = 817 [M]$^+$.

Experimental Part

UV/Vis (CH$_2$Cl$_2$): λ_{max} (log ε) [nm] = 526.5 (4.99), 490.5 (4.79), 458.5 (4.38).

IR (diamond, RT): $\tilde{v}$ [cm^{-1}] = 2925, 2855, 2360, 2342, 1730, 1697, 1655, 1594, 1438, 1403, 1343, 1275, 1258, 1178, 1152, 1023, 810, 747.

Perylene-Porphyrin-dyad (152)
The dyad could be synthesized using GOP XI with perylene derivative **151** (17 mg, 0.02 mmol), DMAP (3 mg, 0.02 mmol), 1-HOBT (3 mg, 0.02 mmol), DCC (9 mg, 0.04 mmol) and zinc porphyrin **147** (23 mg, 0.03 mmol). The reaction mixture was stirred for 72 h. Column chromatography on silica gel (CH$_2$Cl$_2$/EtOAc 8:1) led to the product. Yield: 26 mg (75 %)

^{1}H NMR (CDCl$_3$, 400 MHz, RT):
δ [ppm] = 8.92 (m, 6 H, pyr$H$), 8.93 (d, ^3J = 4.45 Hz, 2H, pyr$H$), 8.38 (m, 8H, PhH), 8.11 (d, ^3J = 7.91 Hz, 1H, PhH), 7.98 (m, 6H, PhH), 7.74 (m, 6H, PhH), 7.62 (t, ^3J = 8.30 Hz, 1H, PhH), 7.57 (d, ^3J = 8.10 Hz, 1H, PhH), 7.22 (d, ^3J = 7.91 Hz, 1H, PhH), 5.12 (m, 1H, CH), 4.58 (m, 2H, NCH_2), 4.50 (m, 4H, OCH_2), 4.09 (t, ^3J = 7.20 Hz, 2H, PhOCH_2), 3.65 (t, ^3J = 7.20 Hz, 2H, OCH_2), 3.43 (m, 2H, OCH_2), 3.02 (s, 2H, COCH_2CO), 2.47 (m, 4H, CH$_2$), 2.21 (m, 4H, CH$_2$), 2.05 (m, 4H, CH$_2$), 1.63 (s, 27H, CH$_3$), 1.23 (m, 16H, CH$_2$), 0.83 (t, ^3J = 8.36 Hz, 6H, CH$_3$).

^{13}C NMR (CDCl$_3$, 100.5 MHz, RT):
δ [ppm] = 172.84 (1C, C=O), 166.00 (1C, C=O), 165.97 (1C, C=O), 162.05 (4C, C=O), 156.89 (1C, PhC), 150.14 (2C, PhC), 144.33 (1C, PhC), 139.60 (3C, PhC), 139.55 (1C, PhC), 134.78 (4C, PhC), 134.69 (2C, PhC), 133.31 (6C, PhC), 132.05

(2C, Ph*C*), 131.84 (1C, Ph*C*), 131.62 (8C, α-pyr*C*), 130.75 (2C, β-pyr*C*), 130.28 (2C, Ph*C*), 128.97 (2C, Ph*C*), 128.42 (6C, β-pyr*C*), 127.37 (1C, Ph*C*), 125.43 (2C, Ph*C*), 122.53 (2C, Ph*C*), 123.48 (2C, Ph*C*), 123.30 (2C, Ph*C*), 122.57 (6 C, Ph*C*), 121.00 (3C, meso*C*), 120.82 (1C, meso*C*), 113.83 (2C, Ph*C*), 66.26 (1C, O*C*H$_2$), 64.08 (1C, O*C*H$_2$), 63.02 (1C, O*C*H$_2$), 61.94 (1C, PhO*C*H$_2$), 54.68 (1C, N*C*H$_2$), 52.57 (1C, *C*H), 41.00 (1C, CO*C*H$_2$CO), 38.01 (2C, *C*H$_2$), 34.85 (3C, *C*(CH$_3$)$_3$), 34.83 (1C, *C*H$_2$), 32.36 (2C, *C*H$_2$), 31.80 (1C, *C*H$_2$), 31.70 (9C, *C*H$_3$), 29.52 (2C, *C*H$_2$), 29.23 (2C, *C*H$_2$), 26.97 (2C, *C*H$_2$), 22.61 (2C, *C*H$_2$), 14.06 (2C, *C*H$_3$).

MS (MALDI, DCTB): m/z = 1705 [M]$^+$.

UV/Vis (CH$_2$Cl$_2$): λ_{max} (log ε) [nm] = 529.5 (4.99), 493.0 (4.76), 421.5 (5.73).

IR (diamond, RT): $\tilde{v}$ [cm^{-1}] = 2955, 2929, 2908, 2361, 2342, 1734, 1697, 1657, 1639, 1620, 1576, 1441, 1341, 1251, 1109, 1068, 957, 854, 797, 746.

9,9-Dihexyl-2-formyl-7-ethynylmethanol-fluorene (154)

Fluorene compound **154** was prepared according to GOP IV with Pd(PPh$_3$)$_2$Cl$_2$ (5 mg, 0.01 mmol), CuI (4mg, 0.02 mmol), 9,9 dihexyl-2-formyl-7-bromo-fluorene **38** (300 mg, 0.68 mmol) and propargylic alcohol **153** (48 mg, 50µL, 0.88 mmol) in a mixture of THF (10mL) and NEt$_3$ (5 mL). The reaction mixture was heated at 90 °C for 3 d. The occurring brownish residue was cleaned up by column chromatography (CH$_2$Cl$_2$) on silica gel.
Yield: 198 mg (70 %).

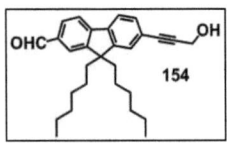

^{1}H NMR (CDCl$_3$, 400 MHz, RT):

δ [ppm] = 10.02 (s, 1H, *H*C=O), 7.98 (m, 3H, Ph*H*), 7.66 (d, ^3J = 8.08 Hz, 1H, Ph*H*), 7.39 (s, 2H, Ph*H*), 4.49 (s, 2H, C*H*$_2$OH), 2.15 (m, 4H, C*H*$_2$), 1.18 (s, 1H, O*H*), 1.03 (m, 16H, hexylC*H*$_2$), 0.70 (t, ^3J = 7.15 Hz, 6H, C*H*$_3$).

¹³C NMR (CDCl₃, 100.5 MHz, RT): δ [ppm] = 192.75 (1C, HC=O), 152.53 (1C, PhC), 152.19 (1C, PhC), 147.00 (1C, PhC), 140.31 (1C, PhC), 135.98 (1C, PhC), 131.31 (1C, PhC), 130.92 (1C, PhC), 126.75 (1C, PhC), 123.50 (1C, PhC), 122.99 (1C, PhC), 121.28 (1C, PhC), 120.74 (1C, PhC), 88.53 (1C, ethynyl$C$), 86.68 (ethynyl$C$), 55.74 (1C, quart$C$), 52.13 (1C, CH_2OH), 40.57 (2C, CH_2), 31.87 (2C, CH_2), 29.97 (2C, CH_2), 24.11 (2C, CH_2), 22.94 (2C, CH_2), 14.36 (2C, CH_3).

MS (MALDI, DCTB): m/z = 418 [M]⁺.

UV/Vis (CH₂Cl₂): λ$_{max}$ (log ε) [nm] = 342.0 (4.58), 260.5 (4.52), 216.5 (4.50).

IR (diamond, RT): $\tilde{\nu}$ [cm⁻¹] = 2956, 2927, 2856, 2361, 2342, 1694, 1606, 1465, 1276, 1261, 1031, 822, 764, 750.

Fluorene derivative (155)

2-((3-(*tert*-butoxycarbonyl)propoxy)carbonyl)acetic acid **148** (13 mg, 0.05 mmol), DMAP (7 mg, 0.06 mmol), 1-HOBT (9 mg, 0.06 mmol), DCC (22 mg, 0.11 mmol) and fluorene compound **154** (27 mg, 0.07 mmol) were used according to GOP XI. Column chromatography on silica gel (CH₂Cl₂) was used to isolate fluorene **155** from byproducts.
Yield: 27 mg (65 %).

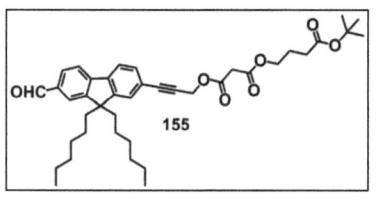

¹H NMR (CDCl₃, 400 MHz, RT):

δ [ppm] = 10.07 (s, 1H, HC=O), 7.87 (m, 3H, PhH), 7.74 (d, ³J = 8.29 Hz, 1H, PhH), 7.47 (m, 2H, PhH), 5.04 (s, 2H, CH_2O), 4.22 (t, ³J = 6.41 Hz, 2H, OCH_2), 3.49 (s, 2H, COCH_2CO), 2.33 (t, ³J = 7.35 Hz, 2H, CH_2), 1.98 (m, 2H, CH_2), 1.45 (s, 9H, C(CH_3)₃), 1.02 (m, 20H, hexylCH_2), 0.75 (t, ³J = 6.97 Hz, 6H, CH_3).

¹³C NMR (CDCl₃, 100.5 MHz, RT):

δ [ppm] = 192.69 (1C, HC=O), 172.39 (1C, C=O), 166.50 (1C, C=O), 166.31 (1C, C=O), 152.52 (1C, PhC), 152.22 (1C, PhC), 146.88 (1C, PhC), 140.68 (1C, PhC), 136.06 (1C, PhC), 131.62 (1C, PhC), 130.89 (1C, PhC), 126.91 (1C, PhC), 123.50

(1C, Ph*C*), 122.35 (1C, Ph*C*), 121.27 (1C, Ph*C*), 120.81 (1C, Ph*C*), 88.96 (1C, ethynyl*C*), 83.44 (ethynyl*C*), 80.98 (1C, *C*(CH$_3$)$_3$), 65.16 (1C, O*C*H$_2$), 55.78 (1C, quart*C*), 54.72 (1C, *C*H$_2$O), 41.64 (1C, CO*C*H$_2$CO), 40.56 (2C, *C*H$_2$), 32.15 (1C, *C*H$_2$), 31.86 (2C, *C*H$_2$), 30.09 (3C, C(*C*H$_3$)$_3$), 29.96 (2C, *C*H$_2$), 28.48 (1C, *C*H$_2$), 24.10 (2C, *C*H$_2$), 22.93 (2C, *C*H$_2$), 14.35 (2C, *C*H$_3$).

MS (MALDI, OM): m/z = 644 [M]$^+$.

UV/Vis (CH$_2$Cl$_2$): λ_{max} (log ε) [nm] = 342.5 (4.66), 260.0 (4.55), 216.5 (4.51).

IR (diamond, RT): $\tilde{v}$ [cm^{-1}] = 2927, 2360, 2342, 1733, 1697, 1607, 1368, 1328, 1276, 1261, 1152, 764, 750.

Fluorene derivative (156)
Ester cleavage of fluorene derivative **155** (24 mg, 0.04 mmol) was done by GOP VIb. Yield: 23 mg (100 %).

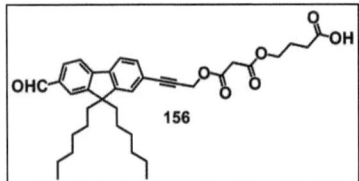

^{1}H-NMR (CDCl$_3$, 400 MHz, RT):
δ [ppm] = 10.07 (s, 1H, *H*C=O), 9.77 (bs, 1C, COO*H*), 7.92 (m, 3H, Ph*H*), 7.74 (d, ^3J = 8.27 Hz, 1H, Ph*H*), 7.48 (m, 2H, Ph*H*), 5.04 (s, 2H, C*H*$_2$O), 4.26 (t, ^3J = 6.42 Hz, 2H, OC*H*$_2$), 3.50 (s, 2H, COC*H*$_2$CO), 2.50 (t, ^3J = 7.34 Hz, 2H, C*H*$_2$), 2.02 (m, 2H, C*H*$_2$), 1.02 (m, 20H, hexylC*H*$_2$), 0.75 (t, ^3J = 6.95 Hz, 6H, C*H*$_3$).

^{13}C NMR (CDCl$_3$, 100.5 MHz, RT):
δ [ppm] = 192.81 (1C, *H*C=O), 170.43 (1C, *C*=O), 166.50 (1C, *C*=O), 166.32 (1C, *C*=O), 152.53 (1C, Ph*C*), 152.22 (1C, Ph*C*), 146.92 (1C, Ph*C*), 140.69 (1C, Ph*C*), 136.03 (1C, Ph*C*), 131.63 (1C, Ph*C*), 130.92 (1C, Ph*C*), 126.91 (1C, Ph*C*), 123.78 (1C, Ph*C*), 122.78 (1C, Ph*C*), 121.58 (1C, Ph*C*), 120.83 (1C, Ph*C*), 87.99 (1C, ethynyl*C*), 83.40 (ethynyl*C*), 64.86 (1C, O*C*H$_2$), 56.06 (1C, quart*C*), 55.78 (1C, *C*H$_2$O), 41.60 (1C, CO*C*H$_2$CO), 40.55 (2C, *C*H$_2$), 31.87 (1C, *C*H$_2$), 31.83 (2C, *C*H$_2$),

29.96 (2C, CH$_2$), 29.90 (1C, CH$_2$), 24.11 (2C, CH$_2$), 22.90 (2C, CH$_2$), 14.36 (2C, CH$_3$).

MS (MALDI, DCTB): m/z = 611 [M-Na]$^+$.

UV/Vis (CH$_2$Cl$_2$): λ_{max} (log ε) [nm] = 342.0 (4.65), 260.5 (4.53), 216.5 (4.49).

IR (diamond, RT): $\tilde{v}$ [cm^{-1}] = 2929, 2856, 1736, 1696, 1606, 1465, 1437, 1415, 1378, 1330, 1271, 1199, 1152, 1101, 1030, 1004, 893, 825, 798, 755.

Fluorene-Porphyrin-dyad (157)
The dyad was synthesized according to GOP XI using fluorene **156** (20 mg, 0.03 mmol), DMAP (5 mg, 0.04 mmol), 1-HOBT (6 mg, 0.04 mmol), DCC (16 mg, 0.08 mmol) and porphyrin alcohol **147** (43 mg, 0.05 mmol) The reaction mixture was stirred for 18 h. Purification could be achieved using column chromatography on silica gel (hexanes/EtOAc 3:1).
Yield: 24 mg (55%)

^{1}H NMR (CDCl$_3$, 400 MHz, RT):
δ [ppm] = 9.96 (s, 1H, HC=O), 9.00 (m, 6H, pyrH), 8.97 (d, ^3J = 4.48 Hz, 2H, pyrH), 8.15 (d, ^3J = 8.30 Hz, 6H, PhH), 7.89 (d, ^3J = 7.81 Hz, 1H, PhH), 7.76 (d, ^3J = 8.30 Hz, 6H, PhH), 7.73 (d, ^3J = 8.27 Hz, 1H, PhH), 7.57 (d, ^3J = 7.06 Hz, 1H, PhH), 7.41 (m, 5H, PhH), 7.36 (m, 1H, PhH), 7.34 (d, ^3J = 7.15 Hz, 1H, PhH), 4.91 (s, 2H, CH$_2$O), 4.52 (t, ^3J = 6.42 Hz, 2H, OCH$_2$), 3.37 (t, ^3J = 7.22 Hz, 2H, PhOCH$_2$), 4.16 (t, ^3J = 7.25 Hz, 2H, OCH$_2$), 3.32 (s, 2H, COCH$_2$CO), 2.47 (t, ^3J = 7.32 Hz, 2H, CH$_2$), 1.98 (m, 2H, CH$_2$), 1.63 (s, 27 H, CH$_3$), 0.98 (m, 20H, hexylCH$_2$), 0.75 (t, ^3J = 6.94 Hz, 6H, CH$_3$).

¹³C NMR (CDCl₃, 100.5 MHz, RT):

δ [ppm] = 192.24 (1C, H*C*=O), 172.63 (1C, *C*=O), 165.94 (1C, *C*=O), 165.74 (1C, *C*=O), 155.53 (1C, Ph*C*), 152.03 (1C, Ph*C*), 151.74 (1C, Ph*C*), 149.84 (2C, Ph*C*), 146.37 (1C, Ph*C*), 144.40 (1C, Ph*C*), 140.14 (1C, Ph*C*), 139.73 (3C, Ph*C*), 135.54 (1C, Ph*C*), 134.27 (1C, Ph*C*), 132.11 (6C, Ph*C*), 131.58 (1C, Ph*C*), 131.12 (1C, Ph*C*), 130.42 (8C, α-pyr*C*), 128.08 (1C, Ph*C*), 127.44 (2C, β-pyr*C*), 126.40 (6C, β-pyr*C*), 123.45 (1C, Ph*C*), 122.99 (1C, Ph*C*), 121.85 (2C, Ph*C*), 121.42 (1C, Ph*C*), 121.25 (6 C, Ph*C*), 120.87 (2C, Ph*C*), 120.75 (1C, Ph*C*), 120.29 (3C, meso*C*), 120.16 (1C, meso*C*), 87.48 (1C, ethynyl*C*), 82.95 (ethynyl*C*), 66.10 (1C, O*C*H₂), 64.49 (1C, O*C*H₂), 63.06 (1C, PhO*C*H₂), 55.28 (1C, quart*C*), 53.77 (1C, *C*H₂O), 41.03 (1C, CO*C*H₂CO), 40.06 (2C, *C*H₂), 34.87 (3C, *C*(CH₃)₃), 31.70 (1C, *C*H₂), 31.42 (2C, *C*H₂), 30.40 (9C, *C*H₃), 29.51 (2C, *C*H₂), 23.73 (1C, *C*H₂), 23.65 (2C, *C*H₂), 22.49 (2C, *C*H₂), 13.92 (2C, *C*H₃).

MS (MALDI, DCTB): m/z = 1477 [M]⁺.

UV/Vis (CH₂Cl₂): λ_{max} (log ε) [nm] = 548.5 (4.23), 421.0 (5.50), 341.5 (4.61), 260 (4.58), 216 (4.50).

IR (diamond, RT): $\tilde{\nu}$ [cm⁻¹] = 2957, 2930, 2860, 2360, 2341, 1740, 1698, 1605, 1338, 1266, 1204, 1166, 998, 810, 797, 749.

Pyrene malonyl ester (159)

Pyrene **159** was accessible *via* GOP XI using 2-((3-(*tert*-butoxycarbonyl)propoxy)carbonyl)acetic acid **148** (200 mg, 0.81 mmol), DMAP (109 mg, 0.89 mmol), 1-HOBT (120 mg, 0.89 mmol), DCC (334 mg, 1.62 mmol) and 1-pyrene butanol **158** (267 mg, 0.98 mmol). The reaction mixture was stirred over night allowing reaching room temperature. The desired pyrene could be isolated by column chromatography (silica gel, hexanes/EtOAc 4:1) of the crude product.
Yield: 240 mg (60 %).

¹H NMR (CDCl₃, 400 MHz, RT):

δ [ppm] = 8.19 (m, 3H, Ph*H*), 8.09 (m, 1H, Ph*H*), 8.06 (d, ³*J* = 7.52 Hz, 1H, Ph*H*), 8.00 (m, 2H, Ph*H*), 7.98 (m, 1H, Ph*H*), 7.82 (d, ³*J* = 7.91 Hz, 1H, Ph*H*), 4.18 (m, 4H,

OCH_2), 3.41 (s, 2H, COCH_2CO), 3.30 (t, ^3J = 7.46 Hz, 2H, CH_2), 2.29 (t, ^3J = 7.27 Hz, 2H, CH_2), 1.93 (m, 6H, CH_2), 1.48 (s, 9H, C(CH_3)$_3$).

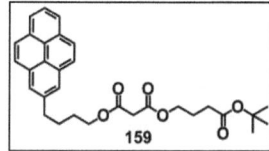

^{13}C NMR (CDCl$_3$, 100.5 MHz, RT):

δ [ppm] = 172.44 (1C, C=O), 167.03 (1C, C=O), 166.97 (1C, C=O), 135.17 (1C, PhC), 136.56 (1C, PhC), 131.82 (1C, PhC), 131.28 (1C, PhC), 130.26 (1C, PhC), 128.97 (1C, PhC), 127.92 (1C, PhC), 127.70 (1C, PhC), 127.59 (1C, PhC), 127.49 (1C, PhC), 127.06 (1C, PhC), 126.25 (1C, PhC), 125.46 (1C, PhC), 125.39 (1C, PhC), 125.22 (1C, PhC), 123.65 (1C, PhC), 80.89 (1C, C(CH$_3$)$_3$), 65.80 (1C, OCH$_2$), 64.98 (1C, OCH$_2$), 41.96 (1C, COCH$_2$CO), 33.30 (2C, CH$_2$), 33.20 (1C, CH$_2$), 30.19 (3C, C(CH$_3$)$_3$), 28.84 (1C, CH$_2$), 28.29 (1C, CH$_2$), 24.47 (1C, CH$_2$).

MS (MALDI, DCTB): m/z = 502 [M]$^+$.

UV/Vis (CH$_2$Cl$_2$): λ$_{max}$ (log ε) [nm] = 344.5 (4.58), 328.5 (4.27), 314.0 (4.10), 277.5 (4.63), 267.0 (4.05), 244.5 (4.81), 235.5 (4.78).

IR (diamond, RT): ṽ [cm^{-1}] = 1750, 1729, 1603, 1457, 1416, 1392, 1367, 1329, 1260, 1150, 1033, 966, 896, 845, 763, 750.

EA: calculated for C$_{31}$H$_{34}$O$_6$ (502.60): C 74.08, H 6.82; found: C 74.13, H 6.81.

Pyrene malonyl acid (160)

Cleavage of the *tert*-butyl ester **159** (211 mg, 0.42 mmol) was achieved using GOP VIb.

Yield: 190 mg (99 %).

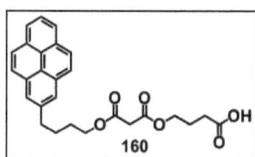

^{1}H NMR (CDCl$_3$, 400 MHz, RT):

δ [ppm] = 9.97 (bs, 1H, COOH), 8.23 (m, 3H, PhH), 8.16 (m, 1H, PhH), 8.12 (d, ^3J = 7.52 Hz, 1H, PhH), 8.03 (m, 2H, PhH), 8.00 (m, 1H, PhH), 7.86 (d, ^3J = 7.91 Hz, 1H,

Ph*H*), 4.23 (t, ^3J = 6.31 Hz, 2H, OC*H*$_2$), 4.16 (t, ^3J = 6.22 Hz, 2H, OC*H*$_2$), 3.40 (s, 2H, COC*H*$_2$CO), 3.35 (t, ^3J = 7.44 Hz, 2H, C*H*$_2$), 2.39 (t, ^3J = 7.25 Hz, 2H, C*H*$_2$), 1.93 (m, 4H, C*H*$_2$), 1.84 (m, 2H, C*H*$_2$).

^{13}C NMR (CDCl$_3$, 100.5 MHz, RT):

δ [ppm] = 179.16 (1C, *C*=O), 167.10 (1C, *C*=O), 166.99 (1C, *C*=O), 136.58 (1C, Ph*C*), 131.82 (1C, Ph*C*), 131.28 (1C, Ph*C*), 130.28 (1C, Ph*C*), 128.99 (1C, Ph*C*), 127.92 (1C, Ph*C*), 127.73 (1C, Ph*C*), 127.63 (1C, Ph*C*), 127.08 (1C, Ph*C*), 126.27 (1C, Ph*C*), 125.47 (1C, Ph*C*), 125.39 (1C, Ph*C*), 125.35 (1C, Ph*C*), 125.23 (1C, Ph*C*), 125.16 (1C, Ph*C*), 123.66 (1C, Ph*C*), 65.94 (1C, O*C*H$_2$), 64.68 (1C, O*C*H$_2$), 41.90 (1C, CO*C*H$_2$CO), 33.67 (1C, *C*H$_2$), 33.33 (1C, *C*H$_2$), 30.67 (1C, *C*H$_2$), 28.84 (1C, *C*H$_2$), 23.93 (1C, *C*H$_2$).

MS (MALDI, DCTB): m/z = 447 [M]$^+$.

UV/Vis (CH$_2$Cl$_2$): λ_{max} (log ε) [nm] = 344.5 (4.59), 328.0 (4.44), 314.0 (4.16), 277.5 (4.70), 266.5 (4.46), 244.5 (4.90), 235.5 (4.78).

IR (diamond, RT): $\tilde{\nu}$ [cm^{-1}] = 2937, 2366, 1732, 1476, 1476, 1414, 1337, 1278, 1185, 1152, 1034, 843.

EA: calculated for C$_{27}$H$_{26}$O$_6$ (446.49): C 72.63, H 5.87; found: C 71.98, H 6.21.

Pyrene-Poprhyrin dyad (161)

Dyad **161** was prepared according to GOP XI using pyrene derivative **160** (113 mg, 0.35 mmol), DMAP (34 mg, 0.28 mmol), 1-HOBT (37 mg, 0.28 mmol), DCC (104 mg, 0.50 mmol) and porphyrin derivative **147** (274 mg, 0.30 mmol). The reaction mixture was stirred for six days. Purification was achieved by column chromatography on silica gel (hexanes/EtOAc 3:1).
Yield: 217 mg (65 %).

^{1}H-NMR (CDCl$_3$, 400 MHz, RT): δ [ppm] = 8.99 (m, 6 H, pyr*H*), 8.95 (d, ^3J = 4.45 Hz, 2H, pyr*H*), 8.16 (m, 10H, Ph*H*), 8.04 (d, ^3J = 9.23 Hz, 1H, Ph*H*), 7.77 (m, 9H, Ph*H*), 7.74 (m, 1H, Ph*H*), 7.66 (d, ^3J = 7.91 Hz, 1H, Ph*H*), 7.64 (t, ^3J = 8.29 Hz , 1H, Ph*H*),

7.32 (d, ^3J = 7.06 Hz, 1H, PhH), 7.29 (d, ^3J = 7.15 Hz, 1H, PhH), 4.42 (t, ^3J = 6.30 Hz, 2H, OCH_2), 4.27 (t, ^3J = 6.28 Hz, 2H, OCH_2), 4.07 (m, 4H, OCH_2), 3.19 (m, 2H, OCH_2), 3.17 (s, 2H, COCH_2CO), 2.75 (m, 2H, CH_2), 2.36 (t, ^3J = 7.25 Hz, 2H, CH_2), 1.88 (m, 4H, CH_2), 1.64 (s, 27 H, CH_3), 1.54 (m, 2H, CH_2).

^{13}C NMR (CDCl$_3$, 100.5 MHz, RT):

δ [ppm] = 173.02 (1C, C=O), 166.81 (1C, C=O), 166.75 (1C, C=O), 157.04 (1C, PhC), 150.65 (1C, PhC), 150.54 (2C, PhC), 150.20 (1 C, PhC), 140.33 (6C, PhC), 134.78 (3C, PhC), 132.46 (1C, PhC), 132.40 (1C, PhC), 132.34 (6C, PhC), 131.94 (1C, PhC), 127.63 (1C, PhC), 127.46 (8C, α-pyr$C$), 127.41 (2C, β-pyr$C$), 126.78 (1C, PhC), 126.27 (1C, PhC), 125.99 (1C, PhC), 125.04 (1C, PhC), 125.47 (1C, PhC), 125.39 (1C, PhC), 125.35 (1C, PhC), 125.23 (1C, PhC), 125.16 (1C, PhC), 124.98 (1C, PhC), 124.87 (1C, PhC), 123.84 (1C, PhC), 123.66 (1C, PhC), 123.39 (3C, meso C), 122.20 (1C, PhC), 121.98 (1C, meso C), 121.50 (6C, β-pyr$C$), 114.40 (2C, PhC), 66.42 (1C, OCH_2), 65.72 (1C, OCH_2), 64.68 (1C, OCH_2), 63.42 (1C, PhOCH_2), 41.73 (1C, COCH_2CO), 35.30 (1C, CH_2), 33.82 (1C, CH_2), 33.19 (3C, C(CH$_3$)$_3$), 32.15 (1C, CH_2), 30.73 (9C, CH_3), 28.71 (1C, CH_2), 24.09 (1C, CH_2).

MS (MALDI, DCTB): m/z = 1335 [M]$^+$.

UV/Vis (CH$_2$Cl$_2$): λ$_{max}$ (log ε) [nm] = 550.5 (4.32), 421.0 (5.60), 344.5 (4.75), 328.0 (4.69), 313.5 (3.60), 277.5 (4.82), 266.5 (3.72), 244.0 (5.06), 231.5 (5.33).

IR (diamond, RT): $\tilde{v}$ [cm^{-1}] = 2959, 2869, 1737, 1626, 1577, 1481, 1460, 1396, 1362, 1339, 1271, 1246, 1169, 1110, 1069, 999, 962, 846, 813, 798.

EA: calculated for C$_{85}$H$_{80}$N$_4$O$_7$Zn * CHCl$_3$ (1454.35): C 71.02, H 5.61, N 3.85; found: C 71.53, H 6.01, N 4.12.

Pyrene-Porphyrin-Fullerene-triad (162)

C_{60} (56 mg, 0.08 mmol), porphyrin derivative **161** (80 mg, 0.06 mmol), CBr_4 (22 mg, 0.07 mmol) and DBU (18 µL, 0.12 mmol) were used under the conditions described in GOP XIV. The reaction mixture was stirred for 18 h. Purification was done using column chromatography on silica gel (toluene→toluene/EtOAc 10:1). Yield: 50 mg (40 %).

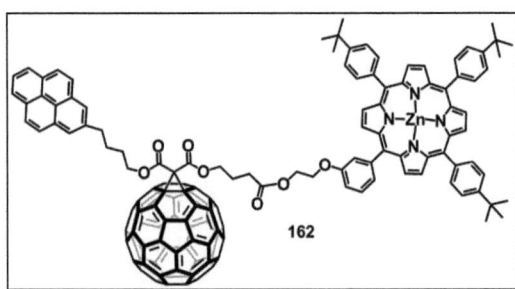

^{1}H NMR (CDCl$_3$, 400 MHz, RT):

δ [ppm] = 9.00 (m, 4H, pyr*H*), 8.89 (m, 4H, pyr*H*), 8.30 (m, 1H, Ph*H*), 8.23 (m, 2H, Ph*H*), 8.13 (m, 2H, Ph*H*), 8.05 (m, 2H, Ph*H*), 7.96 (m, 6H, Ph*H*), 7.91 (s, 3H, Ph*H*), 7.83 (m, 2H, Ph*H*), 7.76 (m, 5H, Ph*H*), 7.65 (t, ^3J = 7.83 Hz, 1H, Ph*H*), 7.43 (t, ^3J = 7.56 Hz , 1H, Ph*H*), 4.43 (m, 2H, OC*H$_2$*), 4.28 (t, ^3J = 5.68 Hz, 2H, OC*H$_2$*), 4.16 (t, ^3J = 5.31 Hz, 2H, OC*H$_2$*), 3.30 (m, 2H, OC*H$_2$*), 2. 28 (t, ^3J = 7.07 Hz, 2H, C*H$_2$*), 1.89 (m, 6H, C*H$_2$*), 1.67 (s, 9H, C*H$_3$*), 1.63 (s, 18H, C*H$_3$*), 0.90 (t, ^3J = 7.07 Hz, 2H, C*H$_2$*).

^{13}C NMR (CDCl$_3$, 100.5 MHz, RT):

δ [ppm] = 172.28 (1C, *C*=O), 163.29 (1C, *C*=O), 163.24 (1C, *C*=O), 156.44 (1C, Ph*C*), 150.43 (1C, Ph*C*), 150.41 (2C, Ph*C*), 150.32 (1 C, Ph*C*), 150.22, 150.15, 149.85, 144.53, 144.47, 144.07, 143.92, 143.81, 143.79, 143.46, 143.34, 143.31, 143.19, 142.94, 142.79, 142.68, 142.52, 142.32, 141.99, 141.87, 141.80, 141.70, 141.42, 141.27, 141.21, 141.02, 140.87, 140.64, 140.58, 140.48, 140.18, 140.06, 139.77, 139.69, 139.64, 139.50, 139.18, 138.26, 137.60 (64C, sp^2*C* C_{60}, Ph*C*), 135.89 (3C, Ph*C*), 134.17 (1C, Ph*C*), 134.07 (1C, Ph*C*), 132.07 (6C, Ph*C*), 131.99 (1C, Ph*C*), 129.76 (1C, Ph*C*), 128.37 (8C, α-pyr*C*), 127.38 (2C, β-pyr*C*), 127.63 (1C, Ph*C*), 127.38 (1C, Ph*C*), 127.30 (1C, Ph*C*), 127.07 (1C, Ph*C*), 126.61 (1C, Ph*C*), 125.78 (1C, Ph*C*), 124.98 (1C, Ph*C*), 124.86 (1C, Ph*C*), 124.69 (1C, Ph*C*), 123.61 (1C, Ph*C*), 123.55 (1C, Ph*C*), 123.48 (1C, Ph*C*), 123.42 (1C, Ph*C*), 123.06 (3C, meso*C*), 121.56 (1C, Ph*C*), 121.32 (1C, meso*C*), 120.69 (6C, β-pyr*C*), 114.11 (2C, Ph*C*), 70.75 (2C, sp^2*C* C_{60}), 67.11 (1C, OC*H$_2$*), 65.95 (1C, OC*H$_2$*), 65.61 (1C, OC*H$_2$*),

62.81 (1C, PhOCH$_2$), 51.83 (1C, COCCO), 34.94 (1C, CH$_2$), 34.88 (1C, CH$_2$), 34.11 (3C, C(CH$_3$)$_3$), 32.87 (1C, CH$_2$), 31.72 (9C, CH$_3$), 22.34 (1C, CH$_2$), 14.07 (1C, CH$_2$).

MS (MALDI, DCTB): m/z = 2053 [M]$^+$.

UV/Vis (CH$_2$Cl$_2$): λ_{max} (log ε) [nm] = 548.0 (4.49), 429.0 (5.35), 344.5 (5.00), 328.0 (5.02), 277.0 (5.15), 264.0 (5.22), 231.5 (5.70).

IR (diamond, RT): $\tilde{v}$ [cm^{-1}] = 2959, 2931, 2359, 2340, 1746, 1275, 1258, 1234, 1208, 1156, 1000, 846, 814, 798, 765, 750, 719.

6.2.7 Porphyrin-Phosphonic Acid and its Precursors

5-(6-Phosponatohexyl-phenyl)-10,15,20-(triphenyl)-porphyrin (10)

Porphyrin phosphonate **170** (30 mg, 0.03 mmol) was dissolved in dry CH$_2$Cl$_2$ (15 mL) under inert conditions and cooled down to 0 °C. Trimethylbromsilane (101 mg, 88 µL, 0.66 mmol) dissolved in dry CH$_2$Cl$_2$ (20 mL) was dropped during 1 h to the rapid stirring solution. After stirring for 3 d (0°C→rt) 5 mL MeOH were added and the mixture was stirred for another 5 h. After distillation of the solvent the crude product was dissolved in CH$_2$Cl$_2$ and precipitated with *n*-pentane.
Yield: 24 mg (92 %).

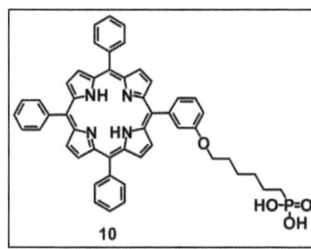

^{1}H NMR (CDCl$_3$/MeOH, 400 MHz, RT):
δ [ppm] = 8.79 (bs, 8H, pyr*H*), 8.10 (bs, 7H, Ph*H*), 7.55 (bs, 11H, Ph*H*), 6.96 (bs, 1H, Ph*H*), 3.76 (bs, 4H, PhOC*H*$_2$, C*H*$_2$P), 1.71 (bs, 2H, C*H*$_2$), 1.46 (bs, 2H, C*H*$_2$), 1.09 (bs, 4H, C*H*$_2$).

^{13}C NMR (CDCl$_3$MeOH, 100.5 MHz, RT):
δ [ppm] = 157.29 (1C, Ph*C*), 149.08 (2C, Ph*C*), 143.04 (1C, Ph*C*), 141.89 (3C, Ph*C*), 134.37 (1C, Ph*C*), 131.09 (6C, Ph*C*), 130.37 (1C, Ph*C*), 127.57 (8C, α-pyr*C*), 127.30 (2C, β-pyr*C*), 126.54 (6C, pyr*C*), 121.09 (3C, Ph*C*), 120.08 (6 C, Ph*C*), 120.02 (3C, meso*C*), 119.81 (1C, meso*C*), 67.97 (1C, PhO*C*H$_2$), 29.41 (1C, *C*H$_2$P), 29.15 (1C, *C*H$_2$), 28.93 (1C, *C*H$_2$), 25.70 (1C, *C*H$_2$), 20.12 (1C, *C*H$_2$).

MS (MALDI, DCTB): m/z = 796 [M]$^+$.

UV/Vis (CH$_2$Cl$_2$): λ_{max} (log ε) [nm] = 648.0 (3.93), 592.5 (3.97), 549.5 (4.22), 516.0 (4.16), 420.0 (5.40).

IR (diamond, RT): $\tilde{\nu}$ [cm^{-1}] = 1597, 1328, 1211, 1178, 1069, 1002, 967, 925, 797, 720.

EA: calculated for C$_{50}$H$_{43}$N$_4$O$_4$P*CH$_2$Cl$_2$*MeOH (911.85): C 68.49, H 5.42, N 6.14; found: C 68.68, H 4.22, N 6.28.

3-(6-Hydroxyhexyloxy)benzaldehyde (165)

3-Hydroxybenzaldeyhde **163** (2 g, 16.4 mmol), 6-bromohexan-1-ol **164** (2.97 g, 16.4 mmol) and 18-crown-6 (43 mg, 0.16 mmol) were dissolved in dry DMF (50 mL) under inert conditions before K$_2$CO$_3$ (2.3 g, 20.5 mmol) was added at once and the reaction mixture was stirred at 80 °C for 24 h. After cooling down to rt the mixture was filtered and CH$_2$Cl$_2$ (50 mL) and H$_2$O (50 mL) were added. The organic layer was separated and washed with HCl (10 %) solution 3 times. Drying over MgSO$_4$ and evaporation of the solvent led to the yellowish crude product which was purified by column chromatography on silica gel (hexanes/EtOAc 1:1).
Yield: 2.3 g (62 %).

^{1}H NMR (CDCl$_3$, 400 MHz, RT): δ [ppm] = 9.97 (s, 1H, HC=O), 7.45 (d, ^3J = 7.39 Hz, 1H, PhH), 7.43 (s, 1H, PhH), 7.38 (d, ^3J = 7.26 Hz, 1H, PhH), 7.18 (m, 1H, PhH), 4.02 (t, ^3J = 6.40 Hz, 2H, OCH$_2$), 3.67 (t, ^3J = 6.50 Hz, 2H, HOCH$_2$), 2.05 (s, 1H, OH), 1.83 (m, 2H, CH$_2$), 1.59 (m, 4H, CH$_2$), 1.48 (m, 2H, CH$_2$).

^{13}C NMR (CDCl$_3$, 100.5 MHz, RT): δ [ppm] = 191.82 (1C, HC=O), 159.17 (1C, PhC), 137.26 (1C, PhC), 129.55 (1C, PhC), 122.94 (1C, PhC), 121.51 (1C, PhC), 112.19 (1C, PhC), 67.64 (1C, OCH$_2$), 62.37 (1C, HOCH$_2$), 32.17 (1C, CH$_2$), 28.61 (1C, CH$_2$), 25.37 (1C, CH$_2$), 25.04 (1C, CH$_2$).

MS (FAB, NBA): m/z = 223 [M]$^+$.

Experimental Part

IR (diamond, RT): ṽ [cm^{-1}] = 2923, 2851, 1686, 1583, 1484, 1446, 1387, 1321, 12256, 1169, 1149, 1076, 1039, 959, 861, 777, 679.

EA: calculated for $C_{13}H_{18}O_3$ * 0.5 EtOAc (266.34): C 67.64, H 8.33; found: C 67.39, H 8.87.

5-(6-Hexoxyloxy-phenyl)-10,15,20-(triphenyl)-porphyrin (167)

Porphyrin synthesis was accomplished using GOP IXb with pyrrole **50** (640 mg, 9 mmol), 3-hydroxyhexyloxy-benzaldehyde **165** (500 mg, 2.3 mmol), benzaldehyde **166** (716 mg, 6.8 mmol), TFA (521 µL, 771 mg, 6.8 mmol), TEA (1.12 mL, 820 mg, 8.12 mmol) and DDQ (1.54 g, 6.8 mmol). The residue was purified by column chromatography with dichloromethane/ethyl acetate 5:1 as eluent.
Yield: 100 mg (7 %).

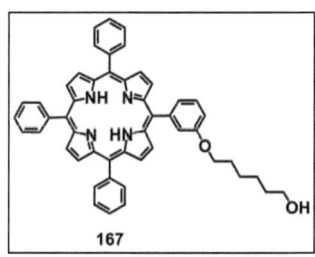

^{1}H NMR (CDCl$_3$, 400 MHz, RT):
δ [ppm] = 8.92 (m, 8 H, pyr$H$), 8.25 (d, ^3J = 8.42 Hz, 6H, PhH), 7.79 (m, 10H, PhH), 7.64 (t, ^3J = 8.10 Hz, 1H, PhH), 7.34 (d, ^3J = 7.06 Hz, 1H, PhH), 7.21 (m, 1H, PhH), 4.15 (t, ^3J = 6.50 Hz, 2H, PhOCH_2), 3.64 (t, ^3J = 6.40 Hz, 2H, HOCH_2), 2.05 (bs, 1H, OH), 1.91 (m, 2H, CH_2), 1.59 (m, 4H, CH_2), 1.28 (m, 2H, CH_2), -2.76 (s, 2H, NH).

^{13}C NMR (CDCl$_3$, 100.5 MHz, RT):
δ [ppm] = 157.40 (2C, PhC), 143.40 (2C, PhC), 142.14 (1C, PhC), 134.53 (1C, PhC), 130.81 (1C, PhC), 127.69 (3C, PhC), 130.81 (16C, pyr$C$), 126.67 (6C, PhC), 121.02 (2C, PhC), 120.09 (6 C, PhC), 119.89 (3C, meso$C$), 114.11 (1C, meso$C$), 112.37 (2C, PhC), 68.05 (1C, HOCH_2), 62.87 (1C, PhOCH_2), 32.65 (1C, CH_2), 29.34 (1C, CH_2), 25.94 (1C, CH_2), 25.55 (1C, CH_2).

MS (FAB, NBA): m/z = 731 [M]$^+$.

UV/Vis (CH$_2$Cl$_2$): λ$_{max}$ (log ε) [nm] = 647.5 (3.92), 590.5 (3.98), 551.5 (4.15), 516.5 (4.39), 419.5 (5.65).

IR (diamond, RT): $\tilde{\nu}$ [cm^{-1}] = 2965, 2360, 2342, 1595, 1438, 1333, 1287, 1262, 1179, 1065, 994, 951, 790, 700.

EA: calculated for C$_{50}$H$_{42}$N$_4$O$_2$ * 2 EtOAc (907.10): C 76.80, H 6.44, N 6.18; found: C 77.32, H 5.88, N 6.69.

5-(6-Hexoxyloxy-phenyl)-10,15,20-(triphenyl)-porphyrinato zinc(II) (168)

Porphyrin **167** (90 mg, 0.12 mmol) and zinc acetate (81 mg, 0.36 mmol) were dissolved in a mixture of CHCl$_3$ and MeOH (40 mL, 3:1) and stirred over night at room temperature. The solvent was removed and the crude product was purified by column chromatography on silica gel (CH$_2$Cl$_2$/EtOAc 5:1).
Yield: 95 mg (97 %).

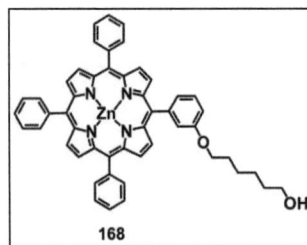

^{1}H-NMR (CDCl$_3$, 400 MHz, RT):

δ [ppm] = 8.97 (m, 8 H, pyr*H*), 8.25 (m , 6H, Ph*H*), 7.82 (d, ^3J = 7.54 Hz, 1H, Ph*H*), 7.76 (m, 10H, Ph*H*), 7.59 (t, ^3J = 7.82 Hz, 1H, Ph*H*), 7.20 (d, ^3J = 8.29 Hz, 1H, Ph*H*), 4.09 (t, ^3J = 7.16 Hz, 2H, PhOC*H$_2$*), 3.96 (t, ^3J = 6.95 Hz, 2H, HOC*H$_2$*), 2.00 (s, 1H, O*H*), 1.62 (m, 2H, C*H$_2$*), 1.26 (m, 2H, C*H$_2$*), 1.16 (m, 2H, C*H$_2$*), 0.72 (m, 2H, C*H$_2$*).

^{13}C NMR (CDCl$_3$, 100.5 MHz, RT):

δ [ppm] = δ [ppm] = 157.51 (1C, Ph*C*), 150.55 (2C, Ph*C*), 144.60 (1C, Ph*C*), 143.37 (3C, Ph*C*), 134.58 (1C, Ph*C*), 132.30 (6C, Ph*C*), 127.92 (1C, Ph*C*), 127.82 (8C, α-pyr*C*), 127.65 (2C, β-pyr*C*), 126.91 (6C, β-pyr*C*), 121.43 (3C, Ph*C*), 121.38 (6 C, Ph*C*), 121.14 (3C, meso*C*), 114.32 (1C, meso*C*), 68.62 (1C, HOC*H$_2$*), 61.97 (1C, PhOC*H$_2$*), 29.46 (1C, C*H$_2$*), 25.69 (1C, C*H$_2$*), 25.21(1C, C*H$_2$*), 21.35 (1C, C*H$_2$*).

MS (FAB, NBA): m/z = 792 [M]$^+$.

UV/Vis (CH$_2$Cl$_2$): λ$_{max}$ (log ε) [nm] = 587.0 (3.84), 548.0 (4.39), 420.5 (5.59).

Experimental Part

IR (diamond, RT): $\tilde{v}$ [cm^{-1}] = 2966, 1338, 1192, 1109, 1071, 1064, 994, 810, 773, 721.

EA: calculated for $C_{50}H_{40}N_4O_2Zn$ * 2 EtOAc (970.48): C 71.78, H 5.82, N 5.77; found: C 71.52, H 5.58, N 5.89.

5-(6-Bromohexoxy-phenyl)-10,15,20-(triphenyl)-porphyrinato zinc(II) (169)

Porphyrin **168** (90 mg, 0.11 mmol) and CBr$_4$ (46 mg, 0.14 mmol) were dissolved in CH$_2$Cl$_2$ (15 mL) before PPh$_3$ (37 mg, 0.14 mmol) was added portionwise. Then the reaction mixture was stirred for 1.5 h at room temperature. After distillation of the solvent the crude product was purified by column chromatography on silica gel (CH$_2$Cl$_2$).
Yield: 78 mg (83 %).

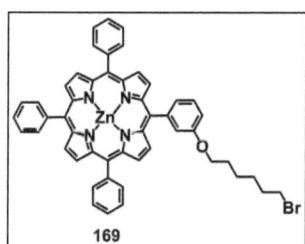

^{1}H-NMR (CDCl$_3$, 400 MHz, RT):
δ [ppm] = 9.00 (m, 8 H, pyr*H*), 8.28 (m, 6H, Ph*H*), 7.86 (m, 10H, Ph*H*), 7.68 (m, 1H, Ph*H*), 7.64 (t, ^3J = 7.85 Hz, 1H, Ph*H*), 7.31 (m, 1H, Ph*H*), 4.11 (t, ^3J = 6.41 Hz, 2H, PhOC*H$_2$*), 3.41 (t, ^3J = 6.78 Hz, 2H, BrC*H$_2$*), 1.90 (m, 2H, C*H$_2$*), 1.53 (m, 2H, C*H$_2$*), 1.31 (m, 2H, C*H$_2$*), 0.93 (m, 2H, C*H$_2$*).

^{13}C NMR (CDCl$_3$, 100.5 MHz, RT):
δ [ppm] = 157.22 (1C, Ph*C*), 150.09 (2C, Ph*C*), 144.62 (1C, Ph*C*), 143.35 (3C, Ph*C*), 134.54 (1C, Ph*C*), 131.95 (6C, Ph*C*), 127.92 (1C, Ph*C*), 127.82 (8C, α-pyr*C*), 127.65 (2C, β-pyr*C*), 126.67 (6C, β-pyr*C*), 121.52 (3C, Ph*C*), 121.32 (6 C, Ph*C*), 121.14 (3C, meso*C*), 114.30 (1C, meso*C*), 67.91 (1C, PhO*C*H$_2$), 33.78 (1C, Br*C*H$_2$), 32.63 (1C, *C*H$_2$), 29.14 (1C, *C*H$_2$), 27.91 (1C, *C*H$_2$), 25.30 (1C, *C*H$_2$).

MS (FAB, NBA): m/z = 857 [M]$^+$.

UV/Vis (CH$_2$Cl$_2$): λ$_{max}$ (log ε) [nm] = 586.5 (3.85), 549.5 (4.04), 419.0 (5.03).

IR (diamond, RT): $\tilde{v}$ [cm^{-1}] = 2952, 1421, 1212, 1150, 1070, 1062, 996, 818, 777, 725.

EA: calculated for $C_{50}H_{39}N_4OZnBr$ * 2 CH_2Cl_2 (1027.05): C 60.81, H 4.22, N 5.46; found: C 59.31, H 4.50, N 4.87.

5-(6-Diethylphosphonatohexyl-phenyl)-10,15,20-(triphenyl)-porphyrinato zinc(II) (170)

Porphyrin **169** (50 mg, 0.06 mmol) was dissolved in triethylphosphite (10 mL) under inert conditions and stirred at 140 °C for 43 h. After cooling down to rt the excessive phosphate was removed in vacuum and the residue was dissolved in CH_2Cl_2. Precipitating with *n*-pentane led to the desired porphyrino phosphonate.
Yield: 53 mg (97 %).

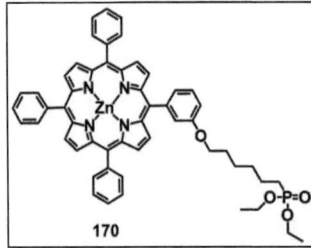

^{1}H NMR (CDCl$_3$, 400 MHz, RT):
δ [ppm] = 8.98 (d, ^3J = 8.20 Hz, 2H, pyr*H*), 8.91 (m, 6H, pyr*H*), 8.22 (m, 6H, Ph*H*), 7.84 (d, ^3J = 7.54 Hz, 1H, Ph*H*), 7.74 (m, 10H, Ph*H*), 7.61 (t, ^3J = 7.91 Hz, 1H, Ph*H*), 7.24 (m , 1H, Ph*H*), 4.05 (t, ^3J = 6.31 Hz, 2H, PhOC*H$_2$*), 3.81 (t, ^3J = 6.52 Hz, 2H, C*H$_2$*P), 2.81 (m, 4H, POC*H$_2$*), 1.74 (m, 2H, C*H$_2$*), 1.30 (m, 2H, C*H$_2$*), 1.20 (m, 2H, C*H$_2$*), 0.90 (m, 2H, C*H$_2$*), 0.66 (t, ^3J = 7.06 Hz, 6H, C*H$_3$*).

^{13}C NMR (CDCl$_3$, 100.5 MHz, RT):
δ [ppm] = 157.55 (1C, Ph*C*), 150.47 (2C, Ph*C*), 144.92 (1C, Ph*C*), 143.64 (3C, Ph*C*), 134.90 (1C, Ph*C*), 132.15 (6C, Ph*C*), 127.91 (1C, Ph*C*), 127.68 (8C, α-pyr*C*), 126.80 (2C, β-pyr*C*), 121.50 (6C, β-pyr*C*), 121.12 (3C, Ph*C*), 120.88 (6 C, Ph*C*), 119.14 (3C, meso*C*), 114.17 (1C, meso*C*), 68.31 (1C, PhO*C*H$_2$), 61.14 (2C, PO*C*H$_2$), 29.41 (1C, *C*H$_2$P), 25.77 (1C, *C*H$_2$), 21.86 (1C, *C*H$_2$), 21.64 (1C, *C*H$_2$), 20.05 (1C, *C*H$_2$), 16.27 (2C, *C*H$_3$).

MS (FAB, NBA): m/z = 913 [M]$^+$.

UV/Vis (CH$_2$Cl$_2$): λ_{max} (log ε) [nm] = 585.0 (3.93), 549.0 (4.41), 419.5 (5.60).

IR (diamond, RT): $\tilde{\nu}$ [cm^{-1}] = 1595, 1439, 1338, 1282, 1187, 1023, 991, 819, 792, 568.

EA: calculated for C$_{54}$H$_{49}$N$_4$O$_4$PZn * 0.5 CH$_2$Cl$_2$ (956.82): C 68.41, H 5.27, N 5.86; found: C 68.85, H 5.04, N 5.43.

7 References

(1) Wackernagel, M.; Schulz, N. B.; Deumling, D.; Linares, A. C.; Jenkins, M.; V. Kapos; Monfreda, C.; Loh, J.; Myers, N.; Norgaard, R.; Randers, J. *Proc. Natl. Acad. Sci.* **2002**, *99*.
(2) Janssens, I. A.; Freibauer, A.; Ciais, P.; Smith, P.; Nabuurs, G. J.; Folberth, G.; Schlamadinger, B.; Hutjes, R. W. A.; Ceulemans, R.; Schulze, E. D.; Valentini, R.; Dolman, A. J. *Science* **2003**, *300*, 1538–1542.
(3) Simmons, M. R. *Twilight in the Desert* Hoboken, NJ **2005**.
(4) Eisenberg, R.; Nocera, D. G. *Inorg. Chem.* **2006**, *45*, 6799–6801.
(5) Science, U. D. o. E. O. o. In *Basic Research Needs for Solar Energy Utilization* 2005, p http://www.er.doe.gov/bes/reports/files/SEU_rpt.pdf.
(6) Ciamician, G. *Science* **1912**, *36*, 385-394.
(7) Kirmaier, C.; Holton, D. *The Photosynthetic Reaction Center (Eds.: Deisenhofer, J.; Norris, J. R.); Academic Press: San Diego, CA* **1993**, Vol. II, 49-70.
(8) Remy, A.; Gerwert, K. *Nature Struct. Biol.* **2003**, *10*, 637-644.
(9) Boxer, S. G. *Annu. Rev. Biophys. Bioeng.* **1990**, *19*, 267.
(10) Fukuzumi, S.; Imahori, H. *Electron Transfer in Chemistry (Ed.: Balzani, V.) Wiley-VCH: Weinheim,* **2001**, Vol. 2, 927-975.
(11) Fukuzumi, S.; Guldi, D. M. *Electron Transfer in Chemistry; (Ed.: Balzani, V.) Wiley-VCH: Weinheim,* **2001**, Vol. 2, 270-337.
(12) Fukuzumi, S. *Org. Biomol. Chem* **2003**, *1*, 609-620.
(13) Guldi, D. M.; Kamat, P. V. *Fullerenes, Chemistry, Physics, and Technology (Eds.: Kadish, K. M., Ruoff, R. S.) Wiley-Interscience: New York* **2000**, 225-281.
(14) Imahori, H.; Tamaki, K.; Yamada, H. Y., K.; Sakata, Y.; Nishimura, Y.; Yamazaki, I.; Fujitsuka, M. O.; Ito, O. *Carbon* **2000**, *38*, 1599-1605.
(15) Fukuzumi, S.; Imahori, H.; Yamada, H.; El-Khouly, M. E.; Fujitsuka, M.; Ito, O.; Guldi, D. M. *J. Am. Chem. Soc.* **2001**, *123*, 2571-2575.
(16) Imahori, H.; Guldi, D. M.; Tamaki, K.; Yoshida, Y.; Luo, C.; Sakata, Y.; Fukuzumi, S. *J. Am. Chem. Soc.* **2001**, *123*, 6617-6628.
(17) Imahori, H.; Tamaki, K.; Guldi, D. M.; Luo, C.; Fujitsuka, M.; Ito, O.; Sakata, Y.; Fukuzumi, S. *J. Am. Chem. Soc.* **2001**, *123*, 2607-2617.
(18) Imahori, H.; Hagiwara, K.; Akiyama, T.; Aoki, M.; Taniguchi, S.; Okada, T.; Shirakawa, M.; Sakata, Y. *Chem. Phys. Lett.* **1996**, *263*, 545-550.
(19) Tkachenko, N. V.; Guenther, C.; Imahori, H.; Tamaki, K.; Sakata, Y.; Fukuzumi, S.; Lemmetyinen, H. *Chem. Phys. Lett.* **2000**, *326*, 344-350.
(20) Vehmanen, V.; Tkachenko, N. V.; Imahori, H.; Fukuzumi, S.; Lemmetyinen, H. *Spectrochim. Acta, Part A* **2001**, *57*, 2229-2244.
(21) Imahori, H. *Org. Biomol. Chem.* **2004**, *2*, 1425-1433.
(22) Imahori, H.; Yamada, H.; Guldi, D. M.; Endo, Y.; Shimomura, A.; Kundu, S.; Yamada, K.; Okada, T.; Sakata, Y.; Fukuzumi, S. *Angew. Chem., Int. Ed.* **2002**, *41*, 2344-2347.
(23) Imahori, H.; Tamaki, K.; Araki, Y.; Sekiguchi, Y.; Ito, O.; Sakata, Y.; Fukuzumi, S. *J. Am. Chem. Soc.* **2002**, *124*, 5165-5174.
(24) Moser, C. C.; Page, C. C.; Dutton, P. L. *Electron Transfer in Chemistry (Ed.: V. Balzani)* **2001**, *3*.
(25) Ritz, T.; A.Damjanovic; Schulten, K. *ChemPhysChem* **2002**, *3*, 243-248.

References

(26) Ort, D. R.; Yocum, C. F. *Oxygenic Photosynthesis: the Light Reactions* Kluwer, Dordrecht, **1996**.
(27) Rhee, K.-H.; Morris, E. P.; Barber, J.; Kühlbrandt, W. *Nature* **1998**, *369*, 283-286.
(28) Renger, T.; Marcus, R. A. *J. Phys. Chem. B* **2002**, *106*, 1809-1819.
(29) Balzani, V.; Venturi, M.; Credi, A. *Molecular Devices and Machines, Wiley-VCH, Weinheim* **2003**.
(30) Dinner, B. A.; Babcock, G. T. *Oxygenic Photosynthesis: the Light Reactions (Eds.: D.R. Ort, C.F. Yocum), Kluwer, Dordrecht*, **1996**.
(31) Deisenhofer, J.; Michel, H. *Angew. Chem. Int. Ed.* **1989**, *28*, 829-847.
(32) Huber, R. *Angew. Chem. Int. Ed.* **1989**, *28*, 848-869.
(33) http://kvhs.nbed.nb.ca/gallant/biology/ps1_2.html.
(34) Adams, D. M.; Brus, L.; Chidsey, C. E. D.; Creager, S.; Creutz, C.; Kagan, C. R.; Kamat, P. V.; Lieberman, M.; Lindsay, S.; Marcus, R. A.; Metzger, R. M.; Michel-Beyerle, M. E.; Miller, J. R.; Newton, M. D.; Rolison, D. R.; Sankey, O.; Schanze, K. S.; Yardley, J.; Zhu, X. *J. Phys. Chem. B.* **2003**, *107*, 6668-6697.
(35) McDermott, G.; Prieee, S. M.; Freer, A. A.; Hawthornthwaite-Lawless, A. M.; Papiz, M. Z.; Cogdell, R. J.; N.W. Isaaacs *Nature Struct. Biol.* **1995**, *374*, 517-521.
(36) Balzani, V.; de Cola, L. *Eds. Supramolecular Chemistry, NATO ASI Series; Kluwer Academic Publishers. Dordrech*, **1992**.
(37) Steed, J. W.; Atwood, J. *Supramolecular Chemistry, Wiley, Chichester* **2000**.
(38) Gust, D.; Moore, T. A.; Moore, A. L. *Acc. Chem. Res.* **2001**, *34*, 40-48.
(39) Imahori, H.; Sakata, Y. *Adv. Mater.* **1997**, *9*, 537-546.
(40) Prato, M. *J. Mater. Chem.* **1997**, *7*, 1097-1109.
(41) Imahori, H.; Sakata, Y. *Eur. J. Org. Chem.* **1999**, *10*, 2445-2457.
(42) Diederich, F.; Gomez-Lopez, M. *Chem. Soc. Rev.* **1999**, *28*, 263-277.
(43) Guldi, D. M. *Chem. Commun.* **2000**, 321-327.
(44) Reed, C. A.; Bolskar, R. D. *Chem. Rev.* **2000**, *100*, 1075-1120.
(45) Gust, D.; Moore, T. A.; Moore, A. L. *J. Photochem. & Photobiol. B* **2000**, *58*, 63-71.
(46) Guldi, D. M.; Martin, N. *J. Mater. Chem.* **2002**, *12*, 1978-1992.
(47) Nierengarten, J. F. *Top. Curr. Chem.* **2003**, *228*, 87-110.
(48) El-Khouly, M. E.; Ito, O.; D'Souza, F. *J. Photochem. Photobiol. C* **2004**, *5*, 79-104.
(49) Guldi, D. M.; Zerbetto, F.; Georgakilas, V.; Prato, M. *Acc. Chem. Res.* **2005**, *38*, 38-43.
(50) Guldi, D. M. *Phys. Chem. Chem. Phys.* **2007**, *12*, 1400-1420.
(51) Gust, D.; Moore, T. A.; Moore, A. L. *Electron Transfer in Chemistry* **2001**, *3*, 272-336.
(52) Guldi, D. M.; Imahori, H.; Tamaki, K.; Kashiwagi, Y.; Yamada, H.; Sakata, Y.; Fukuzumi, S. *J. Phys. Chem. A* **2004**, *108*, 541-548.
(53) Luo, C.; Guldi, D. M.; Imahori, H.; Tamaki, K.; Sakata, Y. *J. Am. Chem. Soc.* **2000**, *122*, 6535-6551.
(54) D'Souza, F.; El-Khouly, M. E.; Gadde, S.; Zandler, M. E.; Zandler, A. L. E.; McCarty, A. L.; Araki, Y.; Ito, O. *Tetrahedron* **2006**, *62*, 1967-1978.
(55) Sessler, J. S.; Wang, B.; Springs, S. L.; T, B. C. *Cromprehensive Supramolecular Chemistry, (Eds: J. L. Atwood, J. E. D. Davies, D. D. MacNicol, F. Vögtle)* **1996**, Chapter 9.
(56) Guldi, D. M. *Chem. Soc. Rev.* **2002**, *31*, 22-36.

(57) Meijer, M. D.; Klink, G. P. M. v.; Koten, G. v. *Coord. Chem. Rev.* **2002**, *230*, 141-163.
(58) Lewis, F. D.; Letsinger, R. I.; Wasielwski, M. R. *Acc. Chem. Res.* **2001**, *34*, 159-170.
(59) Sun, L.; Hammarström, L.; Akermark, B.; Styring, S. *Chem. Soc. Rev.* **2001**, *30*, 36-49.
(60) Balch, A. L.; Olmstead, M. M. *Chem. Rev.* **1998**, *98*, 2123-2165.
(61) Imahori, H.; Fukuzmi, S. *Adv. Func. Mater.* **2004**, *14*, 525-536.
(62) Boyd, P. D.; Reed, C. A. *Acc. Chem. Res.* **2005**, *38*, 235-242.
(63) El-Khouly, M. E.; Ito, O.; Smith, P. M.; D'Souza, F. J. *Photochem. Photobiol.* **2004**, *5*, 79-104.
(64) D'Souza, F.; Ito, O. *Coord. Chem. Rev.* **2005**, *249*, 1410-1422.
(65) Verhoeven, J. *Adv. Chem. Phys.* **1999**, *106*, 603-644.
(66) Piotrowiak, P. *Chem. Soc. Rev.* **1999**, *28*, 143-150.
(67) Martin, N.; Sanchez, L.; Illescas, B.; Perez, I. *Chem. Rev.* **1998**, *98*, 2527-2548.
(68) Huang, C.-H.; McClenaghan, N. D.; Kuhn, A.; Hofstraat, J. W.; Bassani, D. M. *Org. Lett* **2005**, *7*, 3409-3412.
(69) Hong, H.; Davidov, D.; Kallinger, C.; Lemmer, H.; Feldmann, J.; Harth, E.; Gugel, A.; Müllen, K. *Synth. Met.* **1999**, *102*, 1537-.
(70) Haino, T.; Araki, H.; Yamanaka, Y.; Fukazawa, Y. *Tetrahedron Lett.* **2001**, *42*, 3203-3206.
(71) Georgakilas, V.; Pellarini, F.; Prato, M.; Guldi, D. M.; Melle-Franco, M.; Zerbetto, F. *Proc. Natl. Acad. Sci.* **2002**, *99*, 5075-5080.
(72) Schubert, U. S.; Weidl, C. H.; Rapta, P.; Harth, E.; Müllen, K. *Chem. Lett.* **1999**, *9*, 949-950.
(73) Liu, Y.; Wang, H.; Liang, P.; Zhang, H.-Y. *Angew. Chem. Int. Ed.* **2004**, *43*, 2690-2694.
(74) Rispens, M. T.; Sanchez, L.; Beckers, E. H. A.; Hal, P. H. v.; Schenning, A. P. H. J.; Abdelkrim, E.-G.; Meijer, E. W.; Janssen, R. A. J.; Hummelen, J. C. *Synth. Met.* **2003**, *153*, 801-803.
(75) Grimm, F.; Hartnagel, K.; Wessendorf, F.; Hirsch, A. *Chem.Comm.* **2009**, 1331-1333.
(76) Champs, X.; Dietel, E.; Hirsch, A.; Pyo, S.; Echegoyen, L.; Hackbarth, S.; Röder, B. *Chem. Eur. J.* **1999**, *5*, 2362-2373.
(77) Chukharev, V.; Tkachenko, N. V.; Efimov, A.; Guldi, D. M.; Hirsch, A.; Scheloske, M.; Lemmetyinen, H. *J. Phys. Chem. B* **2004**, *108*, 16377-16385.
(78) Sessler, J. L.; Jayawickramarajah, J.; Gouloumis, A.; Torres, T.; Guldi, D. M.; Maldonado, S.; Stevenson, K. J. *Chem. Commun.* **2005**, 1892-1894.
(79) Schuster, D. I.; Li, K.; Guldi, D. M.; Ramey, J. *Org. Lett.* **2004**, *6*, 1919-1922.
(80) Schuster, D. I.; Cheng, P.; Jarowski, P. D.; Guldi, D. M.; Luo, C.; Echegoyen, L.; Pyo, S.; Holzwarth, A. R.; Braslavsky, S. E.; Williams, R. M.; Klihm, G. J. *J. Am. Chem. Soc.* **2004**, *126*, 7257-7270.
(81) D'Souza, F.; Deviprasad, G. R.; Rahman, M. S.; Choi, J.-P. *Inorg. Chem.* **1999**, *38*, 2157-2160.
(82) Ros, T. d.; Prato, M.; Guldi, D. M.; Alessio, E.; Ruzzi, M.; Pasimeni, L. *Chem. Commun.* **1999**, 635-636.
(83) El-Khouly, M. E.; Ito, O.; Gadde, S.; Deviprasad, G. R.; Fujisuka, G. R.; D'Souza, F. *J. Porphyrines Phthalocyanines* **2003**, *7*, 1-7.
(84) Bae, A.-H.; Datano, T.; Sugiyasu, K.; Kishida, T.; Takeuchi, M.; Shinaki, S. *Tetrahedron Lett.* **2005**, *46*, 3169-3173.

(85) Guldi, D. M.; Zilbermann, I.; Anderson, G. A.; Koratos, K.; Prato, M.; Tafuro, R.; Valli, L. *J. Mater. Chem.* **2004**, *14*, 303-309.
(86) D'Souza, F.; Deviprasad, G. R.; Zandler, M. E.; Hoang, V. T.; Arkady, K.; Stipdonk, M. V.; Perera, A.; El-Khouly, M. E.; Fujisuka, M.; Ito, O. *J. Phys. Chem A* **2002**, *106*, 3243-3252.
(87) D'Souza, F.; Deviprasad, G. R.; Zandler, M. E.; El-Khouly, M. E.; Fujisuka, M.; Ito, O. *J. Phys. Chem A* **2003**, *107*, 4801-4807.
(88) Balbinot, D.; Atalick, S.; Guldi, D. M.; Hatzimarinaki, M.; Hirsch, A.; Jux, N. *J. Phys. Chem. B* **2003**, *107*, 13273-13279.
(89) Guldi, D. M.; Zilbermann, I.; Anderson, G.; Li, A.; Balbinot, D.; Jux, N.; Hatzimarinaki, M.; Hirsch, A.; Prato, M. *Chem. Commun.* **2004**, 726-727.
(90) Braun, M.; Atalick, S.; Guldi, D. M.; Lanig, H.; M., B.; Burghardt, S.; Hatzimarinaki, M.; Ravanelli, E.; Prato, M.; van Eldik, R.; Hirsch, A. *Chem. Eur. J.* **2003**, *9*, 5169-5176.
(91) Wessendorf, F.; Gnichwitz, J.-F.; Sarova, G. H.; Hager, K.; Hartnagel, U.; Guldi, D. M.; Hirsch, A. *J. Am. Chem. Soc.* **2007**, *129*, 16057-16071.
(92) Sanchez, L.; Martin, N.; Guldi, D. M. *Angew. Chem.* **2005**, *117*, 5508-5516.
(93) Sanchez, L.; Sierra, M.; Martin, N.; Myles, A. J.; Dale, T. J.; Rebek, J. J.; Seitz, W.; Guldi, D. M. *Angew. Chem. Int. Ed.* **2006**, *45*, 4637-4641.
(94) Gnichwitz, J.-F.; Wielopolski, M.; Hartnagel, K.; Hartnagel, U.; Guldi, D. M.; Hirsch, A. *J. Am. Chem. Soc.* **2008**, *130*, 8491-8501.
(95) D'Souza, F.; Venukadasula, G. M.; Yamanaka, K.-I.; Subbaiyan, N. K.; Zandler, M. E.; Ito, O. *Org. Biomol. Chem.* **2009**, *7*, 1076-1080.
(96) Berl, V.; Schmutz, M.; M.J.Krische; R.G.Khoury; Lehn, J. M. *Chem. Eur. J.* **2002**, *8*, 1227-1244.
(97) Lehn, J.-M. *Science* **2002**, *295*, 2400-2403.
(98) Reinhoudt, D. N.; Crego-Calama, M. *Science* **2002**, *295*, 2403-2407.
(99) Chang, C. J.; Brown, J. D. K.; Chang, M. C. Y.; Baker, E. A.; Nocera, D. G. *in Electron Transfer in Chemistry Vol. 3 (Ed.: V. Balzani), Wiley-VCH, Weinheim* **2001**, *3*.
(100) Lehn, J.-M. *Supramolecular Chemistry, Wiley-VCH, Weinheim* **1995**.
(101) Sun, L. C.; Hammerstrom, L.; Akermark, B.; Styring, S. *Chem. Soc. Rev.* **2001**, *30*, 36-49.
(102) Ghaddar, T. H.; Castner, E. W.; Isied, S. S. *J. Am. Chem. Soc.* **2000**, *122*, 1233-1234.
(103) Hager, K.; Franz, A.; Hirsch, A. *Chem. Eur. J.* **2006**, *12*, 2663-2679.
(104) Baars, M. W. P. L.; Meijer, E. W. *Top. Curr. Chem.* **2000**, *210*, 131-182.
(105) Crooks, R. M.; Lemon, B. I.; Sun, L.; Yueng, L. K.; Zhao, M. *Top. Curr. Chem.* **2001**, *212*, 81-135.
(106) Smith, D. K.; Diederich, F. *Top. Curr.Chem.* **2000**, *210*, 183-227.
(107) Hecht, S.; Frechet, J. M. J. *J. Am. Chem. Soc.* **2001**, *123*, 6959-6960.
(108) A.W.Kleij; Coevering, R. v. d.; Gebbink, R. J. M. K.; Noordman, A.-M.; Spek, A. L.; Koten, G. v. *Chem. Eur. J.* **2001**, *7*, 181-192.
(109) Eckert, J.-F.; Byrne, D.; Nicoud, J.-F.; Oswald, L.; Nierengarten, J.-F.; Numata, M.; Ikeda, A.; Shinaki, S.; Armaroli, N. *New J. Chem* **2000**, *24*, 749-758.
(110) Köhn, F.; Hofkens, J.; Wiesler, U.-M.; Cotlet, M.; Auwerarer, M. v. d.; Müller, K.; Schryver, F. C. d. *Chem. Eur. J.* **2001**, *7*, 4126-4133.
(111) Crooks, R. M.; Zhao, M.; Sun, L.; Chechnik, V.; Yueng, L. K. *Acc. Chem. Res.* **2001**, *34*, 181-190.

(112) Stephan, H.; Spies, H.; Johannsen, B.; Gloe, K.; Gorka, M.; Vögtle, F. *Eur. J. Inorg. Chem.* **2001**, 2957-2963.
(113) Weigl, B. H.; Bardell, B. L.; Cabrera, C. R. *Adv. Drug Delivery Rev.* **2003**, *55*, 349-377.
(114) Chang, S.-K.; Hamilton, A. D. *J. Am. Chem. Soc.* **1988**, *110*, 1318-1319.
(115) Valenta, J. N.; Dixon, R. P.; Hamilton, A. D.; Weber, S. G. *Anal. Chem.* **1994**, *66*, 2397-2403.
(116) Li, S.; Sun, L. F.; Chung, Y. S.; Weber, S. G. *Anal. Chem.* **1999**, *71*, 2146-2151.
(117) Kluger, R.; Tsao, B. *J. Am. Chem. Soc.* **1993**, *115*, 2089-2090.
(118) Tecilla, P.; Jubian, V.; Hamilton, A. D. *Tetrahedron* **1995**, *51*, 435-448.
(119) P.Tecilla; Chang, S.-K.; Hamilton, A. D. *J. Am. Chem. Soc.* **1990**, *112*, 9586-9590.
(120) Wessendorf, F.; Hirsch, A. *Tetrahedron* **2008**, *64*, 11480-11489.
(121) Hager, K.; Hartnagel, U.; Hirsch, A. *Eur. J. Org. Chem* **2007**, *12*, 1942-1956.
(122) Maurer, K.; Hager, K.; Hirsch, A. *Eur. J. Org. Chem.* **2006**, *15*, 3338-3347.
(123) Hirsch, A.; Brettreich, M. *Fullerenes-Chemistry and Reactions, Wiley-VCH, Weinheim* **2005**.
(124) Kroto, H. W.; Allaf, W.; Balm, S. P. *Chem. Rev.* **1991**, *91*, 1213-1235.
(125) Kroto, H. W. *Nature* **1987**, *329*, 529-531.
(126) Schmalz, T. G.; Seitz, W. A.; Klein, D. J.; Hite, G. E. *Chem. Phy. Lett.* **1986**, *130*, 203-207.
(127) Ajie, H.; Alvarez, M. M.; Anz, S. J.; Beck, R. D.; Diederich, F.; Fostiropoulos, K.; Huffmann, D. R.; Kretschma, W.; Rubin, Y. *J. Phys. Chem. A* **1990**, *94*, 8630-8633.
(128) Beckhaus, H. D.; Ruechardt, C.; Kao, M.; Diederich, F.; Foote, C. S. *Angew. Chem.* **1992**, *104*, 69-70.
(129) Hirsch, A. *Angew. Chem.* **1993**, *105*, 1189-1192.
(130) Hirsch, A. *Chem. Uns. Zeit* **1994**, *28*, 79-87.
(131) Echegoyen, L.; Diederich, F.; Echegoyen, L. E. *Full.: Chem. Phys. Technol.* **2000**, 1-51.
(132) Hirsch, A. *Top. Curr. Chem.* **1999**, *199*, 1-65.
(133) Wudl, F.; Hirsch, A.; Khemani, K. C.; Suzuki, T.; Allemand, P. M.; Kosch, A.; Eckert, H.; Srdanov, G.; Webb, H. M. *ACS Symp. Ser.* **1992**, *481*, 161-175.
(134) Hirsch, A.; Soi, A.; Karfunkel, H. R. *Angew. Chem.* **1992**, *104*, 808-810.
(135) Murata, Y.; Komatsu, K.; Wan, T. S. M. *Tetrahedron Lett.* **1996**, *37*, 7061-7064.
(136) Nagashima, H.; Terasaki, H.; Saito, Y.; Jinno, K.; Itoh, K. *J. Org. Chem.* **1995**, *60*, 4966-4967.
(137) Bingel, C. *Chem. Ber.* **1993**, *126*, 1957-1959.
(138) Lamparth, I.; Hirsch, A. *Chem. Commun.* **1994**, 1727-1730.
(139) Nierengarten, J.-F.; Gramlich, V.; Cardullo, F.; Diederich, F. *Angew. Chem. Int. Ed.* **1996**, *35*, 2101-2103.
(140) Camps, X.; Hirsch, A. *J. Chem. Soc. Perkin Trans. 1* **1997**, 1595-1596.
(141) Habicher, T.; Nierengarten, J.-F.; Gramlich, V.; Diederich, F. *Angew. Chem. Int. Ed.* **1998**, *37*, 1916-1919.
(142) Camps, X.; Hirsch, A. *J. Chem. Soc., Perkin Trans. 1* **1997**, 1595-1596.
(143) Hirsch, A.; Li, Q.; Wudl, F. *Angew. Chem.* **1991**, *103*, 1339-1341.
(144) Li, J.; Takeuchi, A.; Ozawa, M.; Li, X.; Saigo, K.; Kitazawa, K. *Chem. Commun.* **1993**, 1784-1785.

(145) Avent, A. G.; Birkett, P. R.; Darwish, A. D.; Houlton, S.; Taylor, R.; Thomson, K. S. T.; Wei, X.-W. *J. Chem. Soc. Perkin Trans. 2* **2001**, 782-786.
(146) Yamago, S.; Yanagaxa, M.; Mukai, H.; Nakamura, E. *Tetrahedron* **1996**, *52*, 5091-5102.
(147) Kusuwaka, T.; Ando, W. *Angew. Chem. Int. Ed.* **1996**, *35*, 1315-1317.
(148) Kusuwaka, T.; Ando, W. *J. Organomet. Chem.* **1998**, *561*, 109-122.
(149) Prato, M. *Top. Curr. Chem.* **1999**, *199*, 173-187.
(150) Yurovsakaya, M. A.; Trushkov, I. V. *Russ. Chem. Bull.* **2002**, *51*, 367-443.
(151) Wilson, R.; Schuster, D. I.; Nuber, B.; Meier, M. S.; Maggini, M.; Prato, M.; Taylor, R. *Full.: Chem., Phys. Technol.* **2000**, 91-176.
(152) Kräutler, B.; Maynollo, J. *Tetrahedron* **1996**, *52*, 5033-5042.
(153) Wilson, S. R.; Lu, Q. *Tetrahedron Lett.* **1993**, *34*, 8043-8046.
(154) Langa, F.; Cruz, P. d. l.; Espildora, E.; Garcia, J. J.; Perez, M. C.; Hoz, A. d. l. *Carbon* **2000**, *38*, 1641-1646.
(155) Prato, M.; Maggini, M. *Acc. Chem. Res.* **1998**, *31*, 519-526.
(156) Isaacs, L.; A. Wehrsig; Diederich, F. *Helv. Chim. Acta* **1993**, *76*, 1231-1250.
(157) Skiebe, A.; Hirsch, A. *Chem. Commun.* **1994**, 335-336.
(158) Nuber, B.; Hampel, F.; Hirsch, A. *Chem. Commun.* **1996**, 1799-1800.
(159) Illescas, B. M.; Martin, N. *J. Org. Chem.* **2000**, *65*, 5986-5995.
(160) Jagerovic, N.; Elguero, J.; Aubagnac, J.-L. *J. Chem. Soc., Perkin Trans. 1* **1996**, 499.
(161) Nakamura, Y.; Takano, N.; Nishimura, T.; Yashima, E.; Sato, M.; Kudo, T.; Nishimura, J. *Org. Lett.* **2001**, *3*, 1193-1196.
(162) Wilson, S. R.; Kaprinidis, N.; Wu, Y.; Schuster, D. I. *J. Am. Chem. Soc.* **1993**, *115*, 8495-8496.
(163) Zhang, X.; Romero, A.; Foot, C. S. *J. Am. Chem. Soc.* **1993**, *115*, 11024-11025.
(164) Zhang, X.; Fan, A.; Foote, C. S. *J. Org. Chem.* **1996**, *61*, 5456-5461.
(165) Matsui, S.; Kinabara, K.; Saigo, K. *Tetrahedron Lett.* **1999**, *40*, 899-902.
(166) Löffler, G.; Petrides, P. E.; Heinrich, P. C. *Biochemie und Pathobiochemie*, Springer Verlag, Berlin **2006**.
(167) Chelikani, P.; Fita, I.; C. Loewen, P. *Cell. Mol. Life Sci.* **2004**, *61*, 192-208.
(168) Putnam, C. D.; Arvai, A. S.; Bourne, Y.; Tainer, J. A. *J. Mol. Biol.* **2000**, *296*, 295-309.
(169) Guengerich, P. F. *Cytochrome P450, 3rd Ed.* Kluwer Academic/Plenum Publisher, New York **2005**.
(170) Küster, W. Z. *Z. Physiol. Chem.* **1913**, *82*, 463-483.
(171) Fischer, H.; Zeile, K. *Liebigs Ann. Chem.* **1929**, *468*, 98-116.
(172) Fischer, H.; Orth, H. *Die Chemie des Pyrrols, Vol II.1*, Akademische Verlagsgesellschaft Leipzig **1937**.
(173) Fischer, H.; Stern, A. *Die Chemie des Pyrrols, Vol II.2*, Akademische Verlagsgesellschaft Leipzig **1940**.
(174) Moss, G. P. *Pure Appl. Chem.* **1987**, *59*, 799-832.
(175) Moss, G. P. *Eur. J. Biochem.* **1988**, *178*, 277-573.
(176) www.chem.qmul.ac.uk/iupac/tetrapyrrole/.
(177) Jackman, L. M.; Sondheimer, F.; Amiel, Y.; Ben-Efraim, D. A.; Gaoni, Y.; Wolovsky, R.; Bothner-By, a. A. *J. Am. Chem. Soc.* **1962**, *84*, 4307-4312.
(178) Vogel, E.; Pretzer, W.; Böll, W. A. *Tetrahedron Lett.* **1965**, *6*, 3613-3617.
(179) Vogel, E. *Pure Appl. Chem.* **1993**, *65*, 143-152.
(180) Cyranski, M. K.; Krygowski, T. M.; Wisiorowski, M.; Hommes, N. J. R. v. E.; Schleyer, P. v. R. *Angew. Chem. Int. Ed.* **1998**, *37*, 177-180.

(181) Juselius, J.; Sundholm, D. *Phys. Chem. Chem. Phys.* **2000**, *2*, 2145-2151.
(182) Medforth, C. J. *in The Porphyrin Handbook*, K. M. Kadish, K. M. Smith, R. Guilard (Eds.), Vol.5, Ch. 35, Academic Press **2000**.
(183) Gouterman, M. *J. Mol. Spectroscopy* **1961**, *6*, 138-163.
(184) Gouterman, M.; Wagniere, G. H. *J. Mol. Spectroscopy* **1963**, *11*, 108-127.
(185) Perkampus, H.-H. *UV/Vis Atlas of Organic Compounds*, 2nd edition, Wiley-VCH, Weinheim **1992**.
(186) Roeder, B. *Einführung in die molekulare Photophysik*, Teubner, Stuttgart-Leipzig **1999**.
(187) Goldsmith, R. H.; Sinks, L. E.; Kelley, R. F.; Betzen, L. J.; Liu, W.; Weiss, E. A.; Ratner, M. A.; Wasielewski, M. R. *Proc. Natl. Acad. Sci. U.S.A.* **2005**, *102*, 3540–3545.
(188) Weiss, E. A.; Tauber, M. J.; Kelley, R. F.; Ahrens, M. J.; Ratner, M. A.; Wasielewski, M. R. *J. Am.Chem. Soc.* **2005**, *127*, 11842–11850.
(189) Tauber, M. J.; Kelley, R. F.; Giaimo, J. M.; Rybtchinski, B.; Wasielewski, M. R. *J. Am. Chem. Soc.* **2006**, *128*, 1782–1783.
(190) Goldsmith, R. H.; Wasielewski, M. R.; Ratner, M. A. *J. Phys. Chem. A* **2006**, *110*, 20258 –20262.
(191) Muellen, K.; Wegner, G. *Electronic Materials: The Oligomer Approach*, Wiley-VCH, Weinheim, Germany, **1998**.
(192) Atienza-Castellanos, C.; Wielopolski, M.; Guldi, D. M.; van der Pol, C.; Bryce, M. R.; Filippone, S.; Martin, N. *Chem. Commun.* **2007**, *48*, 5164-5166.
(193) Redmore, N. P.; Rubstov, I. V.; Therien, M. J. *J. Am. Chem. Soc.* **2003**, *125*, 8769-8778.
(194) Screen, T. O.; Thorne, J. R. G.; Denning, R. G.; Bucknall, D. G.; Anderson, H. L. *J. Mater. Chem.* **2003**, *13*, 2796 –2808.
(195) Thomas, K. G.; Biju, V.; Guldi, D. M.; Kamat, P. V.; George, M. V. *J. Phys. Chem. B* **1999**, *103*, 8864–8869.
(196) Guldi, D. M.; Swartz, A.; Luo, C.; Gomez, R.; Segura, J. L.; Martin, N. *J. Am. Chem. Soc.* **2002**, *124*, 10875 –10886.
(197) Dong, T.-Y.; Chang, S.-W.; Kin, S.-F.; Lin, M.-C.; Wen, Y.-S.; Lee, L. *Organometallics* **2006**, *25*, 2018–2024.
(198) Wielopolski, M.; Atienza, C.; Clark, T.; Guldi, D. M.; Martin, N. *Chem. Eur. J.* **2008**, *14*, 6379–6390.
(199) Boyd, A. S. F.; Carroll, J. B.; Cooke, G.; Garety, J. F.; Jordan, B. J.; Mabruk, S.; Rosaira, G.; Rotellob, V. M. *Chem. Commun.* **2005**, 2468-2470.
(200) Nagy, A.; Novak, Z.; Kotschy, A. *J. Organomet. Chem.* **2005**, *69*, 4453-4461.
(201) Tykwinski, R. R. *Angew. Chem. Int. Ed.* **2003**, *42*, 1566-1568.
(202) Maggini, M.; G., S.; Prato, M. *J. Am. Chem. Soc.* **1993**, *115*, 9798-9799.
(203) Heck, R. F.; Nolley, J. P. *J. Org. Chem* **1972**, *37*, 2320–2322.
(204) Miyaura, N.; Yamada, K.; Suzuki, A. *Tetrahedron Lett.* **1979**, *36*, 3437-3440.
(205) Suzuki, A. *Pure Appl. Chem.* **1991**, *63*, 419-422.
(206) Suzuki, A. *J. Organomet. Chem.* **1999**, *576*, 147-168.
(207) Van der Pol, C.; Bryce, M. R.; Wielopolski, M.; Atienza-Castellanos, C.; Guldi, D. M.; Filippone, S.; Martin, N. *J. Org. Chem.* **2007**, *72*, 6662-6671.
(208) Littler, B. J.; Ciringh, Y.; Lindsey, J. S. *J. Org. Chem.* **1999**, *64*, 2864-2872.
(209) Lindsey, J. S.; MacCrum, K. A.; Thyonas, J. S.; Chuang, Y.-Y. *J. Org. Chem.* **1994**, *59*, 579-587.
(210) Lindsey, J. S.; Schreiman, I. C.; Hsu, H. C.; Kearney, P. C.; Maguerettaz, A. M. *J. Org. Chem.* **1987**, *52*, 827-836.

(211) Geier III, R. G.; Lindsey, J. S. *J. Chem. Soc., Perkin Trans. 2* **2001**, *2*, 677-686.
(212) Ehli, C.; Rahman, G. M. A.; Jux, N.; Balbinot, G., D. M.; Paolucci, F.; Marcaccio, M.; Paolucci, D.; Melle-Franco, M.; Zerbetto, F.; Campidelli, S.; Prato, M. *J. Am. Chem. Soc.* **2006**, *128*, 11222-11231.
(213) Sarova, G. H.; Hartnagel, U.; Balbinot, D.; Sali, S.; Jux, N.; Hirsch, A.; Guldi, D. M. *Chem. Eur. J.* **2008**, *14*, 3137-3145.
(214) Sgobba, V.; Rahman, G. M. A.; Guldi, D. M.; Jux, N.; Campidelli, S.; Prato, M. *Adv. Mater.* **2006**, *18*, 2264-2269.
(215) Wang, Z.; Wang, L.; Zhang, X.; Shen, J. *Macromol. Chem. Phys.* **1997**, *198*, 573-579.
(216) Plater, M. J.; Sinclair, J. P.; Aiken, S.; Gelbrich, T.; Hursthouse, M. B. *Tetrahedron* **2004**, *60*, 6385-6394.
(217) Amann, A. *Dissertation* **2004**, Ulm.
(218) Amann, A.; Rang, A.; Schalley, C. A.; Bäuerle, P. *Eur. J. Org. Chem.* **2006**, *8*, 1940-1948.
(219) Höfle, G.; Steglich, W. *Synthesis* **1972**, 619-621.
(220) Neises, B.; Steglich, W. *Angew. Chem. Int. Ed.* **1978**, *17*, 522-524.
(221) Neises, B.; Steglich, W. *Angew. Chem.* **1978**, *90*, 556-558.
(222) Mahmud, I. M.; Zhou, N.; Wang, L.; Zhao, Y. *Tetrahedron* **2008**, *64*, 11420-11432.
(223) Berget, P. E.; Teixeira, J. M.; Jacobsen, J. L.; Schore, N. E. *Tetrahedron Lett.* **2007**, *48*, 8101-8103.
(224) Buschhaus, B.; Bauer, W.; Hirsch, A. *Tetrahedron* **2003**, *59*, 3899-3915.
(225) Newkome, G. R.; Behera, R. K.; Moorefield, C. N.; Baker, G. R. *J. Org. Chem.* **1991**, *56*, 7162-7167.
(226) Perez, L.; Lenoble, J.; Barbera, J.; Cruz, P. d. l.; Deschenaux, R.; Langa, F. *Chem. Commun.* **2008**, 4590-4592.
(227) Franz, A.; Bauer, W.; Hirsch, A. *Angew. Chem.* **2005**, *117*, 1588-1561.
(228) Solov'ev, V. P.; Baulin, V. E.; Strahova, N. N.; Kazachenko, V. P.; Belsky, V. K.; Varnek, A. A.; Volkova, T. A.; Wipff, G. *J. Chem. Soc. Perkin Trans. 1* **1998**, 1489-1498.
(229) Solov'ev, V. P.; Vnuk, E. A.; Strakhova, N. N.; Reavsky, O. A. *VINITI, Moscow* **1991**.
(230) Valeur, B. *Molecular Fluorescence, Wiley-VCH, Weinheim* **2002**.
(231) Antony, J.; Medvedev, D. M.; Stuchebrukhov, A. A. *J. Am. Chem. Soc.* **2000**, *122*, 1057-1065.
(232) Dirksen, A.; Kleverlaan, C. J.; Reek, J. N. H.; Cola, L. D. *J. Phys. Chem. A* **2005**, *109*, 5248-5256.
(233) Walsh, C. *Acc. Chem. Res.* **1980**, *13*, 148-155.
(234) Massey, V. *Biochem. Soc. Trans* **2000**, *28*, 283-296.
(235) Sun, M.; Moore, T. A.; Song, P.-S. *J. Am. Chem. Soc.* **1972**, *94*, 1730-1740.
(236) Fukuzumi, S.; Tanaka, T. *Photoinduced Electron Transfer (Eds: M. A. Fox, M. Chanon) Elsevier, Amsterdam* **1988**, 636-687.
(237) Fukuzumi, S.; Kuroda, S.; Tanaka, T. *J. Am. Chem. Soc.* **1985**, *107*, 3020-3027.
(238) Fukuzumi, S.; Yasui, K.; Suenobu, T.; Ohkubo, K.; Fujitsuka, M.; Ito, O. *J. Phys. Chem. A* **2001**, *105*, 10501-10510.
(239) Hoeben, F. J. M.; Jonkheijm, P.; Meijer, E. W.; Schenning, A. P. H. J. *Chem. Rev.* **2005**, *105*, 1491-1546.

(240) Escosura, A. d. l.; Martínez-Díanz, M. V.; Guldi, D. M.; Torres, T. *J. Am. Chem. Soc.* **2006**, *128*, 4112-4118.
(241) Konishi, T.; Ikeda, A.; Shinkai, S. *Tetrahedron* **2005**, *61*, 4881-4899.
(242) Bonnett, R.; Martínez, G. *Tetrahedron* **2001**, *57*, 9513-9547.
(243) Hasobe, T.; Kamat, P. V.; Absalom, M. A.; Kashiwagi, Y.; Sly, J.; Crossley, M. J.; Hosomizu, K.; Imahori, H.; Fukuzumi, S. *J. Phys. Chem. B* **2004**, *108*, 12865-12872.
(244) Murakami, M.; Ohkubo, K.; Fukuzumi, S. *J. Am. Chem. Soc.* **2009**, submitted.
(245) Guldi, D. M.; Asmus, K. D. *J. Am. Chem. Soc.* **1997**, *119*, 5744-5745.
(246) Fukuzumi, S.; Ohkubo, K.; Imahori, H.; Guldi, D. M. *Chem.-Eur. J.* **2003**, *9*, 1585-1593.
(247) Franz, A.; Bauer, W.; Hirsch, A. *Angew. Chem. Int. Ed.* **2005**, *44*, 1564-1567.
(248) Likussar, W.; Boltz, D. F. *Anal. Chem.* **1971**, *43*, 1265-1272.
(249) Fukuzumi, S.; Kondo, Y.; Mochizuki, S.; Tanaka, T. *J. Chem. Soc., Perkin Trans. 2* **1989**, 1753-1761.
(250) Stewart, J. J. P. *J. Comput. Chem.* **1989**, *10*, 209-220.
(251) Stewart, J. J. P. *J. Comput. Chem.* **1989**, *10*, 221-264.
(252) Hasobe, T.; Imahori, H.; Kamat, P. V.; Ahn, T. K.; Kim, S. K.; Kim, D.; Fujimoto, A.; Hirakawa, T.; Fukuzumi, S. *J. Am. Chem. Soc.* **2005**, *127*, 1216-1228.
(253) Kamat, P. V.; Barazzouk, S.; Thomas, K. G.; Hotchandani, S. *J. Phys. Chem. B* **2000**, *104*, 4014-4017.
(254) Sudeep, P. K.; Ipe, B. I.; Thomas, K. G.; George, M. V.; Barazzouk, S.; Hotchandani, S.; Kamat, P. V. *Nano Lett.* **2002**, *2*, 29-35.
(255) Kamat, P. V.; Barazzouk, S.; Hotchandani, S.; Thomas, K. G. *Chem.-Eur. J.* **2000**, *6*, 3914-3921.
(256) Barazzouk, S.; Hotchandani, S.; Vinodgopal, K.; Kamat, P. V. *J. Phys. Chem. B* **2004**, *108*, 17015-17018.
(257) Pagona, G.; Sandanayaka, A. S. D.; Hasobe, T.; Charalambidis, G.; Coutsolelos, A. G.; Yudasaka, M.; Iijima, S.; Tagmatarchis, N. *J. Phys. Chem. C* **2002**, *112*, 15735-15741.
(258) Hasobe, T.; Imahori, H.; Fukuzumi, S.; Kamat, P. V. *J. Phys. Chem. B* **2003**, *107*, 12105-12112.
(259) Imahori, H.; Hasobe, T.; Yamada, H.; Kamat, P. V.; Barazzouk, S.; Fujitsuka, M.; Ito, O.; Fukuzumi, S. *Chem. Lett.* **2001**, 784-785.
(260) Hasobe, T.; Kashiwagi, Y.; Absalom, M. A.; Sly, J.; Hosomizu, K.; Crossley, M. J.; Imahori, H.; Kamat, P. V.; Fukuzumi, S. *Adv. Mater.* **2004**, *16*, 975-979.
(261) Fukuzumi, S.; Suenobu, T.; Patz, M.; Hirasaka, T.; Itoh, S.; Fujitsuka, M.; Ito, O. *J. Am. Chem. Soc.* **1998**, *120*, 8060-8068.
(262) Kojima, T.; Nakanishi, T.; Harada, R.; Ohkubo, K.; Yamaguchi, S.; Fukuzumi, S. *Chem.-Eur. J.* **2007**, *13*, 8714-8725.
(263) Plater, M. J.; Sinclair, J. P.; Aiken, S.; Gelbrich, T.; Hursthouse, M. B. *Tetrahedron* **2004**, *30*, 6385-6394
(264) Detert, H.; Schollmeyer, D.; Sugiono, E. *Eur. J. Org. Chem.* **2001**, 2927-2938.
(265) Burroughes, J. H.; Bradley, D. D. C.; Brown, A. R.; Marks, R. N.; Mackay, K.; Friend, R. H.; Burns, P. L.; Holmes, A. B. *Nature* **1990**, *347*, 539-541.
(266) Weder, C.; Wrighton, M. S. *Macromolecules* **1996**, *29*, 5157-5165.
(267) Ohmori, Y.; Uchida, M.; Muro, K.; Katsumi, Y. *Jpn. J. Appl. Phys.* **1991**, *30*, 1938-1940.

(268) Berggren, M.; Inganas, O.; Gustafsson, G.; Rasmusson, J.; Andersson, M. R.; Hjertberg, T.; Wennerström, O. *Nature* **1994**, *372*, 444-446.
(269) Rajca, A.; Rajca, S.; Pink, M.; Miyasaka, M. *Synlett* **2007**, 1799-1822.
(270) Murphy, A. R.; Fréchet, J. M. J. *Chem. Rev.* **2007**, *107*, 1066-1096.
(271) Ajayaghosh, A.; Praveen, V. K. *Acc. Chem. Res.* **2007**, *40*, 644-656.
(272) Tomović, Z.; van Dongen, J.; George, S. J.; Xu, H.; Pisula, W.; Leclere, W.; Smulders, M. M. J.; de Feyter, S.; Meijer, E. W.; Schenning, A. H. P. J. *J. Am. Chem. Soc.* **2007**, *129*, 16190-16196.
(273) Watson, M. D.; Fechtenkötter, A.; Müllen, K. *Chem. Rev.* **2001**, *101*, 1267-1300.
(274) Grimsdale, A. C.; Müllen, K. *Angew. Chem. Int. Ed.* **2005**, *44*, 5592-5629.
(275) Schenning, A. P. H. J.; Meijer, E. W. *Chem. Comm.* **2005**, 3245-3258.
(276) Wu, J.; Pisula, W.; Müllen, K. *Chem. Rev.* **2007**, *107*, 718-747.
(277) Yamamoto, Y.; Fukushima, T.; Suna, Y.; Ishii, N.; Saeki, A.; Seki, S.; Tagawa, S.; Taniguchi, M.; Kawai, T.; Aida, T. *Science* **2006**, *314*, 1761-1764.
(278) Yamamoto, Y.; Fukushima, T.; Jin, W.; Kosaka, A.; Hara, T.; Nakamura, T.; Saeki, A.; Seki, S.; Tagawa, S.; Aida, T. *Adv. Mater.* **2006**, *18*, 1297-1300.
(279) Xiao, S.; Tang, J.; Beetz, T.; Guo, X.; Tremblay, N.; Siegrist, T.; Zhu, Y.; Steigerwald, M.; Nuckolls, C. *J. Am. Chem. Soc.* **2006**, *128*, 10700-10701.
(280) Shklyarevskiy, I. O.; Jonkheijm, P.; Stutzmann, N.; Wasserberg, D.; Wondergem, H. J.; Christianen, P. C. M.; Schenning, A. P. H. J.; de Leeuw, D. M.; Tomović, Z.; Wu, J.; Müllen, K.; Maan, J. C. *J. Am. Chem. Soc.* **2005**, *127*, 16233-16237.
(281) Jonkheijm, P.; Stutzmann, N.; Chen, Z.; de Leeuw, D. M.; Meijer, E. W.; Schenning, A. P. H. J.; Würthner, F. *J. Am. Chem. Soc.* **2006**, *128*, 9535-9540.
(282) Secker, D.; Weber, H. *Phys. Stat. sol. (b)* **2007**, *244*, 4176-4180.
(283) Lörtscher, E.; Weber, H.; Riel, H. *Phys. Rev. Let.* **2007**, *98*, 176807.
(284) Ochs, R.; Secker, D.; Elbing, M.; Mayor, M.; Weber, H. *Faraday Discussions* **2006**, *131*, 281-289.
(285) Elbing, M.; Ochs, R.; Koentopp, M.; Fischer, M.; von Hänisch, C.; Weigend, F.; Evers, F.; Weber, H.; Mayor, M. *Proc. Natl. Acad. Sci. USA* **2005**, *102*, 8815–8820.
(286) van Ruitenbeek, J.; Scheer, E.; Weber, H. *Contacting Individual Molecules Using Mechanically Controllable Break Junctions In: G. Cuniberti; G. Fagas; K. Richter (Eds.): Introducing Molecular Electronics Berlin, Heidelberg: Springer* **2005**, 253–274.
(287) Mas-Torrent, M.; den Boer, D.; Durkut, M.; Hadley, P.; Schenning, A. P. H. J. *Nanotechnology* **2004**, *15*, 265-269.
(288) Akagi, K.; Piao, G.; Kaneko, S.; Sakamaki, K.; Sirakawa, H.; Kyotani, M. *Science* **1998**, *282*, 1683-1686.
(289) Xia, Y.; Yang, P.; Sun, Y.; Wu, Y.; Mayers, B.; Gates, B.; Yin, Y.; Kim, F.; H., Y. *Adv. Mater.* **2003**, *15*, 353-389.
(290) *Acc. Chem. Res.* **2002**, *36*.
(291) Perlmutter-Hayman, B. *Acc. Chem. Res.* **1986**, *19*, 90-96.
(292) Schneider, H.-J.; Yatsimirsky, A. *Principles and Methods in Supramolecular Chemistry, Wiley, Chichester* **2000**.
(293) Guldi, D. M.; Rahman, G. M. A.; Prato, M.; Jux, N.; Qin, S.; Ford, W. *Angew. Chem. Int. Ed.* **2005**, *44*, 2015-2018.

(294) Hirst, A. R.; Coates, I. A.; Boucheteau, T. R.; Miravet, J. F.; Escuder, B.; V., C.; Hamley, I. W.; Smith, D. K. *J. Am. Chem. Soc.* **2008**, *130*, 9113-9121.
(295) Martin, N.; Sanchez, L.; Herranz, M. A.; Illescas, B.; Guldi, D. M. *Acc. Chem. Res.* **2007**, *40*, 1015-1024.
(296) Atienza, C.; Martin, N.; Wielopolski, M.; Haworth, N.; Clark, T.; Guldi, D. M. *Chem. Comm.* **2006**, 3202-3204.
(297) Rauschenbach, S.; Stadler, F. L.; Lunedei, E.; Malinowski, N.; Koltsov, S.; Costantini, G.; Kern, K. *Small* **2006**, *2*, 540-547.
(298) Fenn, J. B.; Mann, M.; Meng, C. K.; Wong, S. F.; Whitehouse, C. M. *Science* **1989**, *246*, 64-71.
(299) Gologan, B.; Takats, Z.; Alvarez, J.; Wiseman, J.; Talaty, N.; Ouyang, Z.; Cooks, R. G. *J. Am. Soc. Mass Spectrom.* **2004**, *15*, 1874-1884.
(300) Bromann, K.; Felix, C.; Brune, H.; Harbich, W.; Monot, R.; Buttet, J.; Kern, K. *Science* **1996**, *274*, 956-958.
(301) Sze, S. M. *Physics of Semiconductor Devices, New York, Wiley-Interscience* **1981**.
(302) Reed, M. A.; Zhou, C.; Muller, C. J.; Burgin, T. P.; Tour, J. M. *Science* **1997**, *278*, 252-254.
(303) Wang, W.; T. Li; Attardo, G. *J. Org. Chem.* **1997**, *62*, 6598–6602.
(304) Wu, C.-W.; Lin, H.-C. *Macromolecules* **2006**, *39*, 7985–7997
(305) Gomez-Escalonilla, M. J.; Langa, F.; Rueff, J.-M.; Oswald, L.; Nierengarten, J.-F. *Tetrahedron* **2002**, *43*, 7507-7511.
(306) Horner, L.; Hoffmann, H. M. R.; Wippel, H. G. *Chem. Ber.* **1958**, *91*, 61-63.
(307) Horner, L.; Hoffmann, H. M. R.; Wippel, H. G.; Klahre, G. *Chem. Ber.* **1959**, *92*, 2499–2505.
(308) Detert, H.; Sugiono, E. *J. Phys. Org. Chem.* **2000**, *13*, 587–590.
(309) Corey, E. J.; Fuchs, P. L. *Tetrahedron Lett.* **1972**, *36*, 3769–3772.
(310) Ruben, M.; Landa, A.; Loertsel, E.; Riel, H.; Mayor, M.; Goerls, H.; Weber, H. B.; Arnold, A.; Evers, F. *Small* **2008**, *4*, 2229-2235.
(311) Lortscher, E.; Elbing, M.; Tschudy, M.; von Hanisch, C.; Weber, H. B.; Mayor, M.; Riel, H. *ChemPhysChem* **2008**, *9*, 2252-2258.
(312) Spaenig, F.; Kovacs, C.; Hauke, F.; Ohkubo, K.; Fukuzumi, S.; Guldi, D. M.; Hirsch, A. *J. Am. Chem. Soc.* **2009**, *131*, 8180-8195.
(313) Guldi, D. M.; Maggini, M.; Scorrano, G.; Prato, M. *J. Am. Chem. Soc.* **1997**, *119*, 974-980.
(314) Guldi, D. M.; Maggini, M.; Scorrano, G.; Prato, M. *Res. Chem. Intermed.* **1997**, *23*, 561-573.
(315) Koepke, J.; Hu, X.; Muenke, C.; Schulten, K.; Michel, H. *Structure* **1996**, *4*, 581-597.
(316) Choi, M.-S.; Yamazaki, T.; Yamazaki, I.; Aida, T. *Angew. Chem.* **2004**, *116*, 152-160.
(317) Dolphin, D. *The Porphyrins, Academic Press* **1978**.
(318) Kadish, K. M.; Smith, K. M.; Guillard, R. *The Porphyrin Handbook, Academic Press* **2000**.
(319) Allemand, P. M.; Koch, A.; Wudl, F.; Rubin, Y.; Diederich, F.; Alvarez, M. M.; Anz, S. J.; Whetten, R. L. *J. Am Chem. Soc.* **1991**, *113*, 1050-1051.
(320) Xie, Q.; Cordero, E. P.; Echegoyen, L. *J. Am. Chem. Soc.* **1992**, *114*, 3978-3980.
(321) Imahori, H.; El-Khouly, M. E.; Fujitsuka, M.; Ito, O.; Sakata, Y.; Fukuzumi, S. *J. Phys. Chem. A* **2001**, *105*, 325-332.

(322) Bell, T. D. M.; Smith, T. A.; Ghiggino, K. P.; Ranasinghe, M. G.; Shephard, M. J.; Row, M. N. P. *Chem. Phys. Lett.* **1997**, *268*, 223-228.
(323) Baran, P. S.; Monaco, R. R.; Khan, A. U.; Schuster, D. I.; Wilson, S. R. *J. Am. Chem. Soc.* **1997**, *119*, 8363-8364.
(324) Higashida, S.; Imahori, H.; Kandea, T.; Sakata, Y. *Chem. Lett.* **1998**, *27*, 605-607.
(325) Tamaki, K.; Imahori, H.; Nishimura, Y.; Yamazaki, I.; Shimomura, A.; Okada, T.; Sakata, Y. *Chem. Lett.* **1999**, *28*, 227-229.
(326) Fong, R.; Schuster, D. I.; Wilson, S. R. *Org. Lett.* **1999**, *1*, 729-732.
(327) Wedel, M.; Montforts, F. P. *Tetrahedron Lett.* **1999**, *40*, 7071-7074.
(328) Armaroli, N.; Marconi, G.; Echegoyen, L.; Bourgeois, J.-P.; Diederich, F. *Chem. Eur. J.* **2000**, *6*, 1629-1645.
(329) Nistri, D.; Dei, D.; Chiti, G.; Cocchi, A.; Fantetti, L.; Roncucci, G. L.; C., M.; dei Filli, A. *J. Porphyrines Phthalocyanines* **2005**, *9*, 290-297.
(330) Hartnagel, U.; Balbinot, D.; Jux, N.; Hirsch, A. *Org. Biomol. Chem.* **2006**, *4*, 1785-1795.
(331) Max-Planck-Gesellschaft *Techmax* **2004**, *3*, 1-4.
(332) Langhals, H. *Chem. Ber.* **1985**, *118*, 4641-4645.
(333) Würthner, F.; Chen, Z.; Dehm, V.; Stepanenko, V. *Chem. Commun.* **2006**, 1188-1190.
(334) Löhmannsröben, H.-G.; Langhals, H. *Appl. Phys.* **1989**, *48*, 449-452.
(335) Ebeid, E. Z.; El-Daly, S. A.; Langhals, H. *J. Phys. Chem. A* **1988**, *92*, 4565-4568.
(336) Sadrai, M.; Bird, G. R. *Opt. Commun.* **1984**, *51*, 62-64.
(337) Würthner, F. *Angew. Chem. Int. Ed.* **2001**, *40*, 1037-1039.
(338) Law, K.-Y. *Chem. Rev.* **1993**, *93*, 449-486.
(339) Schmidt-Mende, L.; Fechtenkötter, A.; Müllen, K.; Moons, E.; Friend, R. H.; MacKenzie, J. D. *Science* **2001**, *293*, 1119-1122.
(340) Yakimov, A.; Forrest, S. R. *Appl. Phys. Lett.* **2002**, *80*, 1667-1669.
(341) Rademacher, A.; Märkle, S.; Langhals, H. *Chem. Ber.* **1981**, *115*, 2927-2934.
(342) Langhals, H.; Jaschke, H.; Bastani-Oskoui, H.; Speckbacher, M. *Eur. J. Org. Chem.* **2005**, *20*, 4313-4321.
(343) Bingel, C. *Chem. Ber.* **1993**, *126*, 1957-1959.
(344) D'Souza, F.; Venukadasula, G. M.; Yamanaka, K.-I.; Subbaiyan, N. K.; Zandler, M. E.; Ito, O. *Org. Biomol. Chem* **2009**, *7*, 1076-1080.
(345) Neher, D. *Macromol. Rapid Commun.* **2001**, *22*, 1365-1385.
(346) Scherf, U.; List, E. J. W. *Adv. Mater.* **2002**, *14*, 477-487.
(347) Perepichka, I. I.; Perepichka, I. F.; Bryce, M. R.; Pålsson, L.-O. *Chem. Commun.* **2005**, 3397-3399.
(348) Hummelen, J. C.; Kroon, J. M.; Inganas, O.; Andersson, M. R. *Adv. Mater.* **2003**, *15*, 988-991.
(349) Zhang, F.; Perzon, E.; Wang, X.; Mammo, W.; Andersson, M. R.; Inganas, O. *Adv. Func. Mater.* **2005**, *15*, 745-750.
(350) Wang, X.; Perzon, E.; Oswald, F.; Langa, F.; Admassie, S.; Andersson, M. R.; Inganas, O. *Adv. Func. Mater.* **2005**, *15*, 1665-1670.
(351) Zhang, F.; Mammo, W.; Andersson, L. M.; Admassie, S.; Andersson, M. R.; Inganas, O. *Adv. Mater.* **2006**, *18*, 2169-2173.
(352) Neiss, B.; Steglich, W. *Angew. Chem.* **1978**, *90*, 556-557.
(353) Neiss, B.; Steglich, W. *Angew. Chem. Int. Ed.* **1978**, *17*, 522-523.
(354) Lakowics, J. R. *Principles of Fluorescence Spectroscopy 2nd Ed.Kluwer Academic/Plenum Publishers New York* **1999**.

(355) Kim, H. M.; Lee, Y. O.; Lim, C. S.; Kim, J. S.; Cho, B. R. *J. Org. Chem.* **2008**, *73*, 5127-5130.
(356) Kim, H. J.; Hong, J.; Hong, A.; Ham, S.; Lee, J. H.; Kim, J. S. *Org. Lett.* **2008**, *10*.
(357) Yang, S.-W.; Elangovan, A.; Hwang, K.-C.; Ho, T.-I. *J. Phys. Chem. B* **2005**, *109*, 16628-16635.
(358) Benniston, A. C.; Harriman, A.; Lawrie, D. J.; A., M. *ChemPhysChem* **2004**, *6*, 51-55.
(359) Benniston, A. C.; Harriman, A.; Lawrie, D. J.; Mayeux, A.; Rafferty, K.; Russell, O. D. *Dalton Trans.* **2003**, 4762-4769.
(360) Leroy, S.; Soujanya, T.; Fages, F. *Tetrahedron Lett.* **1998**, *39*, 1179-1182.
(361) Maeda, H.; Maeda, T.; Mizuno, K.; Fujimoto, K.; Shimizu, H.; Inouye, M. *Chem. Eur. J.* **2006**, *12*, 824-831.
(362) Ji, S.; Yang, J.; Yang, Q.; Liu, S.; M. Chen; Zhao, J. *J. Org. Chem.* **2009**, *74*, 4855-4865.
(363) Halik, M.; Klauk, H.; Zschieschang, U.; Schmid, G.; Ponomarenko, S.; Kirchmeyer, S.; Weber, W. *Adv. Mater.* **2003**, *15*, 917-922.
(364) Halik, M.; Klauk, H.; Zschieschang, U.; Schmid, G.; Dehm, C.; Schütz, M.; Maisch, S.; Effenberger, F. *Nature* **2004**, *431*, 963-966.
(365) Klauk, H.; Zschieschang, U.; Pflaum, J.; Halik, M. *Nature* **2007**, *445*, 745-748.
(366) Halik, M. *Proc. SPIE* **2005**, *5940*, 59400W.
(367) Verma, S.; Kar, P.; Das, A.; Palit, D. K.; Ghosh, H. N. *J. Phys. Chem C* **2008**, *112*, 2918-2926.
(368) Ramakrishnan, G.; Verma, S.; Jose, D. A.; Kumar, D. K.; Das, A.; Palit, D. K.; Ghosh, H. N. *J. Phys. Chem. B* **2006**, *110*, 9012-9021.
(369) Marczak, R.; Werner, F.; Gnichwitz, J. F.; Hirsch, A.; Guldi, D. M.; Peukert, W. *J. Phys. Chem C* **2009**, *113*, 4669-4678.
(370) Kalyanasundaram, K.; Gratzel, M. *Coord. Chem. Rev.* **1998**, *177*, 347-414.
(371) Gallopini, E. *Coord. Chem. Rev.* **2004**, *248*, 1283-1288.
(372) Zschieschang, U.; Klauk, H.; Halik, M.; Schmid, G. *Ger. Offen.* **2005**, 11-13.
(373) Appel, R. *Angew. Chem. Int. Ed.* **1975**, *14*, 801-811.
(374) Arbusow, E. A. *Dissertation, St. Petersburg* **1906**, *38*, 687.
(375) Arbusow, E. A. *Dissertation, Kazan* **1914**, 148-163.
(376) Perrin, D. D.; Amarego, W. L. F. *Purification of Laboratory Chemicals, 3rd ed., Pergamon Press, Oxford,* **1988**.

I want morebooks!

Buy your books fast and straightforward online - at one of the world's fastest growing online book stores! Environmentally sound due to Print-on-Demand technologies.

Buy your books online at

www.get-morebooks.com

Kaufen Sie Ihre Bücher schnell und unkompliziert online – auf einer der am schnellsten wachsenden Buchhandelsplattformen weltweit!
Dank Print-On-Demand umwelt- und ressourcenschonend produziert.

Bücher schneller online kaufen

www.morebooks.de

OmniScriptum Marketing DEU GmbH
Heinrich-Böcking-Str. 6-8
D - 66121 Saarbrücken
Telefax: +49 681 93 81 567-9

info@omniscriptum.com
www.omniscriptum.com

Printed by Books on Demand GmbH, Norderstedt / Germany